2e

SIGNIFICANT CHANGES

to the NEC® 2008 Edition

In Partnership with the **NJATC**

THOMSON

DELMAR LEARNING

Significant Changes to the NEC® 2008 edition
NJATC

**Vice President,
Career Education SBU:**
Gregory L. Clayton

**Product Development
Manager:**
Ed Francis

Product Manager:
Stephanie Kelly

Editorial Assistant:
Jaclyn Ippolito

Director of Marketing:
Beth A. Lutz

Executive Marketing Manager:
Taryn Zlatin

Marketing Specialist:
Marissa Maiella

Director of Technology:
Paul Morris

Technology Project Manager:
Jim Ormsbee

Director of Production:
Patty Stephan

Production Manager:
Andrew Crouth

Content Project Manager:
Andrea Majot

Art/Design Coordinator:
Bethany Casey

Library of Congress Cataloging-in-Publication Data:
Card Number: 2007016982

ISBN-10: 1-4180-6747-4
ISBN-13: 978-1-4180-6747-2

NOTICE TO THE READER

CONTENTS

The National Electrical Code® (NEC®) is in a state of constant change; it is a living document. The revision cycle of the *NEC* never ends. As you begin to read and understand changes to the new 2008 edition, technical committee task groups and individual members are already preparing proposals to develop the 2011 edition. The *NEC* is the most widely used electrical installation standard in the world. It is adopted throughout the United States and is used daily by electricians, contractors, maintainers, inspectors, engineers, and designers. The constant state of change and the three-year revision cycle are necessary to achieve the purpose of the *NEC:* "the practical safeguarding of persons and property from hazards arising from the use of electricity."

The revision process of the *NEC* begins with the proposal stage. Anyone can submit a proposal to change the *NEC*. The code-making panels (CMPs) meet to deliberate on the submitted proposals. The results are printed and distributed in the report on proposals (ROP) and a draft copy of the *NEC* with the ROP changes is also developed. The second stage is the comment stage. Anyone can submit a comment to revise an action taken in the ROP stage. The CMPs meet to deliberate on all submitted comments and the results are printed in the report on comments (ROC). The technical correlating committee has oversight during both the ROP and ROC stages to prevent conflicting actions between CMPs. The next step is at the NFPA annual meeting, where motions may be filed to amend actions taken in the ROP or ROC stages. The final action is a meeting of the standards council. Any appeals made to the council are considered and the new edition of the *NEC* is then final and sent to the printer.

In the front of your *NEC* after the table of contents, all of the technical committees, or code-making panels, are listed with each member's name and classifications. As you review these changes and continue to use the *NEC,* get involved in the revision process by sending in proposals to revise future editions of the *NEC*.

ABOUT THIS BOOK

This text is written to inform electricians, contractors, maintenance personnel, inspectors, engineers, and system designers of the most significant changes to the 2008 *NEC*. This text will notify the reader of all significant changes with valuable information explaining in detail the reason behind the change and how you will be impacted by it.

ABOUT THE AUTHOR

Jim Dollard's vast experience with the Code is evident on each page you're about to read. He is the current Safety Coordinator for IBEW Local 98 in Philadelphia, as well the Chairman of Code Making Panel-10 for the NEC. Jim is also an active member of the NFPA 70E and 90A committees, the UL Electrical Council, and is a Master OSHA 500 Instructor.

ABOUT THE NJATC

Should you decide on a career in the electrical industry, training provided by the International Brotherhood of Electrical Workers and the National Electrical Contractors Association (IBEW-NECA) is the most comprehensive the industry has to offer. If you are accepted into one of their local apprenticeship programs, you'll be trained for one of four career specialties: journeyman lineman, residential wireman, journeyman wireman, or VDV installer/technician. Most importantly, you'll be paid while you learn. To learn more, visit http://www.njatc.org.

ACKNOWLEDGMENTS

The NJATC and Thomson Delmar Learning would like to thank the following technical editors for their valuable contributions to the development of this book:

Rodney D. Belisle, Jacob Benninger, Marc J. Bernsen, J. Ron Caccamese, Paul J. Casparro, Mark E. Christian, Paul Costello, Brian L. Crise, James T. Dollard, Jr., Steve Harper, Palmer Hickman, Jeffrey L. Holmes, Paul Holum, Donald M. King, William F. Laidler, Paul J. LeVasseur, Stephen L. Lipster, Linda J. Little, Ronald Michaelis, Brian Myers, Harold C. Ohde, David R. Quave, Marty L. Riesberg, Rhett A. Roe, John L. Simmons, Jim Weimer, and Andrew White.

As you begin to read through these significant changes consider the following ideas to maximize the benefits of the NJATC *Significant Changes to the NEC® 2008 Edition.*

2008 *National Electric Code (NEC)*

It is not necessary to have a copy of the 2008 *NEC* as you review the NJATC significant changes. The revised code text is provided for you on each change. However, if it is possible, keep your 2008 *NEC* with you as you review this text. As you read and understand the changes, make brief notes in both books to aid you in the future.

Chapter Outline

Read through the outline of changes for each chapter before you begin.

Margin Information

Always read all of the margin information before you review the change. This is essential in helping you to fully understand how to apply the revision. The *NEC* is an extremely organized document; it is arranged in accordance with 90.3:

- Chapters 1 through 4 apply generally.
- Chapters 5, 6, and 7 supplement and/or modify Chapters 1 through 4.
- Chapter 8 stands alone; Chapters 1 through 7 apply only if referenced in a Chapter 8 Article.
- Each chapter is broad in scope, which is addressed through individual articles.
- Articles are, in most cases, separated into parts to logically separate information.
- Parts are separated into sections, which may be subdivided for clarity and usability.
- Sections may contain up to three levels of subdivisions, list items, exceptions, and fine print notes (FPN), which are informational only.

The margin material will always identify the chapter, part (if applicable), section, subdivision, list item, or FPN. Understanding the outline form of the *NEC* will help you to apply the revised code text accurately and quickly.

Change Summary

Read the change summary, which is designed to inform the reader of the change that occurred and the impact on the user of the *NEC* in an abbreviated format.

Revised 2008 *NEC* Text

Read the revised text as written in the 2008 *NEC*. The NJATC text provides the revised text in *legislative format.* This means that new text is <u>underlined</u> and deleted text is noted with a ~~strikethrough~~. This is an extremely useful tool for the user of the *NEC*. It can be extremely difficult to look at a provision in the *NEC* that you may not have dealt with recently and see the change. Through the use of *legislative format* the reader of this text can see how the requirement read in 2005 and the revised text in 2008.

Change Significance

The last step in using this text is to read the change significance. By this point you will have read the margin material, the change summary, and the revised 2008 text. The change summary will bring your understanding of the change to the next level. This part of each change is designed to explain in detail what issues were behind the change, how the electrical industry will be affected, and how this revision must be correlated with the remainder of the *NEC*.

Photographs and Art

Each change is accompanied by at least one photograph or sketch to help you understand the revision with a visual connection to the subject matter. Each photograph or sketch is accompanied with a caption containing relevant information to further aid the reader of this text.

Summary

The *NEC* is revised every three years, which requires code users to constantly update their knowledge and ability to use the document. This text is written for the code user with an emphasis on detail as well as clarity and usability. Each change is explained in depth in an easy-to-read format that explains both the revision as well as the requirement. It is our desire to make the *NEC* the most useful tool in your toolbox or office.

Instructors who plan to use this publication as a reference textbook are encouraged to attend the National Joint Apprenticeship and Training Committee (NJATC) *Significant Changes to the NEC 2008 Train-the-Trainer* (TTT) course. This 2½-day course provides a venue for an in-depth look at the most significant changes in the 2008 edition of the *NEC* with industry members who were directly involved in updating the Code. In addition to the TTT course, a PowerPoint CD containing all of the pictures and illustrations in the textbook is available to assist in the delivery of this course.

SUPPLEMENTAL PACKAGE

The e-Resource is geared to provide instructors with all the tools they need on one convenient CD-ROM. Instructors will find that this resource provides them with a far-reaching teaching partner that includes:

- PowerPoint slides for each book chapter that reinforce key points and feature illustrations and photos from the book,
- the Computerized Test Bank in ExamView format, which allows for test customization for evaluating student comprehension of noteworthy concepts,
- an electronic version of the Instructor's Manual, with supplemental lesson plans and support, and
- the image library, which includes all drawings and photos from the book for the instructor's use to supplement class discussions.

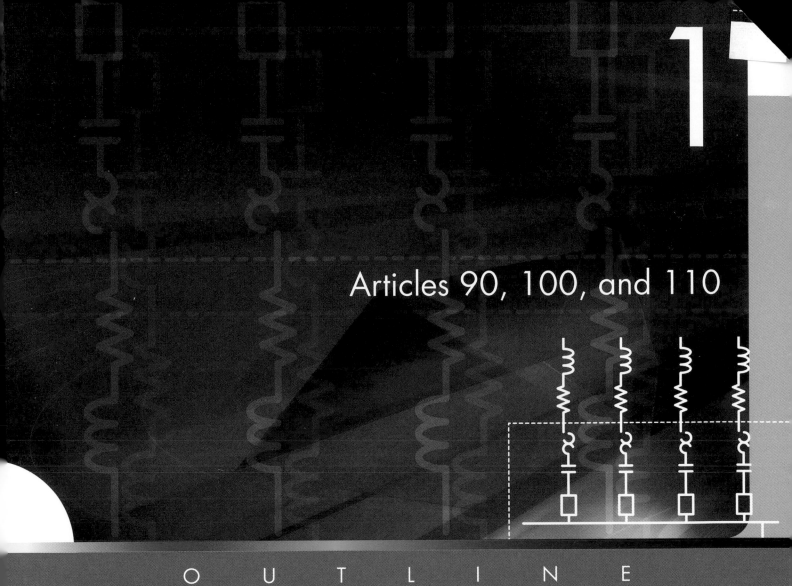

Articles 90, 100, and 110

O U T L I N E

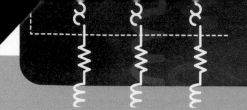

Change Summary:

- The Fine Print Note (FPN) referencing the National Electric Safety Code (NESC) has been deleted from 90.2(A) to eliminate confusion.
- The reference to "or by other agreements" is deleted from 90.2(B)(5). Where utilities perform work outside the right of way or outside easements (such as work on parking lot lighting), the National Electric Code (NEC) now applies.

Revised 2008 NEC Text:

90.2 SCOPE

(A) COVERED: This *Code* covers the installation of electrical conductors, equipment, and raceways; signaling and communications conductors, equipment, and raceways; and optical fiber cables and raceways for the following:

(1), **(3)**, and **(4)** No change.

(2) Yards, lots, parking lots, carnivals, and industrial substations.

~~FPN to (2): For additional information concerning such installations in an industrial or multibuilding complex, see ANSI C2-2002, *National Electrical Safety Code.*~~

(B) NOT COVERED: This *Code* does not cover the following:

(1), **(2)**, **(3)**, and **(4)** No change.

(5) Installations under the exclusive control of an electric utility, where such installations:

    a. Consist of service drops or service laterals, and associated metering, or

    b. Are located in legally established easements <u>or</u> rights-of-way ~~or by other agreements either~~ designated by or recognized by public service commissions, utility commissions, or other regulatory agencies having jurisdiction for such installations, or

    c. Are on property owned or leased by the electric utility for the purpose of communications, metering, generation, control, transformation, transmission, or distribution of electric energy.

FPN to (4) and (5): No change.

Change Significance:

The FPN referencing the NESC is deleted because it creates confusion for the user of the NEC. The NESC contains provisions in conflict with the NEC. Although the NESC may be useful for some engineering design information, its reference as a source from the NEC implies that it is capable of being used without interfering with the use of the NEC.

    90.2(B)(5) lists installations under the exclusive control of the utility that are not covered by the NEC. 90.2(B)(5) now recognizes only legally established easements and right-of-way areas. The deleted reference to "other agreements" permitted utilities to perform work outside their scope as a utility and to ignore the NEC. One example of these situations is parking lot lighting. The NEC will now apply to all installations performed by utilities for private utilization.

90.2
SCOPE

Article 90 Introduction

90.2 Scope

(A) Covered

(B) Not Covered

Change Type: Revision

Installations under the exclusive control of the utility located in legally established easements or rights-of-way are not covered by the NEC.

Article 100
DEFINITION OF BONDED (BONDING)

Chapter 1 General

Article 100 Definitions

Part I General

Change Type: Revision

Change Summary:
- The definition of *Bonding* (*Bonded*) has been revised to apply generally throughout the NEC by describing the purpose and function of bonding.

Typical Pool Bonding

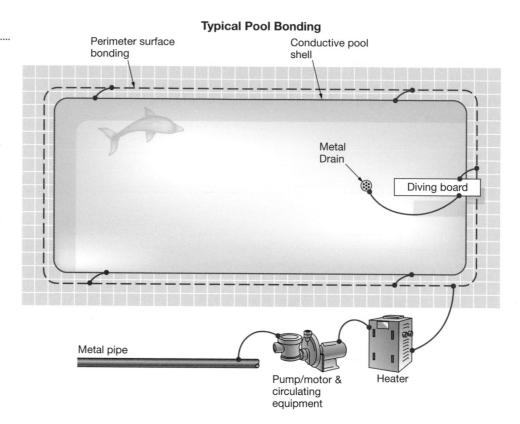

Use of the term *bonded* or *bonding* implies that electrical continuity is established.

Revised 2008 NEC Text:
~~**BONDING (BONDED)**~~ <u>**BONDED (BONDING)**</u>. ~~The permanent joining of metallic parts to form an electrically conductive path that ensures electrical continuity and the capacity to conduct safely any current likely to be imposed.~~ <u>Connected to establish electrical continuity and conductivity.</u>

Change Significance:
The present definition of *bonding* has been rewritten to apply generally throughout the NEC. The revision provides clarity by simply describing the purpose and function of bonding.

As stated in the substantiation for this change, the purpose of bonding is to connect two or more conductive objects together to:

(1) Ensure the electrical continuity of the fault current path, and

(2) Provide the capacity and ability to conduct safely any fault current likely to be imposed, and

(3) Minimize potential differences (voltage) between conductive components.

The intent of the term *bonding* is to convey that normally non–current-carrying conductive materials likely to become energized must be electrically connected to one another and to the supply source in a manner that establishes an effective fault current path. "Normally non–current-carrying conductive materials likely to become energized" include:

(1) Conductive materials enclosing electrical conductors or equipment, or

(2) Forming part of such equipment, or

(3) Other electrically conductive materials and equipment that may present a shock hazard.

There are conditions in the NEC (such as 680.26) where specific bonding is required solely to minimize the difference of potential (voltage) between conductive components.

Notes:

_____

_____

_____

_____

_____

_____

_____

_____

_____

_____

_____

_____

_____

Article 100
DEFINITION OF
CLOTHES CLOSET

Chapter 1 General

Article 100 Definitions

Part I General

Change Type: New

Change Summary:

- New definition to clearly define a clothes closet.
- Based on the definition in the International Building Code.
- Location in Article 100 makes this definition global.
- New definition will aid in the application of 240.24(D), 410.8, and 550.11(A).

Revised 2008 NEC Text:

CLOTHES CLOSET.   A non-habitable room or space intended primarily for storage of garments and apparel.

Change Significance:

The addition of the term *clothes closet* to Article 100 impacts the entire NEC. Users of the NEC now have clear text to identify a clothes closet. In many larger dwelling units, a clothes closet is designed with an entry door and is the size of a small- to medium-sized bedroom or office. When the clothes closet is larger, owners wish to install panelboards and lighting fixtures without restriction. The enforcement community, along with the installer, needs this clarification in the form of a definition for uniform application of the NEC.

The NEC specifically addresses electrical installations in clothes closets. Clothes/garments, boxes used for storage, and many other household items are considered easily ignitable material in the NEC. The term *clothes closet* is used in three locations in the NEC, as follows:

- 240.24(D) requires that overcurrent protective devices shall not be installed in the vicinity of easily ignitable material, such as a clothes closet.

- 410.8 provides a detailed outline of what types of lighting fixtures are permitted in a clothes closet, as well as the location of the lighting fixture.

- 550.11(A) prohibits the distribution panelboard in a mobile or manufactured home from being installed in a bathroom or clothes closet.

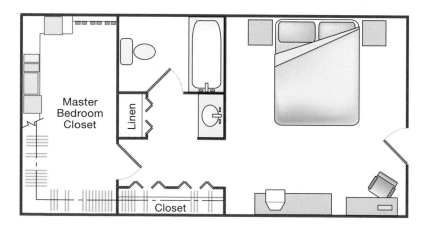

An uninhabitable room or space for storing garments and/or apparel is a clothes closet, regardless of size.

Change Summary:
- The definition of *device* has been editorially revised for usability and clarity. For example, this revision clarifies that a snap switch with a pilot light is a device because its principal function is to carry and control electrical energy.

Revised 2008 NEC Text:
DEVICE. A unit of an electrical system that ~~is intended to~~ carr~~y~~ies or controls ~~but not utilize~~ electric energy <u>as its principal function.</u>

Change Significance:
This revision is intended to clarify that a device has the primary function of carrying and controlling electric energy or current. Snap switches and receptacles carry current and provide a degree of control. The snap switch opens and closes the circuit to a lighting fixture, and a receptacle allows control of appliances as they are plugged in or unplugged. Snap switches may have a pilot light to alert the consumer that the load is connected. Receptacles may have a pilot light to alert the consumer that power is available. Snap switches and receptacles made with pilot lights are "devices." Their primary or principal function is to carry or control electric energy.

Article 100
DEFINITION OF
DEVICE

Chapter 1 General

Article 100 Definitions

Part I General

Change Type: Revision

A pilot-type snap switch is a device and is not considered a luminaire.

Article 100
DEFINITION OF *ELECTRIC POWER PRODUCTION AND DISTRIBUTION NETWORK*

Chapter 1 General

Article 100 Definitions

Part I General

Change Type: New

Change Summary:
* Based on the definition in UL1741.
* Location in Article 100 makes this definition global.
* New definition will aid in the application of Articles 690, 692, and 705.

Revised 2008 NEC Text:
ELECTRIC POWER PRODUCTION AND DISTRIBUTION NETWORK. Power production, distribution, and utilization equipment and facilities, such as electric utility systems that deliver electric power to the connected loads, that are external to and not controlled by an interactive system.

Change Significance:
This new definition (located in Article 100) will now apply globally to the NEC. Article 705 is titled "Interconnected Electric Power Production Sources." This new definition will provide continuity for the application of Article 690 (Solar Photovoltaic Systems) and Article 692 (Fuel Cell Systems) with Article 705 (Interconnected Electric Power Production Sources).

Fuel cell systems may be part of an electrical power production and distribution network. *(Courtesy of S&C Electronic Company.)*

Change Summary:
- The term *equipment* now includes "machinery."
- Industrial machinery falls under the scope of the NEC.

Revised 2008 NEC Text:
EQUIPMENT.   A general term including material, fittings, devices, appliances, luminaries ~~(fixtures)~~, apparatus, <u>machinery</u> and the like used as a part of, or in connection with, an electrical installation.

Change Significance:
This revised definition of *equipment* is intended to more clearly include machinery. The intent of this definition has always included utilization equipment, including machinery. The lack of this term in the definition of equipment created enforcement problems. 90.2(A) outlines what is covered by the NEC.

90.2 clearly requires that the installation of electrical conductors, equipment, and raceways; signaling and communications conductors, equipment, and raceways; and optical fiber cables and raceways are covered in given occupancies. Due to the lack of the term *machinery* being included in 90.2(A) and in the Article 100 definition of equipment, confusion existed as to whether the NEC covered the machinery. The inclusion of the term *machinery* in the revised definition of equipment has eliminated any confusion regarding the application of the NEC.

Article 100
DEFINITION OF
EQUIPMENT

Chapter 1 General

Article 100 Definitions

Part I General

Change Type: Revision

The term *equipment* where used in the NEC now includes "machinery."

Article 100
DEFINITION OF *GROUNDED (GROUNDING)*

Chapter 1 General

Article 100 Definitions

Part I General

Change Type: Revision

Change Summary:
- The term *grounded* has been revised to also define *grounding* by clarifying that where a conductive path such as building steel is used to make a connection to ground the steel serves to connect but does not replace the earth.
- The definition of *grounded effectively* has been deleted. This subjective term is replaced with the terminology *grounded* or *connected to an equipment grounding conductor* to provide clarity and usability.

Revised 2008 NEC Text:
GROUNDED (GROUNDING):   Connected (connecting) to ~~earth~~ ground or to ~~some conducting~~ a conductive body ~~that serves in place of the earth~~ that extends the ground connection.

~~**GROUNDED, EFFECTIVELY.**   Intentionally connected to earth through a ground connection or connections of sufficiently low impedance and having sufficient current-carrying capacity to prevent the buildup of voltages that may result in undue hazards to connected equipment or to persons.~~

Change Significance:
This revision of the term *grounded* now includes the term *grounding* to help clarify the application of these terms as used in the NEC. The concept of using some other body in place of the earth has been deleted because it causes confusion. In the NEC, systems or equipment are essentially either connected to earth or considered ungrounded. If a conducting path is provided to make this connection, such as building steel, it serves as the connection (not in place of the earth). Equipment such as generator frames that are not grounded do not serve in place of the earth. They are actually ungrounded but are bonded to the conductor that provides for an effective fault current path.

The definition of the term *effectively grounded* has been deleted. The use of this term is subjective and without any defined values or parameters for one to judge grounding as either "effective" or "ineffective." "Effective" is described in Section 250.4(A) and (B), but it relates to the effective ground-fault current path as a performance criterion. This term is being replaced this cycle throughout the NEC with the terminology *grounded* or *connected to an equipment grounding conductor* to provide clarity and usability.

The term *grounded* means connected to earth or to a conductive body that extends the earth connection.

Change Summary:

- The term *grounding conductor, equipment* (*EGC*) has been revised to more accurately describe the purpose and function of this conductor.
- Two informational FPNs have been added to inform the code user that an EGC also performs bonding and that types of EGCs are listed in 250.118.

Revised 2008 NEC Text:

GROUNDING CONDUCTOR, EQUIPMENT (EGC). The <u>conductive path installed to connect normally</u> ~~conductor used to connect the~~ non–current-carrying metal parts of equipment, ~~raceways, and other enclosures~~ <u>together and</u> to the system grounded conductor, <u>or to</u> the grounding electrode conductor, or both. ~~, at the service equipment or at the source of a separately derived system.~~

FPN No. 1: <u>It is recognized that the equipment grounding conductor also performs bonding.</u>

FPN No. 2: <u>See 250.118 for a list of acceptable equipment grounding conductors.</u>

Change Significance:

This revision provides clarity by recognizing that the equipment grounding conductor serves in a dual role in bonding equipment and extending the earth connection.

The purposes and functions of the EGC include:

(1) To facilitate the operation of an overcurrent device by providing a low-impedance fault current path between the normally non–current-carrying metal parts of equipment, raceways, and other enclosures to the system grounded conductor and the grounding electrode conductor (or both) at the service equipment, or at a separately derived system, or at a building or structure disconnecting means where supplied by a feeder(s) or branch circuit(s).

(2) To connect all normally non–current-carrying metal parts of equipment and conductive material enclosing electrical conductors so that potential differences (voltage) between equipment enclosures are minimized.

(3) To connect all normally non–current-carrying metal parts of equipment and conductive material enclosing electrical conductors to ground so that potential differences (voltage) to ground are minimized.

The acronym EGC has been introduced into the definition for usability and in accordance with the NEC style manual.

The language "raceways, and other enclosures" within the definition has been removed because types of equipment grounding conductors are listed in 250.118 and therefore this language does not need to be repeated. The language "or both, at the service equipment or at the source of a separately derived system" within the definition is being removed because the installation rules are found elsewhere in the code.

Two informational FPNs have been added to inform the code user that an EGC also performs bonding and that types of EGCs are listed in 250.118.

Article 100
DEFINITION OF *GROUNDING CONDUCTOR, EQUIPMENT (EGC)*

Chapter 1 General
Article 100 Definitions
Part I General
Change Type: Revision

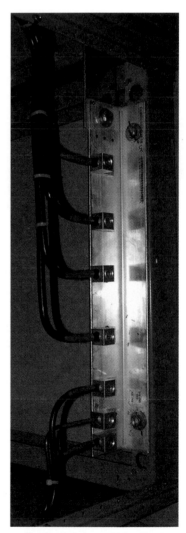

250.118 lists all acceptable methods of establishing an "equipment grounding conductor."

Article 100
DEFINITION OF *GROUNDING ELECTRODE CONDUCTOR*

Chapter 1 General

Article 100 Definitions

Part I General

Change Type: Revision

Change Summary:
- The term *grounding electrode conductor* has been revised to more clearly describe its function, improving clarity and usability.

Revised 2008 NEC Text:
GROUNDING ELECTRODE CONDUCTOR. ~~The~~ A conductor used to connect the <u>system grounded conductor or the equipment to a grounding electrode or to a point on the grounding electrode system.</u> ~~grounding electrode(s) to the equipment grounding conductor, to the grounded conductor, or to both, at the service, at each building or structure where supplied by a feeder(s) or branch-circuit(s), or at the source of a separately derived system.~~

Change Significance:
The revision of the definition of *grounding electrode conductor* simply describes the function of this conductor. The function of the grounding electrode conductor is to connect the grounding electrode(s) to:
- A system conductor for systems intentionally grounded or
- Equipment for systems not intentionally grounded.

The deletion of the components to which a GEC can be connected provides clarity and usability. For example, a conductor is not a "grounded conductor" until the "grounding electrode conductor" is attached. This revision clearly recognizes that the grounding electrode conductor is applied as follows:

Grounded System

From the system grounded conductor to (1) the grounding electrode or (2) a point on the grounding electrode system.

Ungrounded System

From the equipment to (1) the grounding electrode or (2) a point on the grounding electrode system.

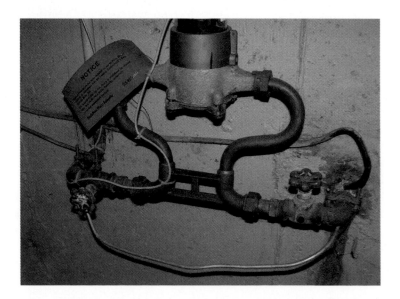

The GEC connects the grounded conductor and/or equipment to the grounding electrode system.

Change Summary:
* The new term *intersystem bonding termination* will aid the code user by clarifying that this term identifies a "device," located as required in 250.94, to accomplish intersystem bonding.

Revised 2008 NEC Text:
INTERSYSTEM BONDING TERMINATION. A device that provides a means for connecting communications system(s) grounding conductor(s) and bonding conductor(s) at the service equipment or at the disconnecting means for buildings or structures supplied by a feeder or branch circuit.

Change Significance:
The new definition of *intersystem bonding termination* has been included in Article 100 because this term is used in more than one NEC article. The NEC requires intersystem bonding of service-/feeder-supplied power systems to metallic raceways, cable sheaths, and/or current-carrying conductors of optical fiber, communications, CATV, and network-powered broadband communications systems. These requirements are located in Articles 250, 770, 800, 820, and 830.

Intersystem bonding is an essential safety-driven requirement designed to prevent occurrences of voltages between communication systems and power systems. The addition of this definition will add clarity by identifying a "device" located as required to accomplish intersystem bonding. This device will be located at the service equipment and at the disconnecting means for any additional buildings or structures and be of a type as listed in 250.94. The intersystem bonding termination is required to be accessible and to have the capacity for connection of not less than three intersystem bonding conductors.

Article 100
DEFINITION OF *INTERSYSTEM BONDING TERMINATION*

Chapter 1 General

Article 100 Definitions

Part I General

Change Type: New

The intersystem boding termination must provide for connection of at least three intersystem bonding conductors.

Article 100
DEFINITION OF
KITCHEN......................

Chapter 1 General

Article 100 Definitions

Part I General

Change Type: New

Change Summary:
- New definition to clearly define a kitchen.
- Based on the description in 210.8(B)(2) in the 2005 NEC.
- Location in Article 100 makes this definition global.
- Requires permanent facilities for food preparation and cooking.

Revised 2008 NEC Text:
KITCHEN. An area with a sink and permanent facilities for food preparation and cooking.

Change Significance:
The need for a definition of the term *kitchen* was created in the 2002 NEC when a new list item (3) was added to 210.8(B) requiring ground-fault circuit interrupter (GFCI) protection for all 125-volt single-phase 15- and 20-amp receptacles located in kitchens in other than dwelling units. Without a clear definition of the term *kitchen,* there was tremendous confusion in the industry. In the 2005 NEC, this problem was addressed by adding a description to illustrate exactly what was meant by the term *kitchen.*

Adding a microwave connected by cord and plug to a countertop with a sink does not constitute a kitchen. Permanent facilities for food preparation and cooking are required. This would include but not be limited to:
- A sink with running water and drainage for cleaning food and utensils
- A table, area, or countertop for preparing food
- A permanently installed range, grill, fryer, cook-top, oven, and so on

The term *kitchen* as now defined in Article 100 is used more than 40 times in the NEC in the following articles:

210	Branch Circuits	**517**	Health Care Facilities
220	Branch Circuit, Feeder, and Service Calculations	**550**	Mobile Homes, Manufactured Homes, and Mobile Home Parks
250	Grounding and Bonding	**551**	Recreational Vehicles, RV Parks, and Park Trailers
422	Appliances		

A new definition of *kitchen* in Article 100 will apply globally throughout the NEC.

Change Summary:

- The parenthetical references to "lighting fixture(s)" and "fixture(s)" after the word *luminaire* are deleted throughout the NEC.
- The definition of *luminaire* is revised to include lighting units without physical protection of the lamp and to clarify that a lampholder is not a luminaire.

Revised 2008 NEC Text:

Delete the parenthetical references to "lighting fixture(s)" or "fixture(s)" after the word "luminaire" throughout the National Electrical Code.

LUMINAIRE. A complete lighting unit consisting of a <u>light source such as a lamp or lamps,</u> together with the parts designed ~~to distribute the light,~~ to position <u>the light source</u> and ~~protect the lamps and ballast (where applicable), and to~~ connect <u>it</u> ~~the lamps~~ to the power supply. <u>It may also include parts to protect the light source or the ballast, or to distribute the light. A lampholder itself is not a luminaire.</u>

Change Significance:

In the 2002 edition of the NEC, the term *luminaire* was inserted to replace the term *lighting fixture.* In an effort to implement this change in a user-friendly fashion throughout the NEC, the term *fixture* was included in parentheses after the term *luminaire.* The terms *lighting fixture* and *fixture* are not defined in the NEC. The parenthetical references to "lighting fixture(s)" or "fixture(s)" after the word *luminaire* are deleted throughout the NEC.

The definition of *luminaire* is revised to include all luminaries. The previous definition inferred that all luminaries were provided with parts designed to distribute the light and to position and protect the lamps. This is not true for all luminaries.

A new last sentence is added to clarify that a lampholder alone is not a luminaire. For example, a lampholder can be a screw shell type to hold a typical incandescent lamp or a tombstone to hold a pin-type fluorescent tube.

Notes:

_____

_____

_____

_____

_____

_____

Article 100
DEFINITION OF
LUMINAIRE

Chapter 1 General

Article 100 Definitions

Part I General

Change Type: Revision

Luminaire

The term *luminaire* will now stand alone because the terms *lighting fixture* and *fixture* are deleted throughout the NEC.

Article 100
DEFINITION OF *METAL-ENCLOSED POWER SWITCHGEAR*

Chapter 1 General

Article 100 Definitions

Part I General

Change Type: Revision

Change Summary:

- A new last sentence in this definition informs the code user that metal-enclosed power switchgear is available in non–arc-resistant and arc-resistant constructions.

Revised 2008 NEC Text:

METAL-ENCLOSED POWER SWITCHGEAR. A switchgear assembly completely enclosed on all sides and top with sheet metal (except for ventilating openings and inspection windows) <u>and</u> containing primary power circuit switching, interrupting devices, or both, with buses and connections. The assembly may include control and auxiliary devices. Access to the interior of the enclosure is provided by doors, removable covers, or both. <u>Metal enclosed power switchgear is available in non-arc resistant or arc resistant constructions.</u>

Change Significance:

Arc-resistant construction of metal-enclosed power switchgear is designed to minimize an arc in a fault and to direct any resulting arcing energy away from persons operating the equipment. The addition of this informative last sentence in the definition is intended to inform the code user of safety-driven equipment options. Arc-resistant switchgear provides multiple safety-driven features, including pressure relief flaps designed to direct energy away from persons, vents and circulating openings at heights above persons, and hinges and gaskets reinforced to prevent the passage of energy.

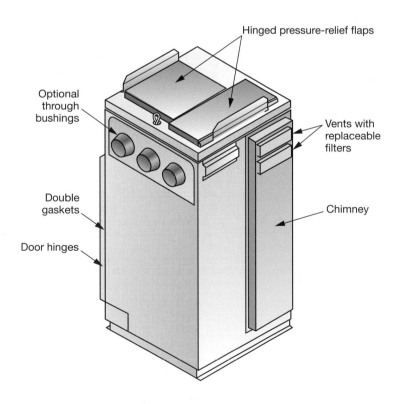

Arc-resistant metal-enclosed switchgear helps to reduce exposure of persons to arc flash hazards. Courtesy of S&C.

Change Summary:

- The new terms *neutral conductor* and *neutral point* have replaced the term *neutral* throughout the NEC. These new definitions throughout the NEC specify to the code user clear and easy-to-apply installation requirements.

Revised 2008 NEC Text:

NEUTRAL CONDUCTOR.   The conductor connected to the neutral point of a system that is intended to carry current under normal conditions.

NEUTRAL POINT.   The common point on a wye-connection in a polyphase system or midpoint on a single-phase, 3-wire system, or midpoint of a single-phase portion of a 3-phase delta system, or a midpoint of a 3-wire, direct current system.

FPN: At the neutral point of the system, the vectorial sum of the nominal voltages from all other phases within the system that utilize the neutral, with respect to the neutral point, is zero potential.

Change Significance:

These new definitions of *neutral conductor* and *neutral point* have been included in Article 100 because these terms are used in more than one NEC article. A global effort occurred this cycle throughout the NEC to replace the term *neutral* with the terms *neutral conductor* and *neutral point* for clarity and usability. The use of the terms *neutral conductor* and *neutral point* (as newly defined in Article 100) throughout the NEC will provide the code user with clear and unambiguous installation requirements.

A *neutral point* is now clearly defined as the common point on the systems identified in the definition. The informational FPN explains that this common point is truly "neutral" when vectorially adding the nominal voltages from all other phases within the system that will utilize a "neutral conductor" connected to the "neutral point."

A *neutral conductor* is simply defined as: "The conductor connected to the neutral point of a system that is intended to carry current under 'normal' conditions." It is important to note that this conductor, connected to the "neutral point," is only a "neutral conductor" when it is installed as an "intentional current-carrying" conductor. This is an extremely important distinction because the "equipment grounding conductor" is connected to the "neutral point" at the service or load side of a separately derived system. The equipment grounding conductor is not intended as a current-carrying conductor under "normal" conditions.

Notes:

_____

_____

Article 100
DEFINITIONS OF *NEUTRAL CONDUCTOR* AND *NEUTRAL POINT*

Chapter 1 General

Article 100 Definitions

Part I General

Change Type: New

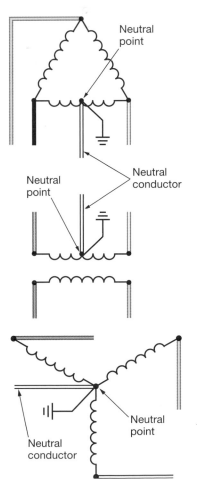

The terms *neutral conductor* and *neutral point* are now defined in Article 100.

Article 100
DEFINITION OF *PREMISES WIRING (SYSTEM)*

Chapter 1 General

Article 100 Definitions

Part I General

Change Type: Revision

Change Summary:
- Revised definition clarifies that all wiring and equipment downstream of the service point (including batteries, solar photovoltaic systems, generators, transformers, and so on) are part of premises wiring.
- Clarification: Where no service point exists, all sources, wiring, and equipment are premises wiring.

Revised 2008 NEC Text:
PREMISES WIRING (SYSTEM). ~~That i~~ Interior and exterior wiring, including power, lighting, control, and signal circuit wiring together with all their associated hardware, fittings, and wiring devices, both permanently and temporarily installed~~., that extends from the service point or source of power, such as a battery, a solar photovoltaic system, or a generator, transformer, or converter windings, to the outlet(s).~~ This includes: (a) wiring from the service point or power source to the outlets or (b) wiring from and including the power source to the outlets where there is no service point.

Such wiring does not include wiring internal to appliances, luminaires ~~(fixtures)~~, motors, controllers, motor control centers, and similar equipment.

(continued)

No utility service. The generator used as a power source is included in the "premises wiring"

An onsite generator is part of the premises wiring system.

Change Significance:

The definition of *premises wiring* has been revised to clearly illustrate that premises wiring (such as supply conductors originating from another system) can and does exist on the supply side of a separately derived system.

Where a service point does not exist (i.e., where no electric utility serves the building or structure), all wiring, sources, and equipment are premises wiring.

The list of examples of supply sources other than service (which must originate from the utility) has been deleted.

Battery, solar, wind, and fuel cell systems are part of the premises wiring system unless they are owned by the electric utility and are in use for generation and transmission of power upstream of the service point.

Generators, transformers, and converter windings are also part of the premises wiring system unless they are owned by the electric utility and are in use for generation and transmission of power upstream of the service point.

This photograph depicts a service lateral on the left that enters a meter housing. All wiring on the load side of the meter housing into the enclosure on the right and downstream is *premises wiring*.

Article 100
DEFINITION OF
QUALIFIED PERSON

Chapter 1 General

Article 100 Definitions

Part I General

Change Type: Revision

Change Summary:
- Revised definition to clarify safety training requirements.
- Qualified persons must be trained to recognize and avoid electrical hazards.

Revised 2008 NEC Text:
QUALIFIED PERSON. One who has skills and knowledge related to the construction and operation of the electrical equipment and installations and has received safety training ~~on~~ to recognize and avoid the hazards involved.

FPN: Refer to NFPA 70E-2004, Standard for Electrical Safety in the Workplace, for electrical safety training requirements.

Change Significance:
The meaning of "qualified person" has been revised to clearly illustrate the safety training referred to in the definition. **Recognition** and **avoidance** of electrical hazards are essential for any person to be considered "qualified."

The term *qualified person* is used more than 115 times in the NEC, and the term *unqualified* more than 15 times. The use of the term *qualified person* in the NEC is in most cases in the form of a glorified exception relaxing more stringent requirements in the NEC. For a person to be considered qualified for a specific task the definition of *qualified person* in the NEC demands that the individual:

- Have **skills** and **knowledge** related to the construction and operation of the electrical equipment and installations.
- Have received safety training to **recognize** and **avoid** the hazards involved.
- Recognize the hazards **SHOCK-ARC, FLASH-ARC,** and **BLAST.**
- Avoid accident and injury through **JUSTIFICATION, HAZARD ANALYSIS,** and **PROTECTION.**

Qualified persons must be trained
to **RECOGNIZE** and **AVOID**
electrical hazards

NFPA 70E provides prescriptive requirements to protect persons from the hazards of exposure to live parts and interaction with electrical equipment.

Change Summary:

- The definition of *surge arrester* has been relocated to Article 100 from 280.2.
- The definition of *transient voltage surge protector* (*TVSS*) is deleted in 285.2 and is replaced with a new definition of *surge protective device* (*SPD*) in Article 100.

Article 100
DEFINITIONS OF *SURGE ARRESTER* AND *SURGE PROTECTIVE DEVICE (SPD)*

Chapter 1 General

Article 100 Definitions

Part I General

Change Type: New

The term *transient voltage surge suppressor* (*TVSS*) is deleted and is replaced by the term *surge protective device* (*SPD*).

Revised 2008 NEC Text:

SURGE ARRESTER   (Relocated from 280.2, No other changes)

SURGE PROTECTIVE DEVICE (SPD). ~~Transient Voltage Surge Suppressor (TVSS).~~ A protective device for limiting transient voltages by diverting or limiting surge current; it also prevents continued flow of follow current while remaining capable of repeating these functions <u>and is designated as follows:</u>

- <u>Type 1: Permanently connected SPDs intended for installation between the secondary of the service transformer and the line side of the service disconnect overcurrent device.</u>
- <u>Type 2: Permanently connected SPDs intended for installation on the load side of the service disconnect overcurrent device; including SPDs located at the branch panel.</u>
- <u>Type 3: Point of utilization SPDs.</u>
- <u>Type 4: Component SPDs, including discrete components, as well as assemblies.</u>

(continued)

FPN: For further information on Type 1, Type 2, Type 3, and Type 4 SPDs, see UL 1449, *Standard for Surge Protective Devices.*

Change Significance:

The definition of *surge arrester* is deleted in 280.2 and has been relocated to Article 100 because this term is used in more than one article. No substantive changes occurred in the definition of the term *surge arrester.*

The definition of *transient voltage surge protector* (*TVSS*) is deleted in 285.2 and is replaced with a new definition of *surge protective device* (*SPD*) in Article 100. This new definition includes the specific types of SPDs and their applications. These changes correlate with revisions in Articles 280 and 285 as well as with the combination of low-voltage surge arrestors and SPD/TVSS into a single standard (UL 1449). A new FPN has been added to reference UL 1449.

Notes:

_____

_____

_____

_____

_____

_____

_____

_____

_____

_____

_____

Change Summary:

- Installers and enforcers now have a clear definition of *short-circuit current rating,* removing subjectivity in determining the suitability of equipment or system.
- This new definition is added to Article 100 because the term *short-circuit current rating* is used more than 20 times in the NEC.

(continued)

Article 100
DEFINITION OF *SHORT-CIRCUIT CURRENT RATING*

Chapter 1 General

Article 100 Definitions

Part I General

Change Type: New

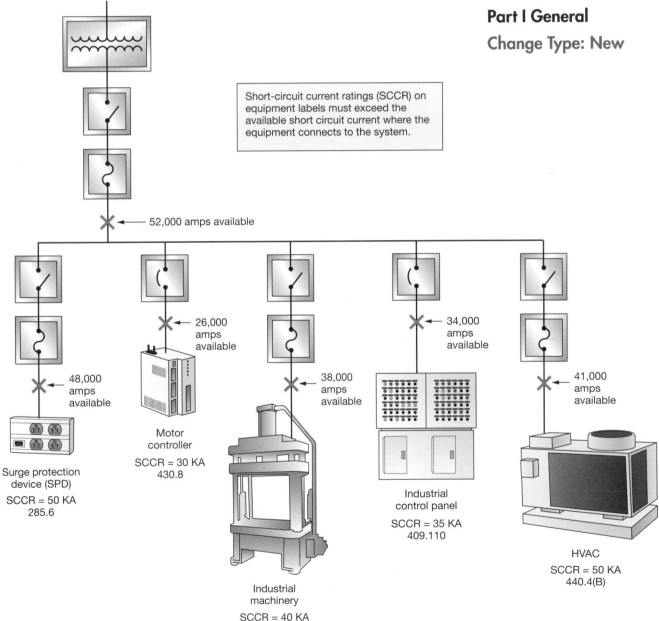

Short-circuit current ratings (SCCR) on equipment labels must exceed the available short circuit current where the equipment connects to the system.

52,000 amps available

48,000 amps available

26,000 amps available

38,000 amps available

34,000 amps available

41,000 amps available

Surge protection device (SPD)

SCCR = 50 KA
285.6

Motor controller

SCCR = 30 KA
430.8

Industrial machinery

SCCR = 40 KA
670.3(A)

Industrial control panel

SCCR = 35 KA
409.110

HVAC

SCCR = 50 KA
440.4(B)

The term *short-circuit current rating* is now defined and applies globally throughout the NEC.

Revised 2008 NEC Text:

SHORT-CIRCUIT CURRENT RATING. <u>The prospective symmetrical fault current at a nominal voltage to which an apparatus or system is able to be connected without sustaining damage exceeding defined acceptance criteria.</u>

Change Significance:

The term *short-circuit current rating* is used more than 20 times in the NEC. Prior to addition of this new definition, the application of the term *short-circuit current rating* was subjective to the individual installer or enforcer.

Short-circuit current ratings are marked on equipment and assemblies (such as industrial control panels)—as required in 409.110(3)—to provide adequate protection of the equipment or assembly from damage in the event of a short-circuit situation.

Installers and enforcers are required by 110.10 to ensure that all over-current devices intended to interrupt current at fault levels have an adequate interrupting rating sufficient for the nominal operating voltage and current available at the line terminals of the equipment.

This new definition describes the determination of "short-circuit current rating" for equipment.

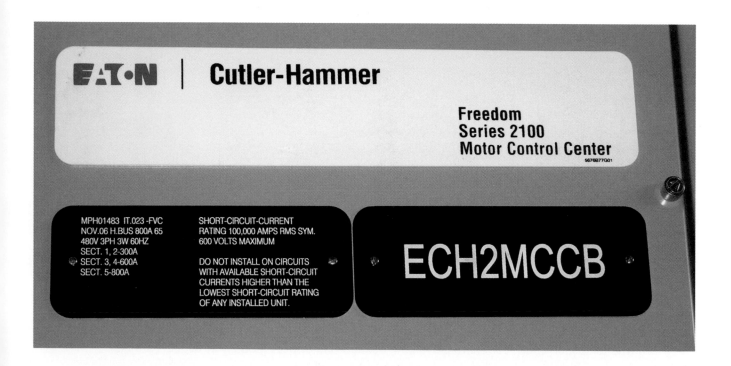

The National Electrical Code contains requirements for equipment to be marked with a *short-circuit current rating*.

Change Summary:

- The definition of *interactive system* is relocated to Article 100 from Article 705 as it applies in Articles 690, 692, and 705. A new definition of *utility-interactive inverter* is added to Article 100 as it applies in Articles 690, 692, and 705.

Revised 2008 NEC Text:

UTILITY-INTERACTIVE INVERTER.   An inverter intended for use in parallel with an electric utility to supply common loads that may deliver power to the utility.

INTERACTIVE SYSTEM.   An electric power production system that is operating in parallel with and capable of delivering energy to an electric primary source supply system.

Change Significance:

These definitions are added to Article 100 because these terms are used in more than one article. The definition of *interactive system* is relocated from Article 705. Similar definitions of *interactive system* existed in Articles 690 and 692 in 2005. The relocation of this definition to Article 100 is part of a reorganization effort to place the interconnection issues for alternative energy sources in Article 705.

A new definition of *utility-interactive inverter* is added to illustrate that this inverter can be used to directly supply loads from an alternative energy source or to supply power to the utility from an alternative energy source.

Article 100
DEFINITIONS OF *UTILITY-INTERACTIVE INVERTER* AND *INTERACTIVE SYSTEM*

Chapter 1 General

Article 100 Definitions

Part I General

Change Type: New

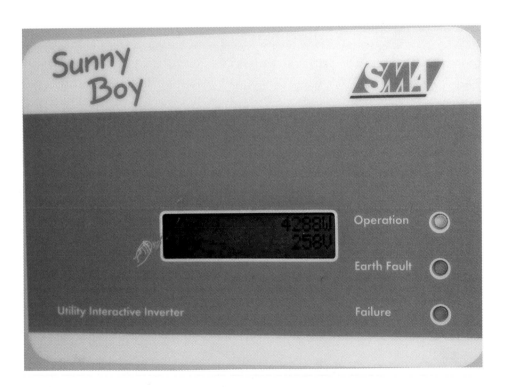

Utility interactive inverters employed in a solar photovoltaic installation, together with utility-supplied service equipment, represent an "interactive system."

110.11
DETERIORATING AGENTS

Chapter 1 General

Article 110 Requirements for Electrical Installations

Part I General

110.11 Deteriorating Agents

Change Type: Revision

Change Summary:

- Expansion of types of equipment requiring protection from weather.
- New FPN to reference National Electrical Manufacturer's Association (NEMA) markings.
- New 110.20 lists NEMA markings.

Revised 2008 NEC Text:

110.11 DETERIORATING AGENTS.   Unless identified for use in the operating environment, no conductors or equipment shall be located in damp or wet locations; where exposed to gases, fumes, vapors, liquids, or other agents that have a deteriorating effect on the conductors or equipment; or where exposed to excessive temperatures.

FPN No. 1: See 300.6 for protection against corrosion.

FPN No. 2: Some cleaning and lubricating compounds can cause severe deterioration of many plastic materials used for insulating and structural applications in equipment.

Equipment <u>not identified for outdoor use and equipment identified only for indoor use, such</u> as "dry locations," ~~"Type 1," or~~ "indoor use only," <u>"damp locations," or enclosure Types 1, 2, 5, 12, 12K and/or 13,</u> shall be protected against permanent damage from the weather during building construction.

<u>FPN No. 3: See Table 110.20 for appropriate enclosure-type designations.</u>

Change Significance:

The revised text now clarifies the types of equipment that must be protected from weather during building construction. All equipment not identified as being suitable for "outdoor use" will require some form of protection from weather during the course of construction. This protection can be provided by covering the equipment with tarps or plastic, enclosing it in plywood, or otherwise protecting it with an equivalent material. The building under construction may constitute adequate protection, provided the roof is complete and the equipment is not exposed to windblown rain or snow.

Equipment not marked "outdoor use" requires protection during construction.

NEMA types 1, 2, 5, 12, 12K, and 13 require protection during construction.

Electrical equipment must be identified for the environment in which it is installed.

Change Summary:

- List items of intended openings deleted. The term *equipment* is all-inclusive.
- All unused openings included, not just cable and raceway.
- Openings for mounting of equipment excluded.
- Openings necessary for operation (ventilation) excluded.

Revised 2008 NEC Text:

110.12(A) UNUSED OPENINGS. Unused ~~cable or raceway~~ openings, ~~in boxes, raceways, auxiliary gutters, cabinets, cutout boxes, meter socket enclosures, equipment cases or housings~~ <u>other than those intended for the operation of equipment, those intended for mounting purposes, or those permitted as part of the design for listed equipment,</u> shall be ~~effectively~~ closed to afford protection substantially equivalent to the wall of the equipment. Where metallic plugs or plates are used with nonmetallic enclosures, they shall be recessed at least 6 mm (¼ in.) from the outer surface of the enclosure.

Change Significance:

The revised text deletes the reference to openings for only "cables and raceways." There are many other types of openings that must be covered when no longer in use. These openings would include those used for devices and meters. The list of equipment has also been deleted. A list used in the NEC is always non-inclusive due to the inability to predict the use of new equipment, installation methods, or types of abuse. The term *equipment* as used in the revised text is all-inclusive.

Openings required for the installation of the equipment (such as mounting holes and ventilation openings) are now specifically permitted.

Unused openings created by the removal of circuit breakers or switches in panel boards and switchboards are required to be closed by text (new in the 2005 NEC) in Section 408.7.

110.12
MECHANICAL EXECUTION OF WORK, (A) UNUSED OPENINGS

Chapter 1 General

Article 110 Requirements for Electrical Installations

Part I General

110.12 Mechanical Execution of Work

(A) Unused Openings

Change Type: Revision

All unused openings in electrical equipment are required to be closed.

110.16
FLASH PROTECTION

Chapter 1 General

Article 110 Requirements for Electrical Installations

Part I General

110.16 Flash Protection

Change Type: Revision

Change Summary:
- Addition of the text "Electrical equipment such as" broadens the scope of labeling requirements.
- CMP-1 states that bus plugs, enclosed disconnect switches, enclosed circuit breakers, and transfer switches are now addressed.

Revised 2008 NEC Text:
110.16 FLASH PROTECTION. ~~Electrical equipment such as s~~Switchboards, panelboards, industrial control panels, meter socket enclosures, and motor control centers that are in other than dwelling occupancies and are likely to require examination, adjustment, servicing, or maintenance while energized shall be field marked to warn qualified persons of potential electric arc flash hazards. The marking shall be located so as to be clearly visible to qualified persons before examination, adjustment, servicing, or maintenance of the equipment.

   FPN No. 1: NFPA 70E-2004, *Standard for Electrical Safety in the Workplace,* provides assistance in determining severity of potential exposure, planning safe work practices, and selecting personal protective equipment.

   FPN No. 2: ANSI Z535.4-1998, *Product Safety Signs and Labels,* provides guidelines for the design of safety signs and labels for application to products.

Change Significance:
The addition of the text "Electrical equipment such as" clarifies and clearly broadens the scope of 110.16. This new text now clearly includes *all* electrical equipment likely to require examination, adjustment, servicing, or maintenance while energized in its requirement that all such equipment be field marked with arc flash warnings. CMP-1 stated in its action to "accept in principle" proposal 1-87 that the inclusion of the text "Electrical equipment such as" met the intent of the submitter to include bus plugs, enclosed disconnect switches, enclosed circuit breakers, and transfer switches in the scope of 110.16.

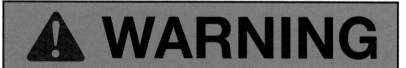

⚠ WARNING

Arc Flash Hazard.
Appropriate PPE Required.
Failure To Comply Can Result in Death or Injury.
Refer to NFPA 70 E.

THOMAS BETTS LB94913

All electrical equipment, in other than dwellings, in which equipment may be accessed in an energized state is required to be marked with this warning.

Change Summary:

- New selection chart for enclosures in all locations other than hazardous.
- Clarification of types of enclosures increases usability.
- Enclosure types listed require a number type marking.

Revised 2008 NEC Text:

110.20 ENCLOSURE TYPES. Enclosures (other than surrounding fences or walls) of switchboards, panelboards, industrial control panels, motor control centers, meter sockets, and motor controllers, rated not over 600 volts nominal and intended for such locations, shall be marked with an enclosure type number as shown in Table 110.20.

Table 110.20 shall be used for selecting these enclosures for use in specific locations other than hazardous (classified) locations. The enclosures are not intended to protect against conditions such as condensation, icing, corrosion, or contamination that may occur within the enclosure or enter via the conduit or unsealed openings.

110.20
ENCLOSURE TYPES

Chapter 1 General

**Article 110
Requirements for
Electrical Installations**

Part I General

110.20 Enclosure Types

Change Type: New

Table 110.20 Enclosure Selection

Provides a Degree of Protection Against the Following Environmental Conditions	For Outdoor Use									
	Enclosure Type Number									
	3	3R	3S	3X	3RX	3SX	4	4X	6	6P
Incidental contact with the enclosed equipment	X	X	X	X	X	X	X	X	X	X
Rain, snow, and sleet	X	X	X	X	X	X	X	X	X	X
Sleet*	=	=	X	=	=	X	=	=	=	=
Windblown dust	X	=	X	X	=	X	X	X	X	X
Hosedown	=	=	=	=	=	=	X	X	X	X
Corrosive agents	=	=	=	X	X	X	=	X	=	X
Temporary submersion	=	=	=	=	=	=	=	=	X	X
Prolonged submersion	=	=	=	=	=	=	=	=	=	X

(continued)

Table 110.20 Enclosure Selection—*cont'd*

	For Indoor Use									
Provides a Degree of Protection Against the Following Environmental Conditions	Enclosure Type Number									
	1	*2*	*4*	*4X*	*5*	*6*	*6P*	*12*	*12K*	*13*
Incidental contact with the enclosed equipment	X	X	X	X	X	X	X	X	X	X
Falling dirt	X	X	X	X	X	X	X	X	X	X
Falling liquids and light splashing	—	X	X	X	X	X	X	X	X	X
Circulating dust, lint, fibers, and flyings	—	—	X	X	—	X	X	X	X	X
Settling airborne dust, lint, fibers, and flyings	—	—	X	X	X	X	X	X	X	X
Hosedown and splashing water	—	—	X	X	—	X	X	—	—	—
Oil and coolant seepage	—	—	—	—	—	—	—	X	X	X
Oil or coolant spraying and splashing	—	—	—	—	—	—	—	—	—	X
Corrosive agents	—	—	—	X	—	—	X	—	—	—
Temporary submersion	—	—	—	—	—	X	X	—	—	—
Prolonged submersion	—	—	—	—	—	—	X	—	—	—

*Mechanism shall be operable when ice covered.

FPN: The term *raintight* is typically used in conjunction with Enclosure Types 3, 3S, 3SX, 3X, 4, 4X, 6, 6P. The term *rainproof* is typically used in conjunction with Enclosure Type 3R, 3RX. The term watertight is typically used in conjunction with Enclosure Types 4, 4X, 6, 6P. The term *driptight* is typically used in conjunction with Enclosure Types 2, 5, 12, 12K and 13. The term *dusttight* is typically used in conjunction with Enclosure Types 3, 3S, 3SX, 3X, 5, 12, 12K and 13.

(continued)

Change Significance:

This table is relocated from 430.91 in Chapter 4 to a new section in Article 110 of Chapter 1. Modified text now applies this table to the following enclosures (rated not over 600 volts nominal):

- Switchboards
- Panelboards
- Industrial control panels
- Motor control centers
- Meter sockets
- Motor controllers

All of the previously listed enclosure types are currently required by industry standards to use a "type-number" marking. This marking of enclosures is now referenced in Chapter 1 to clarify permitted equipment in locations other than hazardous.

This new section and table provide a baseline for installers and enforcers to determine the proper enclosure rating for all locations except "hazardous locations."

It is important to note that the application of these type-number markings in given locations is not intended to protect against conditions such as condensation, icing, corrosion, or contamination that may occur within the enclosure or enter via the conduit or unsealed openings. Relocation of the enclosure, sealing, heating, or other methods may be required to prevent these conditions.

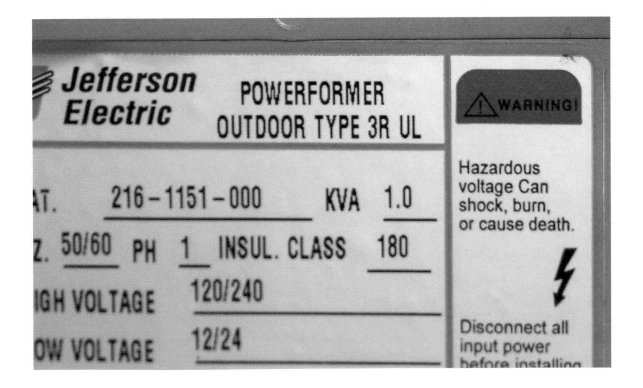

All electrical enclosures are to be marked with an enclosure-type number or NEMA rating. Table 110.20 provides the code user with permitted applications of enclosures by referencing the enclosure type number.

110.22

IDENTIFICATION OF DISCONNECTING MEANS

Chapter 1 General

Article 110 Requirements for Electrical Installations

Part I General

110.22 Identification of Disconnecting Means
(A) General
(B) Engineered Series Combination Systems
(C) Tested Series Combination Systems

Change Type: Revision

Change Summary:
- Series-rated systems selected under engineering supervision are now recognized in Chapter 1 for marking requirements.

Revised 2008 NEC Text:

110.22 IDENTIFICATION OF DISCONNECTING MEANS.

(A) GENERAL. Each disconnecting means shall be legibly marked to indicate its purpose unless located and arranged so the purpose is evident. The marking shall be of sufficient durability to withstand the environment involved.

(B) ENGINEERED SERIES COMBINATION SYSTEMS. Where circuit breakers or fuses are applied in compliance with series combination ratings selected under engineering supervision and marked on the equipment as directed by the engineer, the equipment enclosure(s) shall be legibly marked in the field to indicate the equipment has been applied with a series combination rating. The marking shall be readily visible and state the following: CAUTION—ENGINEERED SERIES COMBINATION SYSTEM RATED _____ AMPERES IDENTIFIED REPLACEMENT COMPONENTS REQUIRED.

FPN: See 240.86(A) for engineered series combination systems.

(continued)

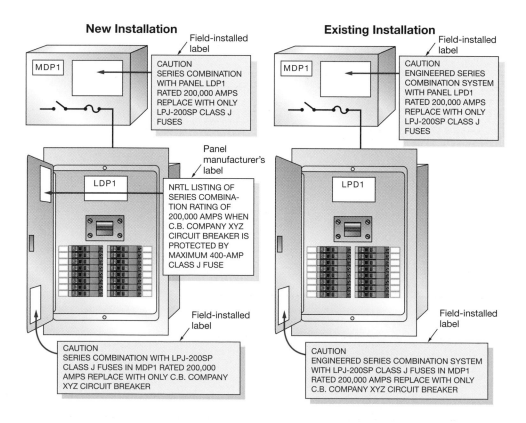

ESC systems permitted in 240.86(A) are required to have field markings applied in accordance with 110.22(B).

(C) TESTED SERIES COMBINATION SYSTEMS.   Where circuit breakers or fuses are applied in compliance with the series combination ratings marked on the equipment by the manufacturer, the equipment enclosure(s) shall be legibly marked in the field to indicate the equipment has been applied with a series combination rating. The marking shall be readily visible and state the following:

CAUTION—SERIES COMBINATION SYSTEM RATED _____ AMPERES IDENTIFIED REPLACEMENT COMPONENTS REQUIRED.

FPN: See 240.86(B) for tested series combination systems. ~~interrupting rating marking for end-use equipment.~~

Change Significance:

110.22 requires marking of all disconnecting means and provides for marking requirements for series-rated systems. This section is now separated into two first-level subdivisions for clarity. It is imperative that replaced overcurrent devices in a series-rated system be identified for such use. Tested combinations are covered by the existing marking requirements. The new marking requirement is included to identify a series-rated system selected under engineering supervision.

Notes:

_____

_____

_____

_____

_____

_____

_____

_____

_____

110.26
SPACES ABOUT ELECTRICAL EQUIPMENT, (C) ENTRANCE TO <u>AND EGRESS FROM</u> WORKING SPACE

Chapter 1 General

Article 110 Requirements for Electrical Installations

Part II 600 Volts, Nominal or Less

110.26 Spaces About Electrical Equipment

(C) Entrance to and Egress from Working Space

(1) Minimum Required
(2) Large Equipment
(3) Personnel Doors

Change Type: Revision

Change Summary:
- There are three significant changes to 110.26(C)(2):
  - **(1)** Addition of the text "and Egress from" illustrates the safety-driven reasons for the requirements of 110.26(C).
  - **(2)** Panic hardware is now required for doors within 25 feet of the working space footprint.
  - **(3)** Large equipment addressed in (C)(2) must be more than 6 feet wide.

Revised 2008 NEC Text:
110.26 SPACES ABOUT ELECTRICAL EQUIPMENT
(C) ENTRANCE TO <u>AND EGRESS FROM</u> WORKING SPACE
(1) MINIMUM REQUIRED.   At least one entrance of sufficient area shall be provided to give access to <u>and egress from</u> working space about electrical equipment.

(2) LARGE EQUIPMENT.   For equipment rated 1200 amperes or more <u>and over 1.8 m (6 ft) wide</u> that contains overcurrent devices, switching devices, or control devices, there shall be one entrance to and egress from the required working space not less than 610 mm (24 in.) wide and 2.0 m (6½ ft) high at each end of the working space.

A single entrance to <u>and egress from</u> the required working space shall be permitted where either of the conditions in 110.26(C)(2)(a) or (C)(2)(b) is met.

   (a) *Unobstructed* <u>Egress</u> ~~Exit~~. Where the location permits a continuous and unobstructed way of <u>egress</u> ~~exit~~ travel, a single entrance to the working space shall be permitted.

(continued)

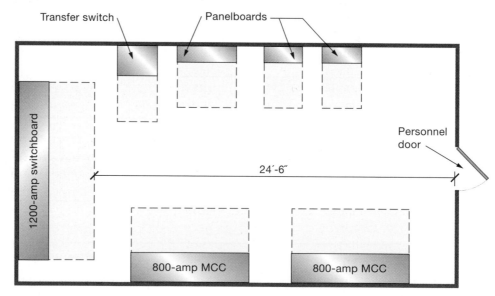

Personnel door shall open in the direction of egress and shall be equipped with panic hardware

Equipment rated at 1200 amps or more will trigger a requirement for doors that open in the direction of egress (with panic hardware) where the door is within 25 feet of the working space.

(b) *Extra Working Space.* Where the depth of the working space is twice that required by 110.26(A)(1), a single entrance shall be permitted. It shall be located so that the distance from the equipment to the nearest edge of the entrance is not less than the minimum clear distance specified in Table 110.26(A)(1) for equipment operating at that voltage and in that condition.

(3) PERSONNEL DOORS. Where equipment rated 1200A or more that contains overcurrent devices, switching devices, or control devices is installed and there is a personnel door(s) intended for entrance to and egress from the working space less than 7.6 m (25 ft) from the nearest edge of the working space, the door(s) shall open in the direction of egress and be equipped with panic bars, pressure plates, or other devices that are normally latched but open under simple pressure.

Change Significance:

The provisions of 110.26(C) supplement the working spaces required in 110.26(A) by requiring a means to access and egress from electrical equipment. These requirements apply to all electrical installations 600 volts nominal or less. Electrical hazards (including shock, arc flash, and arc blast) require that installations of electrical equipment provide adequate space about electrical equipment and safe means of access and egress.

NEC requirements for working space clearances have undergone several changes over the last few cycles in an effort to provide clarity and safe working spaces for installers and maintainers of electrical equipment. The addition of the text "and egress from" provides further clarity that these requirements are essential for the electrical worker not only to enter to the space but to get out quickly if an explosion or arc flash should occur.

There has been much debate over when panic hardware is required for doors of electrical rooms in accordance with 110.26(C)(2). The reason for panic hardware is to facilitate the immediate egress of the electrical equipment room by installers and maintainers in the event of an explosion or arc flash. This safety-driven requirement is essential for all persons accessing the electrical equipment room or area.

There were two areas needing clarification. First, does the size of the room matter? Where electrical equipment is included in a large mechanical room, the doors may be quite a distance from the equipment. Second, there was serious confusion about the location of the door with respect to the working space. Confusion existed because some installers and enforcers believed that if the footprint of the working space was not surrounded by an enclosure the provisions of 110.26(C)(2) did not apply. The requirement for panic hardware in 110.26(C)(2) has been clarified with a specific distance from the working space. If the door comes within 25 feet of the working space footprint, panic hardware is required.

The 6-foot limitation with respect to equipment has been reinserted as part of the limiting criteria for the application of 110.26(C)(2). CMP-1 reversed the position taken in the 2005 cycle of the NEC to remove the 6-foot requirement. The reason for replacing the 6-foot width requirement is due to the negative impact on equipment manufacturers.

110.26(G)

LOCKED ELECTRICAL EQUIPMENT ROOMS OR ENCLOSURES

Chapter 1 General

Article 110 Requirements for Electrical Installations

Part II 600 Volts, Nominal or Less

110.26 Spaces About Electrical Equipment

(G) Locked Electrical Equipment Rooms or Enclosures

Change Type: Revision

Change Summary:
- The last sentence of 110.26 is relocated to a new first-level subdivision for clarity and usability.

Revised 2008 NEC Text:
110.26: SPACES ABOUT ELECTRICAL EQUIPMENT
(G) LOCKED ELECTRICAL EQUIPMENT ROOMS OR ENCLOSURES.
<u>Electrical equipment rooms or</u> enclosures housing electrical apparatus that are controlled by a lock(s) shall be considered accessible to qualified persons.

Change Significance:
This new first-level subdivision provides clarity and usability. 110.26 is titled "Spaces About Electrical Equipment" and this requirement was previously buried in the last sentence of the parent text. Relocation as a new "(G) Locked Electrical Equipment Rooms or Enclosures" provides the code user with quick and easy-to-find access to this requirement. This provision recognizes that a locked door or enclosure is not considered to impede accessibility for qualified persons. In many occupancies, it is necessary to lock electrical equipment rooms to keep out unqualified persons.

Where a lock is used on an electrical equipment room, the room is still considered accessible to qualified persons.

Change Summary:
- FPN referencing the NESC has been deleted.
- High-voltage installations on the load side of the service point are governed by the NEC.

Revised 2008 NEC Text:
110.31: ENCLOSURE FOR ELECTRICAL INSTALLATIONS
(C) OUTDOOR INSTALLATIONS
(1) IN PLACES ACCESSIBLE TO UNQUALIFIED PERSONS. Outdoor electrical installations that are open to unqualified persons shall comply with Parts I, II, and III of Article 225.

~~FPN: For clearances of conductors for system voltages over 600 volts, nominal, see ANSI C2-2002, National Electrical Safety Code.~~

Change Significance:
The reference to the NESC in the FPN to 110.31(C)(1) has been deleted. All electrical installations on the load side of the service point are subject to the requirements of the NEC, not the NESC. This FPN has confused users of the NEC by inferring that clearances of conductors for system voltages greater than 600 volts nominal are located in the NESC. 110.31(C)(1) clearly requires that outdoor electrical installations greater than 600 volts comply with Parts I, II, and III of Article 225.

Section 225.60 provides the user of the NEC with minimum clear distances for conductors greater than 600 volts installed outdoors. In 225.60(C), which addresses "special cases," there is an FPN referring the user to the NESC if they need additional information.

110.31(C)
OUTDOOR INSTALLATIONS

Chapter 1 General

Article 110 Requirements for Electrical Installations

Part III Over 600 Volts, Nominal

110.31 Enclosure for Electrical Installations

(C) Outdoor Installations

(1) In Places Accessible to Unqualified Persons

Change Type: Revision

All electrical installations on the load side of the service point are covered by the NEC, not the NESC.

110.33
ENTRANCE TO ENCLOSURES AND ACCESS TO WORKING SPACE

(A) Entrance

Chapter 1 General

Article 110 Requirements for Electrical Installations

Part III Over 600 Volts, Nominal

110.33 Entrance to Enclosures and Access to Working Space

(A) Entrance

(1) Large Equipment

(2) Guarding

(3) Personnel Doors

Change Type: Revision

Change Summary:

- Title and first sentence of 110.33 are revised to clarify requirements for entering electrical equipment enclosures greater than 600 volts and for accessing working space.

- Panic hardware requirement for greater than 600 volts is clarified. Panic hardware is required where the door is within 25 feet of the working space footprint.

Equipment rated at greater than 600 volts will trigger a requirement for doors that open in the direction of egress (with panic hardware) where the door is within 25 feet of the working space.

Revised 2008 NEC Text:

110.33 ENTRANCE <u>TO ENCLOSURES</u> AND ACCESS TO ~~WORK~~ <u>WORKING</u> SPACE.

(A) Entrance.   At least one entrance <u>to enclosures for electrical installations as described in 110.31</u> not less than 610 mm (24 in.) wide and 2.0 m (6½ ft) high shall be provided to give access to the working space about electric equipment.

(1) LARGE EQUIPMENT.   (No change)

(2) GUARDING.   (No change)

(3) PERSONNEL DOORS.   <u>Where there is a personnel door(s) intended for entrance to and egress from the working space less than 7.6 m (25 ft) from the nearest edge of the working space, the door(s) shall open in the direction of egress and be equipped with panic bars, pressure plates, or other devices that are normally latched but open under simple pressure.</u>

(continued)

Change Significance:

The revision of the title and first sentence clarifies that Section 110.33 addresses two distinct areas: "Entrance to Enclosures" and "Access to Working Space." It has been misconstrued by some for many years that this requirement dealt only with the working space. Confusion existed because some installers and enforcers believed that if the footprint of the working space was not surrounded by an enclosure the provisions of 110.33 did not apply. The new title of this section clarifies that the enclosure must contain an entrance in accordance with 110.33(A), which gives access to the working space required in 110.34.

The reason for panic hardware is to facilitate the immediate egress of electrical enclosure installers and maintainers in the event of an explosion or arc flash. This safety-driven requirement is essential for all persons accessing the electrical enclosure or area.

Relocation of the last sentence of 110.33(A) into a new second-level subdivision (3) clarifies requirements for personnel doors. If the door comes within 25 feet of the working space footprint, panic hardware is required.

Notes:

_____

_____

_____

_____

_____

_____

_____

_____

_____

_____

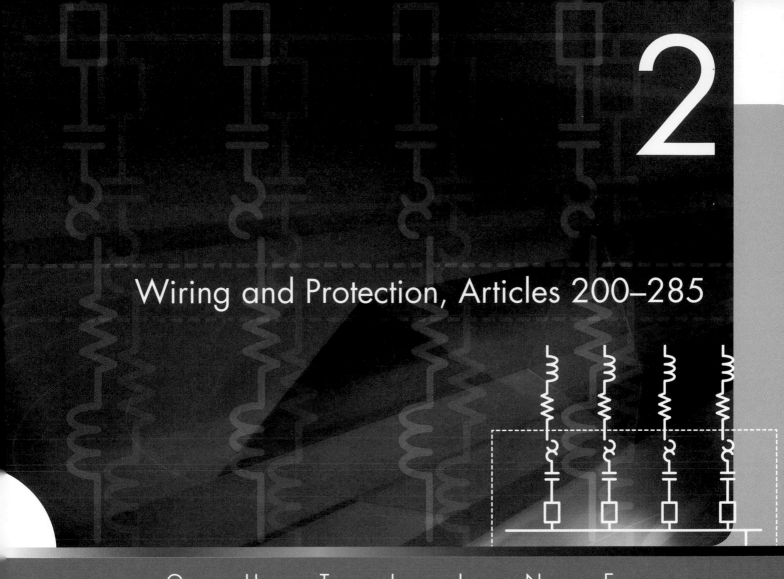

Wiring and Protection, Articles 200–285

OUTLINE

200.2(B)
CONTINUITY

Chapter 2 Wiring and Protection

Article 200 Use and Identification of Grounded Conductors

200.2 General

(A) Insulation

(B) Continuity

Change Type: Revision and New

Change Summary:

- 200.2 is editorially revised into a new first-level subdivision (A) titled "Insulation" to address the existing requirements.

- A new requirement is added in first-level subdivision (B) titled "Continuity" to require that the continuity of grounded conductors does not depend on a connection to a metallic enclosure, raceway, or cable armor.

Revised 2008 NEC Text:

200.2 GENERAL.   All premises wiring systems, other than circuits and systems exempted or prohibited by 210.10, 215.7, 250.21, 250.22, 250.162, 503.155, 517.63, 668.11, 668.21, and 690.41 Exception, shall have a grounded conductor that is identified in accordance with 200.6. The grounded conductor shall comply with (A) and (B).

(A) INSULATION.   The grounded conductor, where insulated, shall have insulation that is (1) suitable, other than color, for any ungrounded conductor of the same circuit on circuits of less than 1000 volts or impedance grounded neutral systems of 1 kV and over, or (2) rated not less than 600 volts for solidly grounded neutral systems of 1 kV and over as described in 250.184(A).

(B) CONTINUITY.   The continuity of a grounded conductor shall not depend on a connection to a metallic enclosure, raceway, or cable armor.

Change Significance:

This section is editorially revised into a new first-level subdivision (A) to contain existing requirements for the insulation of grounded conductors. A new first-level subdivision (B) will now prohibit an enclosure, such as a cabinet enclosing a panelboard, from being utilized as a portion of the grounded conductor circuit. A panelboard installed as service equipment, in accordance with the NEC, permits multiple points of termination for grounded conductors (for example, on separate terminal bars within the panelboard enclosure). Each terminal bar in some cases is then secured to the panelboard enclosure, which becomes part of the circuit. This new requirement will now require that continuity of the grounded conductor not depend on a connection to a metallic enclosure, raceway, or cable armor. For example, where multiple termination points exist they shall be securely connected with a properly sized ground conductor.

The continuity of a grounded conductor is not permitted to be dependent on connection to a metal enclosure, raceway, or cable armor.

Change Summary:

- Clarification: all multiwire branch circuits must originate from the same panelboard or distribution equipment.
- Simultaneous means of disconnect now required for all ungrounded conductors of a multiwire branch circuit.

210.4
MULTIWIRE BRANCH CIRCUITS

Chapter 2 Wiring and Protection

Article 210 Branch Circuits

Part I General Provisions

210.4 Multiwire Branch Circuits

(A) General

(B) Disconnecting Means

Change Type: Revision

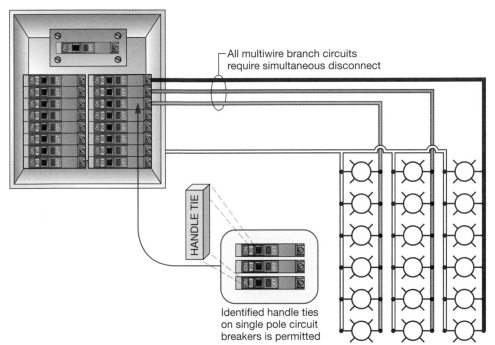

All multiwire branch circuits require simultaneous disconnect

HANDLE TIE

Identified handle ties on single pole circuit breakers is permitted

All multiwire branch circuits now require a simultaneous means of disconnect at the source of supply.

Revised 2008 NEC Text:

210.4 MULTIWIRE BRANCH CIRCUITS

(A) GENERAL.   Branch circuits recognized by this article shall be permitted as multiwire circuits. A multiwire circuit shall be permitted to be considered as multiple circuits. All conductors <u>of a multiwire branch circuit</u> shall originate from the same panelboard or similar distribution equipment.

FPN: A 3-phase, 4-wire, wye-connected power system used to supply power to nonlinear loads may necessitate that the power system design allow for the possibility of high harmonic <u>currents on the</u> neutral-<u>conductor</u> ~~currents~~.

(B) DEVICES OR EQUIPMENT.   ~~Where a multiwire branch circuit supplies more than one device or equipment on the same yoke, a means shall be provided to disconnect simultaneously all ungrounded conductors supplying those devices or equipment at the point where the branch circuit originates.~~

(B) DISCONNECTING MEANS.   <u>Each multiwire branch circuit shall be provided with a means that will simultaneously disconnect all ungrounded conductors at the point where the branch circuit originates.</u>

(continued)

Change Significance:

210.4 addresses multiwire branch circuits. The NEC defines the term *multiwire branch circuit* in Article 100 as follows:

ARTICLE 100
BRANCH CIRCUIT, MULTIWIRE. A branch circuit that consists of two or more ungrounded conductors that have a voltage between them, and a grounded conductor that has equal voltage between it and each ungrounded conductor of the circuit and that is connected to the neutral or grounded conductor of the system.

The last sentence of 210.4(A) is editorially revised to more clearly illustrate that all conductors of a *multiwire branch circuit* are required to originate in the same equipment.

The existing text of first-level subdivision (B)—as well as the title, "Devices or Equipment"—has been deleted. This deleted subdivision previously addressed only "devices or equipment, on the same yoke" supplied by a *multiwire branch circuit.* As written in the 2005 NEC, this deleted text required that where a *multiwire branch circuit* supplied more than one device or equipment on the same yoke a means of simultaneous disconnect was required for the ungrounded conductors. For example, where a duplex receptacle had the tab removed allowing it to be supplied from a *multiwire branch circuit* the two single-pole 20-amp circuit breakers involved required an identified handle tie. This means of simultaneous disconnect was required to ensure that when one of the circuits supplying the device or equipment for maintenance was opened for maintenance all ungrounded conductors supplying the device or equipment were simultaneously opened.

The safety-driven provisions of the previous code text in 210.4(B) have been expanded to require a simultaneous means of disconnect for all ungrounded conductors of all *multiwire branch circuits.* The new title of 210.2(B), "Disconnecting Means," now applies to all *multiwire branch circuits*—not just those supplying more than one device or equipment on the same yoke. The substantiation provided for this change references an occurrence known as an "open neutral" situation. Open neutrals on *multiwire branch circuits* represent a very dangerous situation for electrical workers. For example, an electrician may deenergize a single 20-amp branch circuit supplying a 277-volt lighting fixture to replace a ballast. There would be no indication at the panelboard that the circuit that was deenergized was part of a multiwire branch circuit.

As the electrician begins to work on the now "deenergized" circuit, any action by the electrician himself or by other work in the occupancy to "open a neutral connection" can result in the "neutral conductor" at the deenergized fixture location being energized at line voltage. Many fatalities and serious injuries have occurred to "qualified" persons from open neutrals. This new text is safety driven to ensure that *all* ungrounded conductors of *multiwire branch circuits* are simultaneously disconnected when any circuit of the *multiwire branch circuit* is opened for maintenance.

Change Summary:

- New requirement mandates that all multiwire branch circuits be identified by grouping all associated current-carrying conductors at their source.
- Marking is not required where the grouping is obvious.
- Installer/maintainer will be able to quickly identify grounded and ungrounded conductors that are part of a multiwire branch circuit.

210.4(D)
GROUPING

Chapter 2 Wiring and Protection

Article 210 Branch Circuits

Part I General Provisions

210.4 Multiwire Branch Circuits

(D) Grouping

Change Type: New

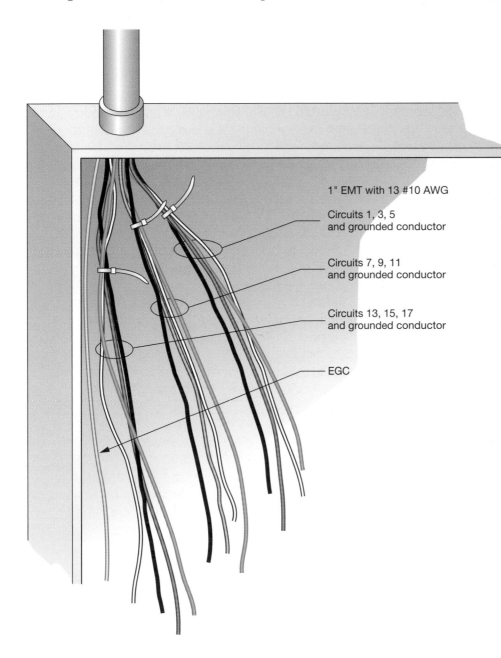

1" EMT with 13 #10 AWG

Circuits 1, 3, 5
and grounded conductor

Circuits 7, 9, 11
and grounded conductor

Circuits 13, 15, 17
and grounded conductor

EGC

Multiwire branch circuits must be grouped in a panelboard or other enclosure to identify the corresponding grounded and ungrounded conductors of the circuit.

(continued)

Revised 2008 NEC Text:
210.4 MULTIWIRE BRANCH CIRCUITS
(D) GROUPING.   The ungrounded and grounded conductors of each multiwire branch circuit shall be grouped by wire ties or similar means in at least one location within the panelboard or other point of origination.

Exception: The requirement for grouping shall not apply if the circuit enters from a cable or raceway unique to the circuit that makes the grouping obvious.

Change Significance:
The revision of 210.4 for multiwire branch circuits is expanded with a new requirement to identify all current-carrying conductors of a multiwire branch circuit at the panelboard or other point of origination. Where a raceway enters a panelboard or other distribution equipment containing many conductors, it can be almost impossible to determine the associated grouping of multiwire branch circuits unless they are grouped when installed.

This new requirement for grouping can be wire ties, ty wraps, tape, or other similar means. The intent of this new requirement is twofold. First, this will now allow for the electrical worker performing maintenance, renovation, or troubleshooting in the panelboard to readily identify multiwire branch circuits. Second, the inspector will be able to trace all mutiwire branch circuits to their source to determine if the provisions for simultaneous disconnection required by 210.4(B) have been met. Where the grouping is obvious—such as a multiwire branch circuit consisting of a single 12/3-type cable assembly, or a raceway containing a single two wire branch circuit—identification is not necessary.

Notes:

_____

_____

_____

_____

_____

_____

_____

_____

Change Summary:
- All ungrounded conductors addressed by 210.5(C) are required to be identified by phase and system.
- This identification shall be at all termination, connection, and splice points.
- Identification means may be documented in a readily available location or posted at each panelboard or distribution point.

Revised 2008 NEC Text:
210.5 IDENTIFICATION FOR BRANCH CIRCUITS
(C) UNGROUNDED CONDUCTORS. Where the premises wiring system has branch circuits supplied from more than one nominal voltage system, each ungrounded conductor of a branch circuit ~~where accessible,~~ shall be identified by <u>phase or line</u> and system <u>at all termination, connection and splice points.</u> The means of identification shall be permitted to be by separate color coding, marking tape, tagging, or other approved means. <u>The method utilized for conductors originating within each branch-circuit panelboard or similar branch-circuit distribution equipment shall be documented in a manner that is readily available or</u> shall be permanently posted at each branch-circuit panelboard or similar branch-circuit distribution equipment.

Change Significance:
During the 2005 NEC cycle, the term *phase* was inadvertently omitted from the identification requirement. The intent of this requirement is to be able to distinguish between the different systems *and* between the different phases of the system. Identification of ungrounded conductors where more than one nominal voltage system exists must be by phase and system to help prevent overloading of grounded conductors when changes are made to the wiring system.

With reference to the required location of the actual marking of ungrounded conductors, the language "where accessible" has been deleted. As written in the 2005 NEC, the marking was required at all "accessible" locations, including conduit fittings and pull boxes. The revised text specifies the locations where the ungrounded branch-circuit conductors shall be identified. Terminations, connections, and splices are the critical locations where the marking is needed.

The means of identification is now permitted to be documented in a readily available location or posted at each panelboard or distribution point. This revision is intended to recognize methods of identifying ungrounded conductors such as cable and conductor labeling/numbering systems used in industrial locations.

Notes:

_____

_____

_____

210.5(C)
UNGROUNDED CONDUCTORS

Chapter 2 Wiring and Protection

Article 210 Branch Circuits

Part I General Provisions

210.5 Identification for Branch Circuits

(C) Ungrounded Conductors

Change Type: Revision

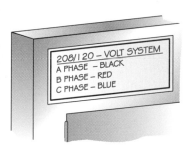

208/120 – VOLT SYSTEM
A PHASE – BLACK
B PHASE – RED
C PHASE – BLUE

Ungrounded branch-circuit conductors must be identified by phase and system at all termination, connection, and splice points where more than one nominal voltage exists in an occupancy.

210.6(D)
600 VOLTS BETWEEN CONDUCTORS

Chapter 2 Wiring and Protection

Article 210 Branch Circuits

Part I General Provisions

210.6 Branch-Circuit Voltage Limitations

(D) 600 Volts Between Conductors

Change Type: New

Change Summary:
- A new list item (3) is added to 210.6(D) permitting luminaires powered from DC systems at voltages above 277 volts to ground and not exceeding 600 volts between conductors.

Revised 2008 NEC Text:
210.6 BRANCH-CIRCUIT VOLTAGE LIMITATIONS.   The nominal voltage of branch circuits shall not exceed the values permitted by 210.6(A) through 210.6(E).

(D) 600 VOLTS BETWEEN CONDUCTORS.   Circuits exceeding 277 volts, nominal, to ground and not exceeding 600 volts, nominal, between conductors shall be permitted to supply the following:
<u>(3) Luminaires powered from direct-current systems where the luminaire contains a listed, dc-rated ballast that provides isolation between the dc power source and the lamp circuit and protection from electric shock when changing lamps.</u>

Change Significance:
This revision will now permit fluorescent lighting in commercial and industrial buildings to be powered directly with DC power or through a DC-assisted lighting system. DC-assisted lighting systems work by supplying rectified AC power to new and existing lighting installations, with additional power provided (directly via diode coupling) from DC sources such as photovoltaics, fuel cells, and wind generators.

DC lighting systems of this type have been researched and have been operating reliably for more than 20 years. The first such system (a photovoltaic-assisted lighting system in a department store in Massachusetts) was installed by the University of Massachusetts and is still operating.

The new text requires that the listed DC-rated ballast must provide isolation between the DC power source and the lamp circuit and protection from electric shock when changing lamps.

Luminaires are permitted to be supplied from a DC system, at voltages up to 600 volts, where installed in accordance with new 210.6(D)(3).

Change Summary:

- All 125-volt single-phase 15- and 20-amp receptacles installed in garages must now be GFCI protected.

- All 125-volt single-phase 15- and 20-amp receptacles installed in unfinished basements, except those supplying a permanently installed fire alarm or burglar alarm systems, must now be GFCI protected.

- New fine print note (FPN) reference requirements in Article 760 prohibiting ground-fault circuit-interrupter/arc-fault circuit-interrupter (GFCI/AFCI) protection of branch circuits supplying fire alarm systems.

210.8(A)
DWELLING UNITS

Chapter 2 Wiring and Protection

Article 210 Branch Circuits

Part I General Provisions

210.8 Ground-Fault Circuit-Interrupter Protection for Personnel

(A) Dwelling Units

(2) Garages

(5) Unfinished basements

Change Type: Revision

Unfinished basement in a dwelling unit

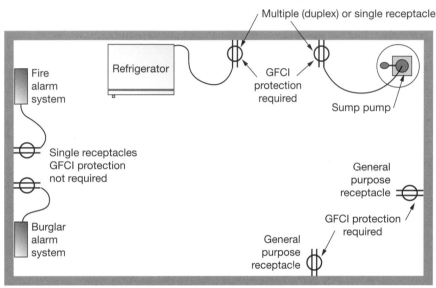

All receptacles installed in dwelling unit basements (unfinished) and garages are now required to be GFCI protected unless they supply a burglar or fire alarm system.

Revised 2008 NEC Text:
210.8 GROUND-FAULT CIRCUIT-INTERRUPTER PROTECTION FOR PERSONNEL

(A) DWELLING UNITS. All 125-volt, single-phase, 15- and 20-ampere receptacles installed in the locations specified in (1) through (8) shall have ground-fault circuit-interrupter protection for personnel.

(2) Garages, and also accessory buildings that have a floor located at or below grade level not intended as habitable rooms and limited to storage areas, work areas, and areas of similar use.

~~Exception No. 1 to (2): Receptacles that are not readily accessible.~~

~~Exception No. 2 to (2): A single receptacle or a duplex receptacle for two appliances located within dedicated space for each appliance that, in normal use, is not easily moved from one place to another and that is cord-and-plug connected in accordance with 400.7(A)(6), (A)(7), or (A)(8).~~

~~Receptacles installed under the exceptions to 210.8(A)(2) shall not be considered as meeting the requirements of 210.52(G).~~

(continued)

(5) Unfinished basements — for purposes of this section, unfinished basements are defined as portions or areas of the basement not intended as habitable rooms and limited to storage areas, work areas, and the like.

~~Exception No. 1 to (5): Receptacles that are not readily accessible.~~

~~Exception No. 2 to (5): A single receptacle or a duplex receptacle for two appliances located within dedicated space for each appliance that, in normal use, is not easily moved from one place to another and that is cord- and plug-connected in accordance with 400.7(A)(6), (A)(7), or (A)(8).~~

Exception ~~No. 3~~ to (5): A receptacle supplying only a permanently installed fire alarm or burglar alarm system shall not be required to have ground-fault circuit-interrupter protection.

FPN: See 760.41(B) and 760.121(B) for power supply requirements for fire alarm systems.

Receptacles installed under the exceptions to 210.8(A)(5) shall not be considered as meeting the requirements of 210.52(G).

Change Significance:

During the 2005 NEC cycle, the revision to expand the provisions of 210.8(A)(7) to include laundry and utility sinks created confusion and contradiction for installers and enforcers. Where receptacle outlets supplying loads such as sump pumps, refrigerators, or freezers were within 6 feet of a laundry or utility sinks the provisions of 210.8(A)(7) seemed to conflict with exceptions numbers 1 and 2 to 210.8(A)(2) and (5). The intent of this change has been reinforced and further clarified with the deletion of the exceptions. All 125-volt single-phase 15- and 20-amp receptacles installed in garages must now be GFCI protected.

All 125-volt single-phase 15- and 20-amp receptacles installed in unfinished basements, except those supplying a permanently installed fire alarm or burglar alarm systems, must now be GFCI protected. The technical committee has recognized that equipment such as sump pumps, refrigerators, freezers, garage door openers, and laundry equipment are compatible with GFCI protection. The permitted leakage current for listed equipment is far below the 4- to 6-milliamp threshold for activation of a GFCI device.

The exceptions for receptacles not readily accessible have also been deleted. The application of the term *readily accessible* for receptacles in basements and garages in dwelling units is subjective. Although a receptacle may not be readily accessible to a person 5 feet tall, it may be readily accessible to a person 6 feet tall. A new FPN to 210.8(A)(5) refers the code user to provisions in Article 760 that prohibit GFCI/AFCI protection of branch circuits supplying fire alarm systems.

Notes:

Change Summary:
- All occupancies—including, commercial, institutional, and nonresidential—with a food preparation area meeting the Article 100 definition of "kitchen" must provide GFCI protection for all 125-volt single-phase 15- and 20-amp receptacles installed in the kitchen.

Revised 2008 NEC Text:
210.8 GROUND-FAULT CIRCUIT-INTERRUPTER PROTECTION FOR PERSONNEL

(B) OTHER THAN DWELLING UNITS. All 125-volt, single-phase, 15- and 20-ampere receptacles installed in the locations specified in (1) through (5) shall have ground-fault circuit-interrupter protection for personnel:

(2) ~~Commercial and institutional~~ Kitchens — ~~for the purposes of this section, a kitchen is an area with a sink and permanent facilities for food preparation and cooking~~

Change Significance:
This new title of "Kitchen" will now clearly include the break rooms in commercial and other nonresidential occupancies where the new Article 100 definition of "Kitchen" is met as follows:

KITCHENS. An area with a sink and permanent facilities for food preparation and cooking.

The 2005 NEC text simply addressed commercial and institutional kitchens. The term *commercial kitchen* is to some degree subjective and interpretations may differ widely depending on the opinion of the installer and the authority having jurisdiction.

A kitchen in a fire house may not be considered a commercial or institutional kitchen due to the fact that food is not sold or prepared for anyone other than the firemen themselves. Break rooms in commercial occupancies may also not be considered commercial or institutional kitchens due to the fact that food is brought by individuals for their own consumption. The shock hazards are the same in any of these occupancies. Where a room or area meets the definition of "Kitchen" as defined in Article 100 GFCI, protection for personnel is now required.

210.8(B)
OTHER THAN DWELLING UNITS (2) KITCHENS

Chapter 2 Wiring and Protection

Article 210 Branch Circuits

Part I General Provisions

210.8 Ground-Fault Circuit-Interrupter Protection for Personnel

(B) Other Than Dwelling Units

(2) Kitchens

Change Type: Revision

All "kitchens" in other than dwelling units must provide GFCI protection for all 125-volt 15- and 20-amp receptacles.

210.8(B)
OTHER THAN DWELLING UNITS (4) OUTDOORS

Chapter 2 Wiring and Protection

Article 210 Branch Circuits

Part I General Provisions

210.8 Ground-Fault Circuit-Interrupter Protection for Personnel

(B) Other Than Dwelling Units

(4) Outdoors

Change Type: Revision

In general, all 125-volt 15- and 20-amp receptacles located outdoors of other than dwelling units require GFCI protection.

Change Summary:
- The requirement for GFCI protection for "other than dwelling units" have been expanded to all outdoor receptacles, not only those in areas for use by or accessible to the general public.
- A new exception exists for GFCI protection of outdoor receptacles where an assured equipment grounding conductor program is implemented in an industrial establishment.

Revised 2008 NEC Text:
210.8 GROUND-FAULT CIRCUIT-INTERRUPTER PROTECTION FOR PERSONNEL

(B) OTHER THAN DWELLING UNITS. All 125-volt, single-phase, 15- and 20-ampere receptacles installed in the locations specified in (1) through (5) shall have ground-fault circuit-interrupter protection for personnel:

(4) Outdoors ~~in public spaces — for the purpose of this section a public space is defined as any space that is for use by, or is accessible to, the public~~

Exception <u>No. 1</u> to (3) and (4): Receptacles that are not readily accessible and are supplied from a dedicated branch circuit for electric snow-melting or deicing equipment shall be permitted to be installed <u>without GFCI protection.</u> ~~in accordance with 426.28.~~

<u>Exception No. 2 to (4): In industrial establishments only, where conditions of maintenance and supervision ensure that only qualified personnel are involved, an assured equipment grounding conductor program as specified in 590.6(B)(2) shall be permitted for only those receptacle outlets used to supply equipment that would create a greater hazard if power is interrupted or having a design that is not compatible with GFCI protection.</u>

~~(5) Outdoors, where installed to comply with 210.63~~

Change Significance:
The requirement for GFCI protection of outdoor receptacles in other than dwelling units has been significantly expanded. The requirement for outdoor receptacles in other than dwelling units was new in the 2005 NEC and was limited to areas accessible by or for use by the general public. This language limited the installation of GFCI-protected receptacles in institutional, commercial, and industrial occupancies where the general public would not be present. The deleted text removes this limitation. The reference to 426.28 for protection of snow-melting or de-icing equipment has been removed from the first exception. This does not, however, permit these receptacles from being installed without GFP protection as required in Article 426.

A new exception has been added to list item "(4) Outdoors" to allow assured equipment grounding conductor programs in lieu of GFCI protection. This exception is limited to industrial establishments where the assured equipment grounding conductor program is implemented in accordance with 590.6(B).

The existing list item (5) is deleted due to the expansion of 210.8(B)(4). The expansion covers all outdoor receptacles, including those installed to meet the requirements of 210.63.

Change Summary:

- All 125-volt single-phase 15- and 20-amp receptacles installed within 6 feet of the outside edge of a nonresidential sink are now required to be provided with GFCI protection.
- Exceptions are provided for industrial laboratories and for patient care areas of health care facilities.

210.8(B)
OTHER THAN DWELLING UNITS (5) SINKS

Chapter 2 Wiring and Protection

Article 210 Branch Circuits

Part I General Provisions

210.8 Ground-Fault Circuit-Interrupter Protection for Personnel

(B) Other Than Dwelling Units

(5) Sinks

Change Type: New

College or high school lab table

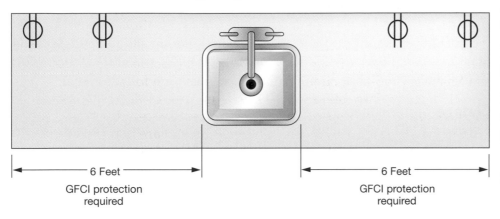

|←——— 6 Feet ———→| |←——— 6 Feet ———→|

GFCI protection required   GFCI protection required

Typical commercial mop closet

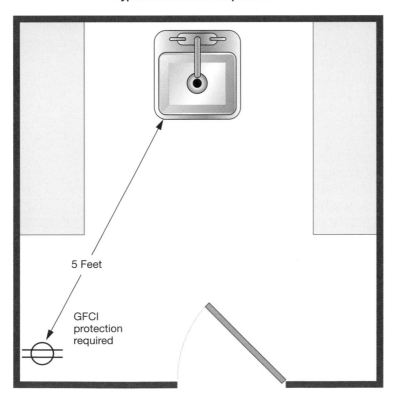

5 Feet

GFCI protection required

In general, all 125-volt 15- and 20-amp receptacles located within 6 feet of the outside edge of a sink in other than dwelling units require GFCI protection.

(continued)

Revised 2008 NEC Text:

210.8 GROUND-FAULT CIRCUIT-INTERRUPTER PROTECTION FOR PERSONNEL

(B) OTHER THAN DWELLING UNITS.   All 125-volt, single-phase, 15- and 20-ampere receptacles installed in the locations specified in (1) through (5) shall have ground-fault circuit-interrupter protection for personnel:

(5) Sinks—where receptacles are installed within 1.8 m (6 ft) of the outside edge of the sink.

Exception No. 1 to (5): In industrial laboratories, receptacles used to supply equipment where removal of power would introduce a greater hazard shall be permitted to be installed without GFCI protection.

Exception No. 2 to (5): For receptacles located in patient care areas of health care facilities other than those covered under 210.8(B)(1), GFCI protection shall not be required.

Change Significance:

This new GFCI requirement for "other than dwelling units" mirrors the requirement for dwelling units in 210.8(A)(7). All 125-volt single-phase 15- and 20-amp receptacles installed within 6 feet of the outside edge of a non-residential sink are now required to be provided with GFCI protection. The safety hazards are the same regardless of occupancy.

Exception No. 1 is provided for industrial laboratory facilities. A potential loss of power to critical lab equipment used to control, cool, heat, or interact with process reactions could result in a greater hazard. Loss of power to such equipment either from a nuisance or intentional GFCI trip may result in fire, explosion, or significant reaction hazards depending on the process.

Exception No. 2 is provided for sinks in a patient care area of a health care facility. Note that this exception does not apply to the bathroom located in a patient care area.

Notes:

_____

_____

_____

_____

_____

_____

Change Summary:

- Most 120-volt single phase 15- and 20-amp branch circuits in dwelling units shall have combination-type AFCI protection.
- Exception No. 1 has been modified allowing metal raceway or steel armored cable to supply an AFCI device at the first outlet.
- Two new FPNs provide informational references for smoke detectors and fire alarm systems.

Revised 2008 NEC Text:

210. 12 ARC-FAULT CIRCUIT-INTERRUPTER PROTECTION

(A) DEFINITION: ARC-FAULT CIRCUIT INTERRUPTER (AFCI). ~~An arc-fault circuit interrupter is a~~ <u>A</u> device intended to provide protection from the effects of arc faults by recognizing characteristics unique to arcing and by functioning to de-energize the circuit when an arc fault is detected.

(B) DWELLING UNITS. ~~BEDROOMS.~~ All 120-volt, single phase, 15- and 20-ampere branch circuits supplying outlets installed in dwelling unit <u>family rooms, dining rooms, living rooms, parlors, libraries, dens,</u> bedrooms<u>, sun rooms, recreation rooms, closets, hallways, or similar rooms or areas</u> shall be protected by a listed arc-fault circuit interrupter, combination type installed to provide protection of the branch circuit.

~~Branch/feeder AFCIs shall be permitted to be used to meet the requirements of 210.12(B) until January 1, 2008.~~

FPN <u>No. 1</u>: For information on types of arc-fault circuit interrupters, see UL 1699-1999, *Standard for Arc-Fault Circuit Interrupters.*

<u>FPN No. 2: See 11.6.3(5) of NFPA 72-2007, *National Fire Alarm Code*® for information related to secondary power supply requirements for smoke alarms installed in dwelling units.</u>

<u>FPN No. 3: See 760.41(B) and 760.121(B) for power supply requirements for fire alarm systems.</u>

~~Exception: The location of the arc-fault circuit interrupter shall be permitted to be at other than the origination of the branch circuit in compliance with (a) and (b):~~

~~(a) The arc-fault circuit interrupter installed within 1.8 m (6 ft) of the branch circuit overcurrent device as measured along the branch circuit conductors.~~

~~(b) The circuit conductors between the branch circuit overcurrent device and the arc-fault circuit interrupter shall be installed in a metal raceway or a cable with a metallic sheath.~~

Exception No. 1: Where RMC, IMC or EMT or steel armored cable, Type AC, meeting the requirements of 250.118 using metal outlet and junction boxes is installed for the portion of the branch circuit between the branch circuit overcurrent device and the first outlet, it shall be permitted to install a combination AFCI at the first outlet to provide protection for the remaining portion of the branch circuit.

Exception No. 2: Where a branch circuit to a fire alarm system installed in accordance with 760.41(B) and 760.121(B) is installed in RMC, IMC, EMT, or steel armored cable, type AC, meeting the requirements of 250.118, with metal outlet and junction boxes, AFCI protection shall be permitted to be omitted.

(continued)

210.12
ARC-FAULT CIRCUIT-INTERRUPTER PROTECTION

Chapter 2 Wiring and Protection

Article 210 Branch Circuits

Part I General Provisions

210.12 Arc-Fault Circuit-Interrupter Protection

(A) Definition Arc-Fault Circuit Interrupter (AFCI)

(B) Dwelling Units

Change Type: Revision

As of January 1, 2008, only "combination-type" AFCIs will be permitted.

Change Significance:

AFCI protection is now required for all 120-volt single-phase 15- and 20-amp branch circuits supplying outlets installed in dwelling-unit family rooms, dining rooms, living rooms, parlors, libraries, dens, bedrooms, sun rooms, recreation rooms, closets, hallways, or similar rooms or areas. This is not a "whole house" requirement. Documented use of AFCIs is only in the bedrooms of dwelling units. This expansion leaves out areas of the dwelling unit in which different types of electrical loads are used such as unfinished basements, garages, bathrooms, outdoors, and kitchens.

Attempts to go "whole house" AFCI in the 2005 NEC cycle made it all the way through the NFPA consensus process, including an appeal to the Standards Council, but ultimately failed. Part of the reluctance on the part of the technical committee was the continuing evolution of AFCI technology. The last sentence of 210.12(B) has been deleted, and now only "combination"-type AFCI protection is permitted. The previous text permitted branch/feeder or combination devices. A combination-type AFCI combines both "branch/feeder" and "outlet circuit" technologies. The UL descriptions of these devices follow.

- *Combination Type AFCI, UL Product Category AWAH*

Arc-Fault Circuit-Interrupters, Combination Type—An AFCI which complies with the requirements for both branch/feeder and outlet circuit AFCIs. It is intended to protect downstream branch circuit wiring and cord sets and power-supply cords.

- *Branch Feeder Type AFCI, UL Product category AVZQ*

Arc-Fault Circuit-Interrupters, Branch/Feeder Type—A device intended to be installed at the origin of a branch circuit or feeder, such as at a panelboard. It is intended to provide protection of the branch circuit wiring, feeder wiring, or both, against unwanted effects of arcing. This device also provides limited protection to branch circuit extension wiring. It may be a circuit-breaker type device or a device in its own enclosure mounted at or near a panelboard.

- *Outlet Circuit Type AFCI, UL Product category AWCG*

Arc-Fault Circuit-Interrupters, Outlet Circuit Type—A device intended to be installed at a branch circuit outlet, such as at an outlet box. It is intended to provide protection of cord sets and power-supply cords connected to it (when provided with receptacle outlets) against the unwanted effects of arcing. This device may provide feed-through protection of the cord sets and power-supply cords connected to downstream receptacles.

AFCI technology has proven itself reliable and effective, with millions of devices in service. The technical committee CMP-2, the Consumer Product Safety Commission, and the National Association of State Fire Marshals, along with many other individuals and organizations, have recognized that implementation of AFCI technology in dwelling units is absolutely necessary to save lives. Although some may argue about the additional cost, no one can put a price on life itself. Lives will be saved and tremendous savings will be realized in reduced fire loss and property damage.

The existing exception is deleted. Two new exceptions are added. Exception No. 1 now permits AFCI protection to be installed at the first outlet where the conductors are installed from the branch circuit overcurrent

(continued)

protective device to the outlet in RMC, IMC, or EMT or in steel-armored cable (type AC) meeting the requirements of 250.118 and using metal outlet and junction boxes.

Exception No. 2 allows for the omission of AFCI protection for branch circuit conductors serving a fire alarm system, provided the conductors are installed in RMC, IMC, EMT, or steel-armored cable (type AC) meeting the requirements of 250.118 and with metal outlet and junction boxes. To meet the requirements of 210.12(B) Exception No. 2 and 760.41(B) or 760.121(B), the branch circuit conductors supplying a fire alarm system must now be installed in a metal raceway or steel-jacketed cable assembly.

Two new FPNs have been added to 210.12 (B). The existing FPN is now listed as No. 1 and references UL 1699-1999 for information on types of AFCIs.

FPN No. 2 references 11.6.3(5) of NFPA 72-2007, National Fire Alarm Code, for information related to secondary power supply requirements for smoke alarms installed in dwelling units.

FPN No. 3 references 760.41(B) and 760.121(B) for power supply requirements for fire alarm systems. Both of these requirements prohibit "fire alarm systems" from being supplied through AFCI- or GFCI-type devices. It is important to note that these are *fire alarm systems* and not single- or multiple-station 120-volt smoke alarms. These are required to have battery backup, and in the event of a power loss due to the operation of a GFCI- or AFCI-type device are also required to provide the necessary level of protection.

The technical committee continues to require that single- and multiple-station smoke alarms be AFCI protected. These devices are powered by 120-volt branch circuits backed up with batteries and are compatible with AFCI-type devices. Smoke detectors powered through a fire alarm system are covered under the requirements of NFPA 72 and Article 760.

Existing dwelling units continue to be excluded from this requirement. The technical committee is concerned that problems will occur due to the wide variety of existing wiring configurations in existing dwelling units. As additional input on the compatibility of these wiring systems with AFCI protection becomes available, existing dwelling units may be included in future NEC cycles.

Notes:

210.25
BRANCH CIRCUITS IN BUILDINGS WITH MORE THAN ONE OCCUPANCY

Chapter 2 Wiring and Protection

Article 210 Branch Circuits

Part II Branch Circuit Ratings

210.25 Branch Circuits in Buildings With More Than One Occupancy

(A) Dwelling Unit Branch Circuits

(B) Common Area Branch Circuits

Change Type: Revision

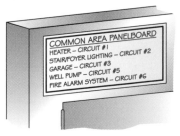

Dwelling unit branch circuits may only supply loads within or associated with that dwelling unit.

Change Summary:

- 210.25 has been revised/expanded to address common-area branch circuits in two or multifamily dwelling units and in multi-occupancy buildings of all types.

Revised 2008 NEC Text:

210.25 ~~COMMON AREA~~ BRANCH CIRCUITS IN <u>BUILDINGS WITH MORE THAN ONE OCCUPANCY</u>

(A) DWELLING UNIT BRANCH CIRCUITS. Branch circuits in <u>each</u> dwelling units shall supply only loads within that dwelling unit or loads associated only with that dwelling unit.

(B) COMMON AREA BRANCH CIRCUITS. Branch circuits required for the purpose of lighting, central alarm, signal, communications, or other needs for public or common areas of a two-family dwelling<u>, or multifamily dwelling, or a multi-occupancy building</u> shall not be supplied from equipment that supplies an individual dwelling unit <u>or tenant space</u>.

Change Significance:

210.25 has been renamed "Branch Circuits in Buildings With More Than One Occupancy" to represent the revised requirements included. Expansion of the requirements of 210.25 required that the text be separated in two first-level subdivisions. First-level subdivision (A), Dwelling Unit Branch Circuits, continues to require that branch circuits supplied from each dwelling unit supply loads within that dwelling unit or loads associated only with that dwelling unit. The word *each* has been added to prohibit grouping of branch circuits within several dwelling units.

First-level subdivision (B), Common Area Branch Circuits, expands the "common-area branch circuit" concept to all multi-occupancy buildings, including dwelling units. As written in the 2005 NEC, a multi-occupancy/tenant retail or commercial building may have branch circuits supplying common-area lighting, central alarm, signal, communications, or other needs for public or common areas connected to a single tenant/occupant meter. If the single tenant/occupant de-energizes the branch circuit(s) supplying the common need branch circuit(s) or leaves the space and the utility owned meter is removed, the power supply is lost. This potential loss of power represents a serious safety concern. The revised text now requires that all multi-occupancy buildings with branch circuits required for a common purpose—including lighting, central alarm, signal, communications, or other needs for public or common areas—supply those circuits from a common or house panel.

Notes:

_____

_____

_____

Change Summary:

- 210.52 has been clarified to require that all receptacles required shall not be:
  - Part of a lighting fixture or appliance
  - Controlled by a wall switch where permitted in 210.70(A)(1) Exception No. 1
  - Located within cabinets or cupboards
  - Located more than 5½ feet above the floor

Revised 2008 NEC Text:

210.52 DWELLING UNIT RECEPTACLE OUTLETS.   This section provides requirements for 125-volt, 15- and 20-ampere receptacle outlets. The ~~R~~ <u>re</u>ceptacle<u>s</u> ~~outlets~~ required by this section shall be in addition to any receptacle that is<u>: (1) P</u>art of a luminaire ~~(lighting fixture)~~ or appliance, <u>or (2) Controlled by a wall switch in accordance with 210.70(A)(1) Exception No. 1, or (3) L</u>ocated within cabinets or cupboards, <u>or (4) L</u>ocated more than 1.7 m (5½ ft) above the floor

Permanently installed electric baseboard heaters equipped with factory-installed receptacle outlets or outlets provided as a separate assembly by the manufacturer shall be permitted as the required outlet or outlets for the wall space utilized by such permanently installed heaters. Such receptacle outlets shall not be connected to the heater circuits.

FPN: Listed baseboard heaters include instructions that may not permit their installation below receptacle outlets.

Change Significance:

The previous text in 210.52 allowed a switched receptacle, such as those permitted in 210.70(A)(1) Exception No. 1 for lighting, to also be included in the receptacle spacing requirements of 210.52(A)(1). It is important to note that the Article 100 definition of receptacle describes it as a "contact" device installed at the outlet. A single receptacle has a single contact device. A duplex-type device is defined as a multiple receptacle with two contact devices.

A duplex-type receptacle with the tab removed—which is essentially two receptacles providing one contact device for general provisions and one contact device switched for lighting as per 210.70(A)(1)—is still permitted. The spacing requirements of 210.52(A)(1) require a "receptacle," not a duplex-type device. This new text provides clarification that any receptacle switched for lighting purposes as permitted in 210.70(A)(1) does not also meet the requirements of 210.52.

210.52
DWELLING UNIT RECEPTACLE OUTLETS

Chapter 2 Wiring and Protection

Article 210 Branch Circuits

Part III Required Outlets

210.52 Dwelling Unit Receptacle Outlets

Change Type: Revision

Receptacles controlled by a wall switch do not fulfill the requirements of 210.52.

210.52
DWELLING UNIT RECEPTACLE OUTLETS
(C) COUNTERTOPS

Chapter 2 Wiring and Protection

Article 210 Branch Circuits

Part III Required Outlets

210.52 Dwelling Unit Receptacle Outlets

(C) Countertops

(1) Wall Countertop Spaces

Exception

(2) Island Countertop Spaces

(3) Peninsular Countertop Spaces

Figure 210.52(C)(1) Determination of Area Behind a Range, or Counter-Mounted Cooking Unit or Sink

Change Type: Revision

Change Summary:

- Figure 210.52 has been modified to clarify that the space directly behind a wall-countertop-mounted cooking unit or sink is exempt from the wall line measurement in 210.52(C)(1).

- The term *rangetop* has been deleted to clarify the intent of referring to a counter-mounted cooking unit, range, or sink.

- New text clarifies that a peninsular countertop with a range-, sink-, or counter-mounted cooking unit installed may create an island countertop.

- The provisions of 210.52(C) are expanded to include countertops in pantries, breakfast rooms, and similar areas in addition to kitchens and dining rooms.

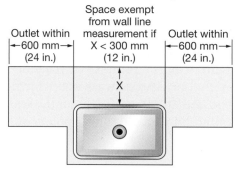

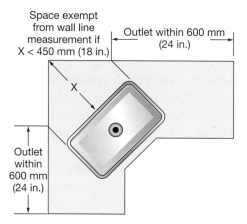

Countertops in kitchens, pantries, breakfast rooms, and similar areas of dwelling units must comply with 210.52(C).

Revised 2008 NEC Text:
210.52 DWELLING UNIT RECEPTACLE OUTLETS

(C) COUNTERTOPS. In kitchens, pantries, breakfast rooms, and dining rooms and similar areas of dwelling units, receptacle outlets for countertop spaces shall be installed in accordance with 210.52(C)(1) through (C)(5).

Where a range, counter-mounted cooking unit or sink is installed in an island or peninsular countertop and the width of the countertop behind the range, counter-mounted cooking unit, or sink is less than 300 mm (12 in.), the range, counter-mounted cooking unit, or sink is considered to divide the the countertop space into two separate countertop spaces as defined in 210.52(C)(4). Each separate countertop space shall comply with the applicable requirements in 210.52(C).

(1) WALL COUNTERTOP SPACES. A receptacle outlet shall be installed at each wall countertop space that is 300 mm (12 in.) or wider. Receptacle outlets shall be installed so that no point along the wall line is more than 600 mm (24 in.) measured horizontally from a receptacle outlet in that space.

(continued)

Exception: Receptacle outlets shall not be required on a wall directly behind a range, <u>counter-mounted cooking unit,</u> or sink in the installation described in Figure 210.52 <u>(C)(1).</u>

(2) ISLAND COUNTER<u>TOP</u> SPACES.   At least one receptacle shall be installed at each island counter<u>top</u> space with a long dimension of 600 mm (24 in.) or greater and a short dimension of 300 mm (12 in.) or greater.

(3) PENINSULAR COUNTER<u>TOP</u> SPACES.   At least one receptacle outlet shall be installed at each peninsular counter<u>top</u> space with a long dimension of 600 mm (24 in.) or greater and a short dimension of 300 mm (12 in.) or greater. A peninsular countertop is measured from the connecting edge.

Change Significance:

The provisions of 210.52(C) are expanded to include countertops in pantries, breakfast rooms, and similar areas in addition to kitchens and dining rooms. Countertop receptacles in these rooms will serve appliances similar to those in a kitchen or dining room. 210.52(B) already requires such outlets to be supplied from the required small-appliance branch circuits.

Figure 210.52 is revised to clarify that the space behind a counter-mounted cooking unit, range, or sink is exempt from the wall line measurement requirements of 210.52(C)(1). The exception further clarifies that receptacle outlets are not required in that space.

The term *rangetop* has been deleted to clarify the intent of referring to a counter-mounted cooking unit, range, or sink and the wall space or countertop space affected.

Peninsular and island countertops are now included in the 210.52(C) requirement for installations of a range, a sink, or a counter-mounted cooking unit. Where a range, counter-mounted cooking unit, or sink is installed in an island or peninsular countertop and the width of the counter behind the range, counter-mounted cooking unit, or sink is less than 300 mm (12 inches), the range, counter-mounted cooking unit, or sink is considered to divide the countertop space into two separate countertop spaces as defined in 210.52(C)(4). This separation of a wall, peninsular, or island countertop space would essentially create another wall, peninsular, or island countertop space.

Notes:

_____

_____

_____

_____

_____

210.52
DWELLING UNIT RECEPTACLE OUTLETS (E) OUTDOOR OUTLETS

Chapter 2 Wiring and Protection

Article 210 Branch Circuits

Part III Required Outlets

210.52 Dwelling Unit Receptacle Outlets

(E) Outdoor Outlets

(1) One-Family and Two-Family Dwellings

(2) Multifamily Dwellings

(3) Balconies, Decks, and Porches

Change Type: Revision

Change Summary:
- 210.52(E) has been editorially revised into three separate second-level subdivisions for the addition of new requirements, for clarity, and for usability.
- Grade-level receptacles are now required to be accessible while standing at grade level.
- In addition to outlets required in 210.52(E)(1) and (2), all balconies, decks, and porches attached to dwelling units accessible from inside the dwelling shall have at least one receptacle outlet installed accessible from the balcony, deck, or porch.

In general, all balconies, decks, and porches accessible from inside a dwelling unit must have an outdoor receptacle installed not higher than 6½ feet.

Revised 2008 NEC Text:
210.52 DWELLING UNIT RECEPTACLE OUTLETS
(E) OUTDOOR OUTLETS.   <u>Outdoor receptacle outlets shall be installed in accordance with (E)(1) through (E)(3).</u>

(continued)

(1) ONE-FAMILY AND TWO-FAMILY DWELLINGS. For a one-family dwelling and each unit of a two-family dwelling that is at grade level, at least one receptacle outlet accessible <u>while standing</u> at grade level and <u>located</u> not more than 2.0 m (6½ ft) above grade shall be installed at the front and back of the dwelling.

(2) MULTI-FAMILY DWELLINGS. For each dwelling unit of a multifamily dwelling where the dwelling unit is located at grade level and provided with individual exterior entrance/egress, at least one receptacle outlet accessible from grade level and not more than 2.0 m (6½ ft) above grade shall be installed. ~~See 210.8(A)(3).~~

(3) BALCONIES, DECKS AND PORCHES. <u>Balconies, decks, and porches that are accessible from inside the dwelling unit shall have at least one receptacle outlet installed within the perimeter of the balcony, deck, or porch. The receptacle shall not be located more than 2.0 m (6½ ft) above the balcony, deck, or porch surface.</u>

   <u>*Exception to (3): Balconies, decks, or porches with a useable area of less than 1.86m² (20 ft²) are not required to have a receptacle installed.*</u>

Change Significance:

This first-level subdivision is broken into three second-level subdivisions for clarity and usability. Outdoor receptacles installed at grade level are now required to be accessible while standing at grade level.

A new 210.52(E)(3) will now require a receptacle outlet on all porches, decks, and balconies accessible from inside the dwelling unit. The technical committee noted that a space accessible only through a window is not a balcony, deck, or porch. These outdoor outlets are required by 210.8(A)(3) to be GFCI protected. An exception is provided for balconies, decks, or porches with a useable area of less than 20 square feet. These smaller balconies, decks, or porches are in most cases for decorative or architectural purposes and are not used in the same manner as a larger area. The reference to 210.8(A)(3) is deleted because it is not necessary to repeat requirements in the NEC.

Previous editions of the NEC required outdoor receptacle outlets for only those dwelling units built at grade level. Dwelling units built at grade level with a porch, deck, or balcony without access to grade did not require a receptacle outlet for the porch or balcony.

Apartments and condominiums also provide porches and balconies built above grade for use by occupants. It is common practice for holiday lighting and electrical appliances of all types to be used on these porches, decks, and balconies. This is accomplished in most cases with the thinnest extension cord available—the indoor two-wire type of cord. Serious safety hazards are created when these cords are plugged into receptacles without GFCI protection. Additional concerns are created when these cords are run through doors and windows that are then closed as weather conditions change.

210.52
DWELLING UNIT RECEPTACLE OUTLETS
(G) BASEMENTS AND GARAGES

Chapter 2 Wiring and Protection

Article 210 Branch Circuits

Part III Required Outlets

210.52 Dwelling Unit Receptacle Outlets

(G) Basements and Garages

Change Type: Revision

Change Summary:
* The intent of 210.52(G) has been clarified to require a receptacle outlet in addition to any installed for specific equipment.

Revised 2008 NEC Text:
210.52 DWELLING UNIT RECEPTACLE OUTLETS (G) BASEMENTS AND GARAGES.   For a one-family dwelling, <u>the following provisions shall apply:</u>

<u>(1)</u> At least one receptacle outlet, in addition to ~~any provided for laundry equipment~~ <u>those for specific equipment</u>, shall be installed in each basement, ~~and~~ in each attached garage, and in each detached garage with electric power ~~See 210.8(A)(2) and (A)(5).~~

<u>(2)</u> Where a portion of the basement is finished into one or more habitable rooms, each separate unfinished portion shall have a receptacle outlet installed in accordance with this section.

Change Significance:
The intent of 210.52(G) has been clarified with an editorial revision to separate the requirements into list items. The text of the 2005 NEC would allow a receptacle outlet installed for a sump pump, laundry, alarm system, furnace, humidifier, and the like to satisfy the requirement of 210.52(G). The only prohibition in previous code text was that the receptacle outlet had to be in addition to an outlet installed for laundry equipment. The reference to a laundry outlet has been deleted. There is no longer a need to specifically exclude a laundry or any other type of outlet due to the clarification provided with the term *specific equipment.* This revision clearly requires the installation of a receptacle outlet intended solely for the general use of the occupant.

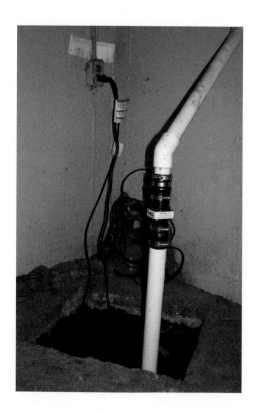

Dwelling unit basements and garages require a receptacle for general use.

Change Summary:

- Sleeping rooms in dormitories must comply with the dwelling unit receptacle outlet spacing requirements of 210.52(A) and (D).

Revised 2008 NEC Text:

210.60 GUEST ROOMS, ~~OR~~ GUEST SUITES, <u>DORMITORIES AND SIMILAR OCCUPANCIES</u>

(A) GENERAL. Guest rooms or guest suites in hotels, motels, <u>sleeping rooms in dormitories</u> and similar occupancies shall have receptacle outlets installed in accordance with 210.52(A) and 210.52(D). Guest rooms or guest suites provided with permanent provisions for cooking shall have receptacle outlets installed in accordance with all of the applicable rules in 210.52.

Change Significance:

Dormitory "sleeping rooms" are now required to meet the spacing requirements for receptacle outlets of 210.52(A) and the requirement for a receptacle outlet within 3 feet of the outside edge of all basins in bathrooms in 210.52(D). Although dormitory rooms do not meet the Article 100 definition of a dwelling unit, they do qualify as a guest room or guest suite. Dormitory rooms, however, will rarely include permanent provisions for cooking and will not be subject to the rules for dwelling unit branch circuits and outlets (as would a guest room/suite) as required in 210.18 and 210.60(A). This revision will ensure adequate receptacle placement and a properly located bathroom receptacle outlet for student "sleeping rooms" in a dormitory room. The use of the term *sleeping rooms* has been identified by this committee as "vague and unenforceable." This change will surely be revisited in the next cycle.

210.60
GUEST ROOMS, GUEST SUITES, DORMITORIES, AND SIMILAR OCCUPANCIES

Chapter 2 Wiring and Protection

Article 210 Branch Circuits

Part III Required Outlets

210.60 Guest Rooms, Guest Suites, Dormitories, and Similar Occupancies

(A) General

Change Type: Revision

Receptacle placement for sleeping rooms and bathrooms in dormitories and similar occupancies are now addressed.

210.62
SHOW WINDOWS

Chapter 2 Wiring and Protection

Article 210 Branch Circuits

Part III Required Outlets

210.62 Show Windows

Change Type: Revision

Change Summary:
- The required receptacle outlets for show windows shall be installed within 18 inches of the top of the show window.

Revised 2008 NEC Text:
210.62 SHOW WINDOWS. At least one receptacle outlet shall be installed ~~directly above~~ within 450 mm (18 in.) of the top of a show window for each 3.7 linear m (12 linear ft) or major fraction thereof of show window area measured horizontally at its maximum width.

Change Significance:
The intent of this revision is to limit the distance of the receptacle outlet required in 210.62 from the top of the show window. The 2005 NEC text required only that a receptacle outlet be installed above a show window. This literally allowed installations in which the outlet was installed as high as 18 feet above the show window. The result in these installations is the use of extension cords creating additional safety concerns.

This revision will now clearly require that the receptacle outlet be installed within 18 inches of the top of the show window, which will eliminate the need for extension cords and result in a safer installation.

Required receptacles for show windows now have placement restrictions.

Change Summary:

- The exception for fire pumps has been deleted. Ground fault protection of equipment (GFPE) for fire pumps is prohibited in 695.6(H)
- Additional text has been added to Exception No. 2 to provide clarity on the location of the GFPE.

215.10
GROUND-FAULT PROTECTION OF EQUIPMENT

Chapter 2 Wiring and Protection

Article 215 Feeders

215.10 Ground-Fault Protection of Equipment

Change Type: Revision

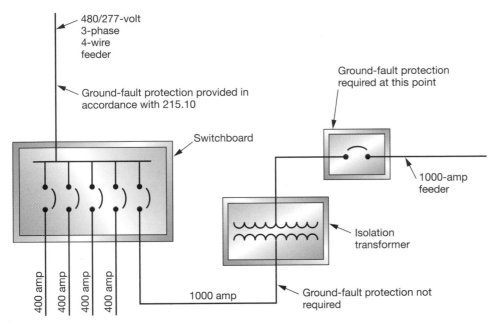

All disconnects rated at 1000 amps or more must be provided with GFPE.

Revised 2008 NEC Text:

215.10 GROUND-FAULT PROTECTION OF EQUIPMENT. Each feeder disconnect rated 1000 amperes or more and installed on solidly grounded wye electrical systems of more than 150 volts to ground, but not exceeding 600 volts phase-to-phase, shall be provided with ground-fault protection of equipment in accordance with the provisions of 230.95.

FPN: For buildings that contain healthcare occupancies, see the requirements of 517.17.

Exception No. 1: The provisions of this section shall not apply to a disconnecting means for a continuous industrial process where a nonorderly shutdown will introduce additional or increased hazards.

~~*Exception No. 2: The provisions of this section shall not apply to fire pumps.*~~

Exception No. ~~3~~ 2: The provisions of this section shall not apply if ground-fault protection of equipment is provided on the supply side of the feeder <u>and on the load side of any transformer supplying the feeder</u>.

Change Significance:

The GFPE exception for fire pumps has been deleted. A new first-level subdivision 695.6(H) in the 2005 NEC prohibits GFPE protection of fire pumps.

(continued)

The action to delete this exception is in conformance with the arrangement of the NEC as outlined in 90.3. Chapters 5, 6, and 7 are "special" and per 90.3 modifies or supplements the rules in Chapters 1 through 4.

Exception No. 2 has been modified to clarify intent. The intent of the exception is to not require GFPE on a feeder disconnect if that feeder has GFPE protection provided upstream. Prior to this revision, some users of the code believed that as written, if GFPE protection was provided upstream of a transformer on the primary, no GFPE was required on the load side (the secondary) of that transformer. GFPE on the primary side of a transformer provides no equipment protection to equipment connected to the secondary because the ground fault on the secondary only returns current to the secondary of the transformer. This revision clarifies that any GFPE protection on the supply side used to exempt the feeder disconnect from GFPE protection must be on the load side of the transformer supplying the feeder.

In general the NEC requires ground fault protection of equipment (GFPE) for all disconnecting means rated at 1,000 amps or more when the system is wye connected with more than 150-volts to ground but not exceeding 600-volts phase to phase.

Change Summary:
- All ungrounded conductors addressed by 215.12(C) are required to be identified by phase or line and system.
- This identification shall be at all termination, connection, and splice points.
- Identification means may be documented in a readily available location or posted at each panelboard or distribution point.

Revised 2008 NEC Text:
215.12 IDENTIFICATION FOR FEEDERS
(C) UNGROUNDED CONDUCTORS. Where the premises wiring system has feeders supplied from more than one nominal voltage system, each ungrounded conductor of a feeder, ~~where accessible~~, shall be identified by <u>phase or line and</u> system <u>at all termination, connection and splice points</u>. The means of identification shall be permitted to be by separate color coding, marking tape, tagging, or other approved means<u>. The method utilized for conductors originating within each feeder panelboard or similar feeder distribution equipment shall be documented in a manner that is readily available or</u> ~~and~~ shall be permanently posted at each feeder panelboard or similar feeder distribution equipment.

Change Significance:
During the 2005 NEC cycle the term *phase* was inadvertently omitted from the identification requirement. The intent of this requirement is to be able to distinguish between the different systems *and* between the different phases of the system. Identification of ungrounded conductors where more than one nominal voltage system exists must be by phase or line and system to help prevent overloading of grounded conductors when changes are made to the wiring system.

With reference to the required location of the actual marking of ungrounded conductors, the language "where accessible" has been deleted. As written in the 2005 NEC, the marking was required at all "accessible" locations, including conduit fittings and pull boxes. The revised text specifies the locations where the ungrounded feeder conductors shall be identified. Terminations, connections, and splices are the critical locations where the marking is needed.

The means of identification is now permitted to be documented in a readily available location or posted at each panelboard or distribution point. This revision is intended to recognize methods of identifying ungrounded conductors such as cable and conductor labeling/numbering systems used in industrial locations.

Notes:

_____

_____

215.12
IDENTIFICATION FOR FEEDERS

Chapter 2 Wiring and Protection

Article 215 Feeders

215.12 Identification for Feeders

(C) Ungrounded Conductors

Change Type: Revision

Ungrounded feeder conductors must be identified by phase or line and system at all termination, connection, and splice points where more than one nominal voltage exists in an occupancy.

220.52

SMALL-APPLIANCE AND LAUNDRY LOADS— DWELLING UNIT

Chapter 2 Wiring and Protection

Article 220 Branch-Circuit, Feeder, and Service Calculations

Part III Feeder and Service Load Calculations

220.52 Small-Appliance and Laundry Loads—Dwelling Unit

(A) Small Appliance Circuit Load

(B) Laundry Circuit Load

Change Type: Revision

Change Summary:

• All small-appliance and laundry branch circuits installed must be included in the feeder/service load calculation.

A dwelling unit can be supplied with as many small-appliance branch circuits desired, but only two are required.

Revised 2008 NEC Text:

220.52 SMALL APPLIANCE AND LAUNDRY LOADS—DWELLING UNIT

(A) SMALL APPLIANCE CIRCUIT LOAD. In each dwelling unit, the load shall be calculated at 1500 volt-amperes for each 2-wire small-appliance branch circuit ~~required~~ as covered by 210.11(C)(1). Where the load is subdivided through two or more feeders, the calculated load for each shall include not less than 1500 volt-amperes for each 2-wire small-appliance branch circuit. These loads shall be permitted to be included with the general lighting load and subjected to the demand factors provided in Table 220.42.

Exception: The individual branch circuit permitted by 210.52(B)(1), Exception No. 2, shall be permitted to be excluded from the calculation required by 220.52.

(B) LAUNDRY CIRCUIT LOAD. A load of not less than 1500 volt-amperes shall be included for each 2-wire laundry branch circuit installed as ~~required~~ as covered by 210.11(C)(2). This load shall be permitted to be included with the general lighting load and subjected to the demand factors provided in Table 220.42.

Change Significance:

This revision to 220.52(A) and (B) now clearly requires that where more than the minimum number of small appliance or laundry branch circuits are included in a dwelling unit each of these branch circuits must be calculated into the feeder or service load calculation.

210.11(C)(1) requires that "two or more 20-ampere small appliance branch circuits" be provided for all receptacle outlets serving the kitchen, pantry, breakfast room, dining room, or similar area of a dwelling unit. Previous text of 220.52(A) required the calculation for only the minimum (two) required. Many larger dwelling units may have four or more 20-amp small-appliance branch circuits installed. All of these small-appliance branch circuits must now be included in the feeder or service load calculation.

210.11(C)(2) requires that "at least one additional 20-ampere branch circuit shall be provided to supply the laundry receptacle outlet(s)" for the laundry of a dwelling unit. Previous text of 220.52(B) required the calculation for only the minimum (one) required. Many larger dwelling units may have two or more 20-amp laundry branch circuits installed. All of these laundry branch circuits must now be included in the feeder or service load calculation.

Notes:

_____

_____

_____

_____

_____

_____

_____

_____

_____

_____

225.18

CLEARANCE FOR OVERHEAD CONDUCTORS AND CABLES

Chapter 2 Wiring and Protection

Article 225 Outside Branch Circuits and Feeders

Part I General

225.18 Clearance for Overhead Conductors and Cables

Change Type: Revision

Change Summary:

*   The term *ground* has been deleted from the title. This change clarifies that the clearance requirements for overhead conductors are measured from the finished or highest grade level.

Revised 2008 NEC Text:

225.18 CLEARANCE ~~FROM GROUND~~ FOR OVERHEAD CONDUCTORS AND CABLES.   Overhead spans of open conductors and open multiconductor cables of not over 600 volts, nominal, shall have a clearance of not less than the following:

(1) 3.0 m (10 ft)—above finished grade, sidewalks, or from any platform or projection from which they might be reached where the voltage does not exceed 150 volts to ground and accessible to pedestrians only

(2) 3.7 m (12 ft)—over residential property and driveways, and those commercial areas not subject to truck traffic where the voltage does not exceed 300 volts to ground

(3) 4.5 m (15 ft)—for those areas listed in the 3.7-m (12-ft) classification where the voltage exceeds 300 volts to ground

(4) 5.5 m (18 ft)—over public streets, alleys, roads, parking areas subject to truck traffic, driveways on other than residential property, and other land traversed by vehicles, such as cultivated, grazing, forest, and orchard

Change Significance:

The global effort within the NEC to revise the application of grounding and bonding terms has impacted this section and others to improve clarity. The term *ground* has been deleted from the title. Use of the term *ground* is subjective and as used in the title of this section may erroneously lead the code user to believe that the clearances required are from all "grounded" objects. Within this section, the term *voltage to ground* is referred to three times—further illustrating the need to revise the title. The revised title now clarifies that the clearance requirements for overhead conductors are measured from the finished or highest grade level.

Overhead conductors and cables must meet minimum clearance requirements.

Change Summary:
- The term *identified* is added to clarify that "handle ties" used on single-pole switches or circuit breakers are intended for that function.

Revised 2008 NEC Text:
225.33 MAXIMUM NUMBER OF DISCONNECTS
(B) SINGLE-POLE UNITS.   Two or three single-pole switches or breakers capable of individual operation shall be permitted on multiwire circuits, one pole for each ungrounded conductor, as one multipole disconnect, provided they are equipped with <u>identified</u> handle ties or a master handle to disconnect all ungrounded conductors with no more than six operations of the hand.

Change Significance:
The addition of the term *identified* clarifies that the handle ties used as permitted in this first-level subdivision are suitable for the specific purpose of simultaneous disconnection. Nails, wire, and other methods of achieving simultaneous disconnect will not be considered "identified" for that purpose. Article 100 contains the following definition:

<u>Identified</u> (as applied to equipment). Recognizable as suitable for the specific purpose, function, use, environment, application, and so forth, where described in a particular Code requirement.

This definition of *identified* in Article 100 includes an informational FPN to illustrate examples of how to determine suitability. Suitability of handle ties for this specific purpose may be determined by investigations performed by qualified testing laboratories (listing and labeling), inspection agencies, or other organizations concerned with product evaluation.

225.33
MAXIMUM NUMBER OF DISCONNECTS

Chapter 2 Wiring and Protection

Article 225 Outside Branch Circuits and Feeders

Part II More Than One Building or Other Structure

225.33 Maximum Number of Disconnects

(B) Single-Pole Units

Change Type: Revision

Wire, finishing nails, and other improvised methods are not acceptable handle ties.

225.39
RATING OF DISCONNECT

Chapter 2 Wiring and Protection

Article 225 Outside Branch Circuits and Feeders

Part II More Than One Building or Other Structure

225.39 Rating of Disconnect

Change Type: Revision

Change Summary:
- The rating of an outside feeder or branch circuit disconnecting means is now permitted to be achieved by combining the ratings of multiple disconnects (not more than six) as permitted in 225.33.

Revised 2008 NEC Text:
225.39 RATING OF DISCONNECT.   The feeder or branch-circuit disconnecting means shall have a rating of not less than the load to be supplied, determined in accordance with Parts I and II of Article 220 for branch circuits, Parts III or IV of Article 220 for feeders, or Part V of Article 220 for farm loads. <u>Where the branch circuit or feeder disconnecting means consists of more than one switch or circuit breaker, as permitted by 225.33, combining the ratings of all the switches or circuit breakers for determining the rating of the disconnecting means shall be permitted.</u> In no case shall the rating be lower than specified in 225.39(A), (B), (C), or (D).

Change Significance:
Article 225 contains requirements for outside feeders and branch circuits run on or between buildings, structures, or poles on the premises; and electric equipment and wiring for the supply of utilization equipment that is located on or attached to the outside of buildings, structures, or poles. These requirements are in many cases parallel to the requirements for service-supplied equipment, buildings, or structures.

For services, 230.80 permits the ratings of multiple disconnects (not more than six per 230.71) to be added to achieve the minimum required rating in 230.79. This new sentence in 225.39 will now permit the rating of an outside feeder or branch circuit disconnecting means, to be achieved by combining the ratings of multiple disconnects (not more than six) as permitted in 225.33.

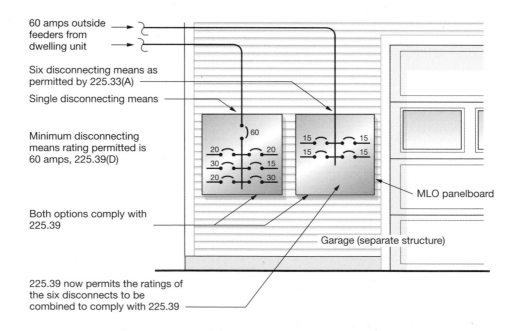

60 amps outside feeders from dwelling unit

Six disconnecting means as permitted by 225.33(A)

Single disconnecting means

Minimum disconnecting means rating permitted is 60 amps, 225.39(D)

Both options comply with 225.39

60

20   20
30   15
20   30

15   15
15   15

MLO panelboard

Garage (separate structure)

225.39 now permits the ratings of the six disconnects to be combined to comply with 225.39

OCPD ratings may be combined to determine compliance with Article 220.

Change Summary:
- Exception No. 1 to 230.40 has been editorially revised to clarify that this exception applies to buildings with more than one occupancy that are supplied with a single service drop or service lateral.

Revised 2008 NEC Text:
230.40 NUMBER OF SERVICE-ENTRANCE CONDUCTOR SETS. Each service drop or lateral shall supply only one set of service-entrance conductors.

Exception No. 1: A building <u>with more than one occupancy</u> shall be permitted to have one set of service-entrance conductors for each service, as defined in 230.2, run to each occupancy or group of occupancies.

Change Significance:
The general rule of 230.40 states that each service drop or lateral shall supply only one set of service-entrance conductors. Service-entrance conductors are generally defined as conductors between the terminals of the service equipment and a point of connection to the service drop or service lateral. There are four exceptions to the general rule in 230.40. Exception No.1 deals specifically with a single building that contains multiple occupancies. This exception now clearly applies to buildings "with more than one occupancy" supplied with a single service drop or service lateral.

230.40
NUMBER OF SERVICE-ENTRANCE CONDUCTOR SETS

Chapter 2 Wiring and Protection

Article 230 Services

Part IV Service-Entrance Conductors

230.40 Number of Service-Entrance Conductor Sets

Exception No. 1

Change Type: Revision

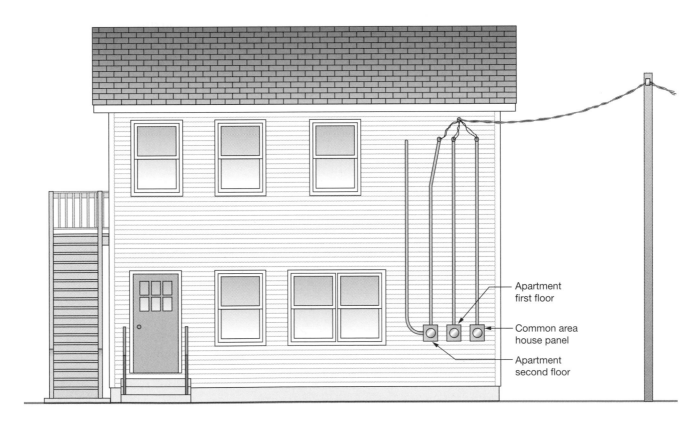

Apartment first floor

Common area house panel

Apartment second floor

Multiple-occupancy buildings may supply a set of service-entrance conductors to each occupancy.

230.44
CABLE TRAYS

Chapter 2 Wiring and Protection

Article 230 Services

Part IV Service-Entrance Conductors

230.44 Cable Trays

Change Type: Revision

Change Summary:
- Where cable trays are permitted to contain both service-entrance and other conductors, the portion of the cable tray containing service-entrance conductors shall be clearly marked "Service Entrance Conductors."

Revised 2008 NEC Text:
230.44 CABLE TRAYS.   Cable tray systems shall be permitted to support service-entrance conductors. Cable trays used to support service-entrance conductors shall contain only service-entrance conductors.

Exception: Conductors other than service-entrance conductors shall be permitted to be installed in a cable tray with service-entrance conductors, provided a solid fixed barrier of a material compatible with the cable tray is installed to separate the service-entrance conductors from other conductors installed in the cable tray. <u>*Cable trays shall be identified with permanently affixed labels with the wording "Service-Entrance Conductors." The labels shall be located so as to be visible after installation and placed so that the service-entrance conductors may be readily traced through the entire length of the cable tray.*</u>

Change Significance:
The general rule in 230.44 clearly prohibits "other conductors" such as branch-circuit or feeder conductors from being installed in a cable tray that contains service-entrance conductors. The exception allows other than service-entrance conductors to be installed in a cable tray with service-entrance conductors, provided a solid fixed barrier of a material compatible with the cable tray is installed to separate the service-entrance conductors from other conductors installed in the cable tray. This revision now requires marking of the sections of cable tray containing service-entrance conductors. This marking is required to be visible and permanent.

Two sentences have been added to this exception that permits "other conductors" to be installed in a cable tray containing service-entrance conductors requiring the following:
- Cable trays containing both service-entrance and "other conductors" shall be identified with permanently affixed labels with the wording "Service-Entrance Conductors."
- The labels shall be located so as to be visible after installation and placed so that the service-entrance conductors may be readily traced through the entire length of the cable tray.

Where service-entrance conductors are installed in a cable tray with other conductors, labeling is required.

Change Summary:

- Raceways enclosing service-entrance conductors are no longer required to be "raintight" but are instead required to be "suitable for use in wet locations."
- Flexible metal conduit is no longer permitted to contain service-entrance conductors where exposed to the weather.

Revised 2008 NEC Text:

230.53 RACEWAYS TO DRAIN.   Where exposed to the weather, raceways enclosing service-entrance conductors shall be ~~raintight~~ <u>suitable for use in wet locations</u> and arranged to drain. Where embedded in masonry, raceways shall be arranged to drain.

> ~~Exception: As permitted in 348.12(1).~~

Change Significance:

Section 230.43 lists 16 wiring methods permitted to contain service-entrance conductors. The previous text of 230.53 required all raceways to be "raintight." The word *raintight* is not appropriate in this section. The "uses permitted" and "uses not permitted" sections in the articles for each raceway type address the provisions for the use of raceways in wet locations. The new text added in this section "suitable for use in wet locations" will now properly correlate this requirement with the provisions of each raceway article. The deletion of the exception in this section now limits the use of flexible metal conduit containing service-entrance conductors to installations not exposed to the weather.

230.53
RACEWAYS TO DRAIN

Chapter 2 Wiring and Protection

Article 230 Services

Part IV Service-Entrance Conductors

230.53 Raceways to Drain

Change Type: Revision

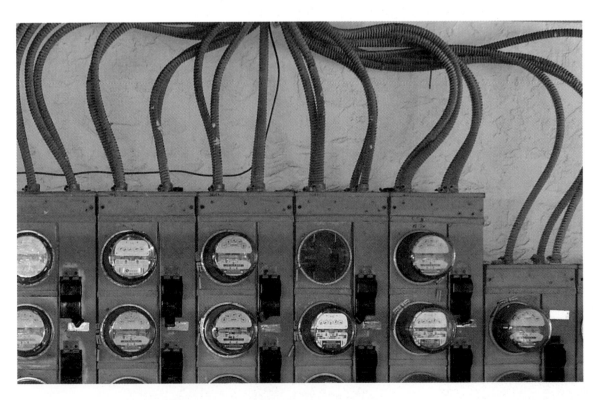

Flexible metal conduit is not permitted in wet locations.

230.54
OVERHEAD SERVICE LOCATIONS

Chapter 2 Wiring and Protection

Article 230 Services

Part IV Service-Entrance Conductors

230.54 Overhead Service Locations

(A) Service Head

(B) Service Cable Equipped With Service Head or Gooseneck

Change Type: Revision

Change Summary:
- The term *raintight* has been deleted from 230.54 and service heads are now required to comply with 314.15.

Revised 2008 NEC Text:
230.54 OVERHEAD SERVICE LOCATIONS

(A) ~~RAINTIGHT~~ SERVICE HEAD.   Service raceways shall be equipped with a ~~raintight~~ service head at the point of connection to service-drop conductors. <u>The service head shall comply with the requirement for fittings in 314.15.</u>

(B) SERVICE CABLE EQUIPPED WITH ~~RAINTIGHT~~ SERVICE HEAD OR GOOSENECK.   Service cables shall be equipped with a ~~raintight~~ service head. <u>The service head shall comply with the requirement for fittings in 314.15.</u>

Exception: Type SE cable shall be permitted to be formed in a gooseneck and taped with a self-sealing weather-resistant thermoplastic.

Change Significance:
The word *raintight* is not appropriate in 230.54(A) or (B). A service head is a fitting and is intended for installation in damp or wet locations. The additional sentence in each first-level subdivision referencing 314.15 provides additional clarity and usability. The use of the term *raintight* created confusion. 314.15 requires that in damp or wet locations these fittings (service head) shall be placed or equipped so as to prevent moisture from entering or accumulating within the fitting (service head) and shall be listed for use in wet locations.

Service heads are no longer required to be "raintight."

Change Summary:

- The language "transient voltage surge suppressors" has been deleted and replaced with type 1 or type 2 "surge protective devices."
- User-friendly reference to applicable parts of Article 250 have been added with reference to meters, meter sockets, and meter disconnect switches.
- Meter disconnect switches are now required to be rated to interrupt the load served.

230.82
EQUIPMENT CONNECTED TO THE SUPPLY SIDE OF SERVICE DISCONNECT

Chapter 2 Wiring and Protection

Article 230 Services

Part VI Service Equipment— Disconnecting Means

230.82 Equipment Connected to the Supply Side of Service Disconnect

Change Type: Revision

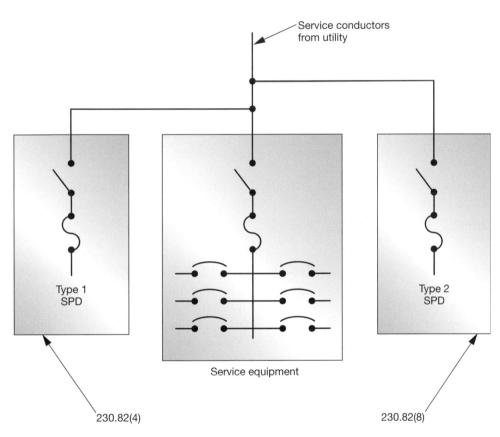

Service conductors from utility

Type 1 SPD

Type 2 SPD

Service equipment

230.82(4)   230.82(8)

230.82 contains a list of equipment which is permitted to be connected on the supply side of the service disconnecting means.

Revised 2008 NEC Text:

230.82 EQUIPMENT CONNECTED TO THE SUPPLY SIDE OF SERVICE DISCONNECT.   Only the following equipment shall be permitted to be connected to the supply side of the service disconnecting means:

(No change to list items (1), (5), (6) & (7))

(2) Meters and meter sockets nominally rated not in excess of 600 volts, provided all metal housings and service enclosures are grounded <u>in accordance with Part VII and bonded in accordance with Part V of Article 250.</u>

(3) Meter disconnect switches nominally rated not in excess of 600 volts that have a short-circuit current rating equal to or greater than the available short circuit current, provided all metal housings and service enclosures are

(continued)

grounded <u>in accordance with Part VII and bonded in accordance with Part V of Article 250. A meter disconnect switch shall be capable of interrupting the load served.</u>

(4) Instrument transformers (current and voltage), impedance shunts, load management devices, ~~and~~ <u>surge</u> arresters <u>and Type 1 surge protective devices</u>

(8) Ground-fault protection systems or <u>Type 2 surge protective devices</u> ~~transient voltage surge suppressors~~, where installed as part of listed equipment, if suitable overcurrent protection and disconnecting means are provided.

Change Significance:

The language "transient voltage surge suppressors" has been deleted and replaced with "surge protective devices." The technology of both low-voltage surge arresters and transient voltage surge suppressors (TVSSs) is now basically the same and is covered under one standard (UL 1449). Designations are provided for surge protective devices with consideration given to the installation location on the line side (type 1) or load side (type 2) of the service disconnect overcurrent protection.

The categories of Surge Arresters (Article 280) and Transient Voltage Surge Suppressors (Article 285) have been combined into one category and one standard (UL 1449), renamed Surge Protective Devices (SPDs). The Surge Arrester designation will only be retained for devices used in circuits of 1 kV and over and evaluated to IEEE C62.11-1999.

List items (2) and (3) have been clarified and updated with user-friendly references to applicable parts of Article 250 for meters, meter sockets, and meter disconnect switches. List item (3) now requires meter disconnect switches to be capable of interrupting the load served. These switches are typically used to interrupt the entire load for meter replacement or repair.

Notes:

_____

_____

_____

_____

_____

_____

Change Summary:

- Prescriptive text has been added to clearly require that the grounded conductor for the solidly grounded wye system, addressed in 230.95, shall be connected directly to ground "through a grounding electrode as specified in 250.50."

Revised 2008 NEC Text:

230.95 GROUND-FAULT PROTECTION OF EQUIPMENT. Ground-fault protection of equipment shall be provided for solidly grounded wye electrical services of more than 150 volts to ground but not exceeding 600 volts phase-to-phase for each service disconnect rated 1000 amperes or more. The grounded conductor for the solidly grounded wye system shall be connected directly to ground <u>through a grounding electrode as specified in 250.50</u> without inserting any resistor or impedance device.

The rating of the service disconnect shall be considered to be the rating of the largest fuse that can be installed or the highest continuous current trip setting for which the actual overcurrent device installed in a circuit breaker is rated or can be adjusted.

Exception No. 1 (No change)

Exception No. 2 (This exception for fire pumps is deleted, see 695.6(G))

Change Significance:

This revision provides clarification and prescriptive text to require that the grounded conductor for the solidly grounded wye system, addressed in 230.95, shall be connected directly to ground "through a grounding electrode as specified in 250.50." The connection to ground is now clearly defined as a solid connection to a grounding electrode in accordance with 250.50.

Exception No. 2, which excluded fire pumps, has been deleted. 695.6(G) specifically prohibits ground-fault protection for fire pump sources. This revision represents part of the global effort in the NEC to clarify references to grounding and bonding.

230.95
GROUND-FAULT PROTECTION OF EQUIPMENT

Chapter 2 Wiring and Protection

Article 230 Services

Part VII Service Equipment—Overcurrent Protection

230.95 Ground-Fault Protection of Equipment

Change Type: Revision

Where GFPE is required for service conductors, the grounded conductor must be connected to ground through a grounding electrode.

240.4(D)
SMALL CONDUCTORS

Chapter 2 Wiring and Protection

Article 240 Overcurrent Protection

Part I General

240.4 Protection of Conductors

(D) Small Conductors

(1) 18 AWG Copper

(2) 16 AWG Copper

(3) 14 AWG Copper

(4) 12 AWG Aluminum and Copper-Clad Aluminum

(5) 12 AWG Copper

(6) 10 AWG Aluminum and Copper-Clad Aluminum

(7) 10 AWG Copper

Change Type: Revision

Change Summary:
* New provisions for protection of 16- and 18-AWG copper conductors have been added to correlate with changes made in NFPA 79, the Electrical Standard for Industrial Machinery.

2005 NEC

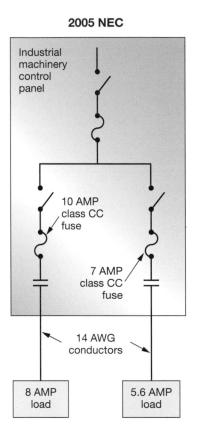

2008 NEC

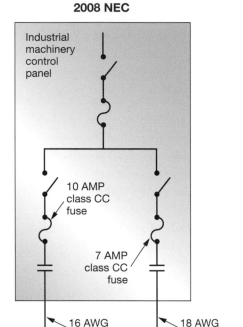

Smallest conductor allowed to supply loads is 14 AWG

Smallest conductor allowed to supply loads is 18 AWG

The provisions for overcurrent protection of small conductors in NFPA 79 for industrial machinery are now recognized in 240.4(D).

Revised 2008 NEC Text:
240.4 PROTECTION OF CONDUCTORS
(D) SMALL CONDUCTORS. Unless specifically permitted in 240.4(E) or (G), the overcurrent protection shall not exceed ~~15 amperes for 14 AWG, 20 amperes for 12 AWG, and 30 amperes for 10 AWG copper; or 15 amperes for 12 AWG and 25 amperes for 10 AWG aluminum and copper-clad aluminum~~ that required by (D)(1) through (D)(7) after any correction factors for ambient temperature and number of conductors have been applied.

(1) 18 AWG COPPER. 7 amperes, provided all the following conditions are met:
(1) Continuous loads do not exceed 5.6 amperes.

(continued)

(2) Overcurrent protection is provided by one of the following:

    a. Branch circuit rated circuit breakers listed and marked for use with 18 AWG copper wire

    b. Branch circuit rated fuses listed and marked for use with 18 AWG copper wire

    c. Class CC, Class J, or Class T fuses

(2) 16 AWG COPPER.   10 amperes, provided all of the following conditions are met:

(1) Continuous loads do not exceed 8 amperes

(2) Overcurrent protection is provided by one of the following:

    a. Branch circuit rated circuit breakers listed and marked for use with 16 AWG copper wire

    b. Branch circuit rated fuses listed and marked for use with 16 AWG copper wire

    c. Class CC, Class J, or Class T fuses

(3) 14 AWG COPPER.   15 amperes

(4) 12 AWG ALUMINUM AND COPPER-CLAD ALUMINUM.   15 amperes

(5) 12 AWG COPPER.   20 amperes

(6) 10 AWG ALUMINUM AND COPPER-CLAD ALUMINUM.   25 amperes

(7) 10 AWG COPPER.   30 amperes

Change Significance:

This change is made to correlate with changes in the NFPA 79 standard. The NFPA 79 technical committee implemented conditions of use for 16- and 18-AWG copper conductors to help United States industrial machinery manufacturers remain competitive in the global marketplace. These changes will allow industrial machinery manufacturers to utilize power circuit conductors that are smaller than 14 AWG, the existing minimum allowed for branch circuits in the NEC. The use of 16- and 18-AWG conductors for other than control loads will now be permitted under the provided conditions of use. The committee noted that this revision provides only overcurrent protection requirements for 16- and 18-AWG copper conductors. Use of these smaller conductors is only permitted where specifically referenced elsewhere in the NEC.

Notes:

240.21
LOCATION IN CIRCUIT

Chapter 2 Wiring and Protection

Article 240 Overcurrent Protection

Part II Location

240.21 Location in Circuit

Change Type: Revision

Change Summary:
- The parent text of 240.21 has been editorially revised to clarify that it is prohibited to "tap a tap."

Revised 2008 NEC Text:
240.21 LOCATION IN CIRCUIT. Overcurrent protection shall be provided in each ungrounded circuit conductor and shall be located at the point where the conductors receive their supply except as specified in 240.21(A) through (G). ~~No c~~ Conductor<u>s</u> supplied under the provisions of 240.21(A) through (G) shall not supply another conductor ~~under those provisions~~, except through an overcurrent protective device meeting the requirements of 240.4.

Change Significance:
This change is editorial in nature and provides clarity and usability for the user of the NEC. The general rule of 240.21 requires that overcurrent protection be provided for all ungrounded conductors at the point the conductors receive their supply. The general rule is then expanded by allowing conductors to be supplied in accordance with the requirements of 210.21(A) through (H), which are sometimes referred to as the tap rules. The last sentence of this section has been modified to more clearly illustrate that it is prohibited to "tap a tap." For example, a 25-foot feeder tap installed in accordance with 240.21(B)(2) or transformer secondary conductors installed in accordance with 240.21(C)(6) must terminate in a single circuit breaker or single set of fuses. It is not permitted to "tap" these conductors in, for example, a wireway to serve multiple loads.

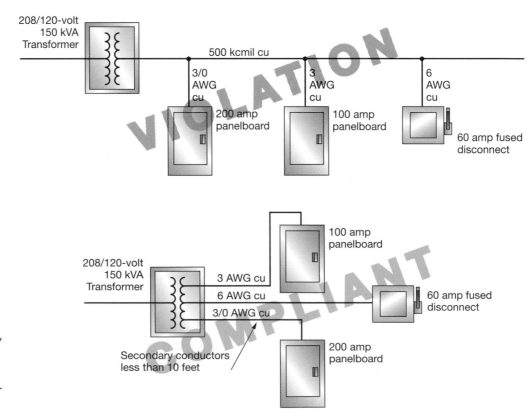

Transformer secondary conductors are not considered protected by the primary overcurrent protective device in delta four-wire and wye-connected systems.

Change Summary:

- Clearer text has been provided to inform the code user that a transformer secondary is permitted to have multiple loads and that each "set" of conductors must meet the provisions of 240.21(C).

Revised 2008 NEC Text:

240.21 LOCATION IN CIRCUIT
(C) TRANSFORMER SECONDARY CONDUCTORS. A set of conductors feeding a single load, or Eeach set of conductors feeding separate loads, shall be permitted to be connected to a transformer secondary, without overcurrent protection at the secondary, as specified in 240.21(C)(1) through (C)(6). The provisions of 240.4(B) shall not be permitted for transformer secondary conductors.

FPN: For overcurrent protection requirements for transformers, see 450.3.

Change Significance:

The parent text of 240.21(C) has been modified to more clearly illustrate the following two key points for code users applying these requirements.

- Transformer secondaries are permitted to serve more than one load.
- Each set of conductors serving a load, regardless of the number of loads served, must be installed in compliance with all rules in 240.21.

For example, a three-phase 208/120-volt transformer may be installed to serve three panelboards and a fused disconnect. 240.21(C)(1) clearly prohibits the protection of the secondary conductors by the primary overcurrent device. This requires application of 240.21(C)(2) for conductors not more than 10 feet long or of 240.21(C)(6) for conductors not more than 25 feet long. A separate set of conductors is required to serve each of these three loads. The parent text of 240.21 prohibits "tapping a tap." Each of these sets of conductors must be installed in accordance with all applicable rules of 240.21.

240.21(C)
TRANSFORMER SECONDARY CONDUCTORS

Chapter 2 Wiring and Protection

Article 240 Overcurrent Protection

Part II Location

240.21 Location in Circuit

(C) Transformer Secondary Conductors

Change Type: Revision

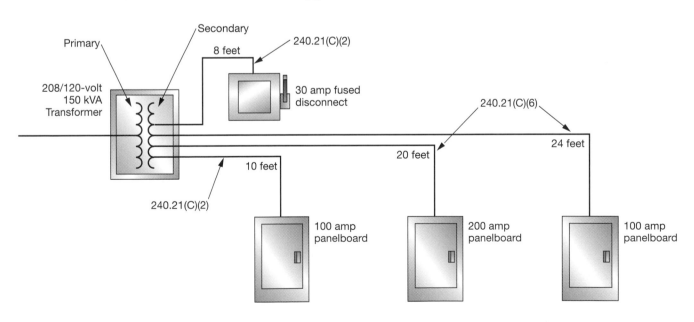

Transformer secondaries may supply more than one load, provided the provisions of 240.21(C) are met.

240.21(C)(2)

TRANSFORMER SECONDARY CONDUCTORS NOT OVER 3 M (10 FT) LONG

Chapter 2 Wiring and Protection

Article 240 Overcurrent Protection

Part II Location

240.21 Location in Circuit

(C) Transformer Secondary Conductors

(2) Transformer Secondary Conductors Not Over 3 m (10 ft) Long

Change Type: Revision

Change Summary:

- 240.21(C)(2) has been editorially revised for usability, and the minimum-size conductor requirements have been clarified to apply only to field-installed conductors that leave the enclosure or vault.

Revised 2008 NEC Text:

240.21 LOCATION IN CIRCUIT
(C) TRANSFORMER SECONDARY CONDUCTORS.
(2) TRANSFORMER SECONDARY CONDUCTORS NOT OVER 3 M (10 FT) LONG. Where the length of secondary conductor does not exceed 3 m (10 ft) and complies with all of the following:

(1) (Delete list item (c))
(2) and (3) No change.
(4) For field installations where the secondary conductors leave the enclosure or vault in which the supply connection is made, the rating of the overcurrent device protecting the primary of the transformer, multiplied by the primary to secondary transformer voltage ratio, shall not exceed 10 times the ampacity of the secondary conductor.

FPN: For overcurrent protection requirements for ~~lighting and appliance branch-circuit~~ panelboards ~~and certain power panelboards~~, see 408.36 ~~(A), (B), and (E)~~.

Change Significance:

The requirements for minimum conductor size from a transformer secondary where the conductors are not more than 10 feet long have been clarified to apply only to "field installation" whereby the secondary conductors leave the enclosure or vault in which the connection is made. This revision will now allow manufacturers or field installations to install smaller conductors, provided those conductors do not leave the enclosure or vault. These installations could include conductors to supply instrumentation equipment or fans. Note that all of the other rules for 240.21(C)(2) would apply.

Relocation and an editorial revision of 240.21(C)(2)(1)(c) into a new list item (4) with this revision provides increased usability. To determine the minimum size of conductor, multiply the size of the primary overcurrent protective device (OCPD) by the primary/secondary ratio and divide that by 10.

Transformer secondary conductors are not considered protected by the primary OCPD unless the provisions of 240.21(C)(1) are met.

Change Summary:
- A new first-level subdivision (H) is added to 240.21 to address the location of overcurrent protection for battery conductors.

Revised 2008 NEC Text:
240.21 LOCATION IN CIRCUIT
(H) BATTERY CONDUCTORS. Overcurrent protection shall be permitted to be installed as close as practicable to the storage battery terminals in a non-hazardous location. Installation of the overcurrent protection within a hazardous location shall also be permitted.

Change Significance:
A new first-level subdivision, (H) Battery Conductors, has been added to 240.21 to provide guidance on the installation and location of overcurrent protective devices for battery installations. In the 2005 NEC, there are no specific requirements for the location of OCPDs for conductors used in battery installations. Batteries were recognized as a separately derived system, but no specific rules existed to protect conductors similar to those in 240.21(C) for transformer secondary conductors or (G) for generators.

When one applied the rules of the 2005 NEC, the conductors according to the parent text in 240.21 had to be protected at their point of supply. That is not practical because we cannot locate an overcurrent protective device at the battery terminals. This new text provides guidance and recognizes practical design for these installations. No conductor length limitation is given (this is determined in the design of the installation). Storage battery areas may constitute a hazardous location due to the accumulation of hydrogen. OCPDs are required to be installed as close as practical in a non-hazardous location. Installation in the hazardous location is permitted. However, the rules for hazardous locations would apply.

240.21(H)
BATTERY CONDUCTORS

Chapter 2 Wiring and Protection

Article 240 Overcurrent Protection

Part II Location

240.21 Location in Circuit

(H) Battery Conductors

Change Type: New

Conductors supplied from batteries are not provided with overcurrent protection at their point of supply.

240.24(B)
OCCUPANCY

Chapter 2 Wiring and Protection

Article 240 Overcurrent Protection

Part II Location

240.24 Location in or on Premises

(B) Occupancy

(1) Service and Feeder Overcurrent Devices

(2) Branch Circuit Overcurrent Devices

Change Type: Revision

Change Summary:
- Where "permanent provisions for cooking" exist in a "guest room" or "guest suite," all branch circuit overcurrent devices shall be accessible to the occupant.

Revised 2008 NEC Text:
240.24 LOCATION IN OR ON PREMISES
(B) OCCUPANCY.   Each occupant shall have ready access to all overcurrent devices protecting the conductors supplying that occupancy, unless otherwise permitted in 240.24(B)(1) and (B)(2).

(1) SERVICE AND FEEDER OVERCURRENT DEVICES.   ~~Exception No. 1:~~ Where electric service and electrical maintenance are provided by the building management and where these are under continuous building management supervision, the service overcurrent devices and feeder overcurrent devices supplying more than one occupancy shall be permitted to be accessible to only authorized management personnel in the following:
(1) Multiple-occupancy buildings
(2) Guest rooms or guest suites ~~of hotels and motels that are intended for transient occupancy~~

(2) BRANCH CIRCUIT OVERCURRENT DEVICES.   ~~Exception No. 2:~~ Where electric service and electrical maintenance are provided by the building management and where these are under continuous building management supervision, the branch circuit overcurrent devices supplying any guest rooms or guest suites without permanent provisions for cooking shall be permitted to be accessible to only authorized management personnel. ~~for guest rooms of hotels and motels that are intended for transient occupancy.~~

Change Significance:
The existing exceptions have been editorially eliminated and rolled into the positive text in two second-level subdivisions. There is no longer a need for the qualifier "intended for transient occupancy" due to the addition of these definitions in Article 100 and qualifying requirements for "guest rooms" and "guest suites" in Article 210.

The qualifier "intended for transient occupancies" has been deleted.

240.24(B)(2) permits the branch-circuit overcurrent devices supplying a guest room or guest suite to be accessible to only authorized management. The addition of new text in 240.24(B)(2) clarifies that where "permanent provisions for cooking" are installed in a guest room or guest suite branch-circuit overcurrent devices shall be accessible to the occupant in the guest room or guest suite.

In essence, the only difference between a "guest room" or "guest suite" and a "dwelling unit" is a requirement for permanent provisions for cooking. Where a "guest room" or "guest suite" is provided with "permanent provisions for cooking," Section 210.18 requires that they be treated the same as dwelling units in regard to the branch-circuit requirements contained in Parts I, II, and III of Article 210. The requirements for accessibility of overcurrent devices for "guest rooms" or "guest suites" provided with "permanent provisions for cooking" have been revised to meet the needs of an "extended stay."

Where permanent provisions for cooking exist in a guest room or guest suite, this extended-stay occupancy is in essence an apartment (which is a dwelling unit).

Change Summary:
- A new first-level subdivision (F) has been added to 240.24 to prohibit the location of overcurrent devices over steps in a stairway.

Revised 2008 NEC Text:
240.24 LOCATION IN OR ON PREMISES
(F) NOT LOCATED OVER STEPS. Overcurrent devices shall not be located over steps of a stairway

Change Significance:
A stairway is a flight of steps and a supporting structure connecting separate levels. This new first-level subdivision now clearly prohibits the installation of overcurrent devices over steps in a stairway. The new text will now provide the enforcement community with a specific requirement to prohibit the installation of overcurrent devices in stairways.

The installation of panelboards or overcurrent protective devices of any type is prohibited in stairways by building codes in commercial and institutional occupancies. Electrical equipment is prohibited in stairways or egress corridors unless that equipment (such as emergency lighting, fire alarm tamper/flow switches, and purge/pressurization fans) directly serves the stair or corridor. The installation of panelboards in the stairway of dwelling units, however, may not be a violation of local building codes.

This new requirement addresses a serious safety issue. No one should be required to stand on different treads on a stairway to repair, replace, maintain, install, troubleshoot, or reset overcurrent devices in a stairway.

240.24(F)
NOT LOCATED OVER STEPS

Chapter 2 Wiring and Protection

Article 240 Overcurrent Protection

Part II Location

240.24 Location in or on Premises

(F) Not Located Over Steps

Change Type: New

Installing overcurrent devices over steps is prohibited.

240.86(A)
SELECTED UNDER ENGINEERING SUPERVISION IN EXISTING INSTALLATIONS

Chapter 2 Wiring and Protection

Article 240 Overcurrent Protection

Part VII Circuit Breakers

240.86 Series Ratings

(A) Selected Under Engineering Supervision in Existing Installations

Change Type: Revision

Change Summary:
- Additional text has been added to 240.86(A) to help clarify the requirement for selecting series-rated combination devices under engineering supervision in existing installations.

Revised 2008 NEC Text:
240.86 SERIES RATINGS
(A) SELECTED UNDER ENGINEERING SUPERVISION IN EXISTING INSTALLATIONS.   The series rated combination devices shall be selected by a licensed professional engineer engaged primarily in the design or maintenance of electrical installations. The selection shall be documented and stamped by the professional engineer. This documentation shall be available to those authorized to design, install, inspect, maintain, and operate the system. This series combination rating, including identification of the upstream device, shall be field marked on the end use equipment.

For calculated applications, the engineer shall ensure that the downstream circuit breaker(s) that are part of the series combination remain passive during the interruption period of the line side fully rated, current-limiting device.

Change Significance:
A new last sentence has been added to 240.86(A) to help clarify the application of an engineered series rating. This new last sentence requires that the licensed professional engineer evaluate downstream circuit breakers. The current limiting fuses or circuit breaker—added to the existing system in an engineered solution to achieve the necessary interrupting rating—must be given time to open before the downstream devices. This text requires that the downstream circuit breakers dynamically affected by the engineered solution be evaluated with respect to passivity. The term *passive* means that the circuit breaker will not begin to open before the upstream current-limiting device opens.

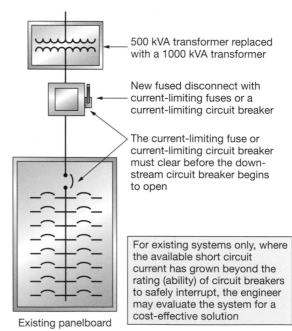

500 kVA transformer replaced with a 1000 kVA transformer

New fused disconnect with current-limiting fuses or a current-limiting circuit breaker

The current-limiting fuse or current-limiting circuit breaker must clear before the downstream circuit breaker begins to open

For existing systems only, where the available short circuit current has grown beyond the rating (ability) of circuit breakers to safely interrupt, the engineer may evaluate the system for a cost-effective solution

Existing panelboard

Series ratings applied under engineering supervision provide an alternative to complete replacement of equipment.

Change Summary:

- New first-level subdivision allowing an engineered calculation of feeder tap conductor size in Part VIII, Supervised Industrial Installations.

Revised 2008 NEC Text:

240.92 LOCATION IN CIRCUIT

(B) FEEDER TAPS.   For feeder taps specified in 240.21(B)(2), (B)(3), and (B)(4), the tap conductors shall be permitted to be sized in accordance with Table 240.92(B).

Table 240.21(B)(2)(1) Tap Conductor Short-Circuit Current Ratings.

Tap conductors are considered to be protected under short-circuit conditions when their short-circuit temperature limit is not exceeded. Conductor heating under short-circuit conditions is determined by (1) or (2):

(1) Short-Circuit Formula for Copper Conductors

$(I^2/A^2) \, t = 0.0297 \log_{10} ((T_2 + 234)(T_1 + 234))$

(2) Short-Circuit Formula for Aluminum Conductors

$(I^2/A^2) \, t = 0.0125 \log_{10} ((T_2 + 228)(T_1 + 228))$

where

I = short-circuit current in amperes

A = conductor area in circular mils

t = time of short-circuit in seconds (for times less than or equal to 10 seconds)

T_1 = initial conductor temperature in degrees Celsius.

T_2 = final conductor temperature in degrees Celsius.

Copper conductor with paper, rubber, varnished cloth insulation $T_2 = 200$

Copper conductor with thermoplastic insulation $T_2 = 150$

Copper conductor with crosslinked polyethylene insulation $T_2 = 250$

Copper conductor with ethylene propylene rubber insulation $T_2 = 250$

Aluminum conductor with paper, rubber, varnished cloth insulation $T_2 = 200$

Aluminum conductor with thermoplastic insulation $T_2 = 150$

Aluminum conductor with crosslinked polyethylene insulation $T_2 = 250$

Aluminum conductor with ethylene propylene rubber insulation $T_2 = 250$

Change Significance:

This change is designed to permit tap conductors to be sized using formulas that have been widely utilized by IEEE, the Canadian Electrical Code, and the IEC. The existing requirements of 240.21(B)(2), (3), and (4) provide safe conductor sizing over all possible tap conductor sizes. In some cases, however, these rules may be conservative and result in a conductor sized larger than is actually required according to the laws of physics. By using formulas that have been widely utilized by IEEE, the Canadian Electrical Code, and the IEC, smaller conductors can be installed. This will provide significant cost savings for industrial electrical distribution systems, allowing North American manufacturers to be more competitive in the global marketplace.

240.92(B)
FEEDER TAPS

Chapter 2 Wiring and Protection

Article 240 Overcurrent Protection

Part VIII Supervised Industrial Installations

240.92 Location in Circuit

(B) Feeder Taps

Change Type: New

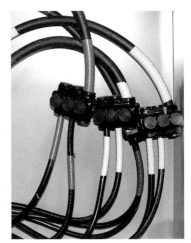

Feeder tap requirements in supervised industrial installations are modified to permit an engineered calculation.

250.4(B)
UNGROUNDED SYSTEMS

Chapter 2 Wiring and Protection

Article 250 Grounding and Bonding

Part I General

250.4 General Requirements for Grounding and Bonding

(B) Ungrounded Systems

(4) Path for Fault Current

Change Type: Revision

Change Summary:
- 250.4(B)(4) has been revised to further explain a typical ground-fault scenario in an ungrounded system.
- The term *permanent* has been removed from this section, as well as from other sections in Article 250, where the ground-fault return path is described.

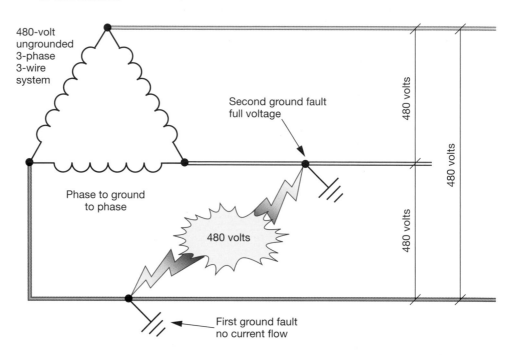

The voltage to ground in an ungrounded system is the maximum voltage between any two-phase conductors.

Revised 2008 NEC Text:
250.4 GENERAL REQUIREMENTS FOR GROUNDING AND BONDING
(B) UNGROUNDED SYSTEMS
(4) PATH FOR FAULT CURRENT.
Electrical equipment, wiring, and other electrically conductive material likely to become energized shall be installed in a manner that creates a ~~permanent,~~ low-impedance circuit from any point on the wiring system to the electrical supply source to facilitate the operation of overcurrent devices should a second <u>ground</u> fault <u>from a different phase</u> occur on the wiring system. The earth shall not be considered as an effective fault-current path.

~~FPN No. 1: A second fault that occurs through the equipment enclosures and bonding is considered a ground fault.~~

~~FPN No. 2: See Figure 250.4 for information on the organization of Article 250.~~

Change Significance:
250.4(B) outlines the general requirements for grounding and bonding of ungrounded systems. This section provides a performance-based outline, which is met with the prescriptive requirements of all 10 parts of Article 250.

(continued)

250.4(B)(4) requires that in an ungrounded system all electrical wiring as well as conductive objects that could become energized be bonded to form a low-impedance path for fault current to "facilitate the operation" of an overcurrent device. This means bond all conductors and conductive materials in such a manner that fault current will flow to trip the breaker or open the fuse.

The text added in 250.4(B)(4) explains how fault current will flow in an ungrounded system. The first phase goes to ground, and the system is now grounded. If a second phase goes to ground, a line voltage fault occurs from first phase through ground to the second phase.

FPN 1 has been deleted due to the addition of new text. The remaining FPN has been revised and relocated to 250.1, the scope of Article 250.

The term *permanent* has been removed from this section, as well as from other sections in Article 250, where the ground-fault return path is described. The use of this term is subjective in this and other requirements and leads to enforcement problems.

Ungrounded systems are commonly used in manufacturing and industrial type facilities.

250.8
CONNECTION OF GROUNDING AND BONDING EQUIPMENT

Chapter 2 Wiring and Protection

Article 250 Grounding and Bonding

Part I General

250.8 Connection of Grounding and Bonding Equipment

(A) Permitted Methods

(B) Methods Not Permitted

Change Type: Revision

Change Summary:
- 250.8 has been revised to clearly outline permissible and nonpermissible methods of connecting grounding and bonding equipment.
- The prohibition of sheet metal-type screws is more clearly addressed with a list of permissible methods.

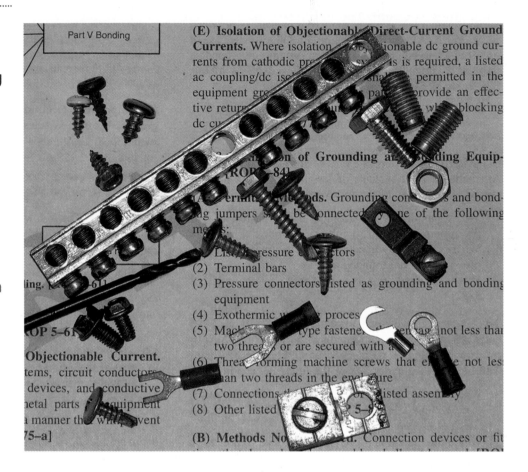

250.8 outlines permitted methods for connection of grounding and bonding equipment.

Revised 2008 NEC Text:
250.8 CONNECTION OF GROUNDING AND BONDING EQUIPMENT
(A) PERMITTED METHODS. Grounding conductors and bonding jumpers shall be connected by <u>one of the following means:</u>

    (1) Listed pressure connectors

    (2) <u>Terminal bars</u>

    (3) Pressure connectors listed <u>as grounding and bonding equipment</u>

    (4) <u>The</u> exothermic welding <u>process</u>

    (5) <u>Machine screw-type fasteners that engage not less than two threads or are secured with a nut</u>

    (6) <u>Thread-forming machine screws that engage not less than two threads in the enclosure</u>

(continued)

(7) Connections that are part of a listed assembly

(8) Other listed means

(B) METHODS NOT PERMITTED.   Connection devices or fittings that depend solely on solder shall not be used. ~~Sheet metal screws shall not be used to connect grounding conductors or connection devices to enclosures.~~

Change Significance:

This revision of 250.8 separates permitted methods for the connection of grounding and bonding equipment and methods not permitted into separate first-level subdivisions for clarity. New first-level subdivision "250.8(A) Permitted Methods" is separated into eight list items to outline all permitted methods for the connection of grounding and bonding equipment.

This list includes evaluated connections that are part of listed assemblies, listed pressure connectors, and other listed means. Also included are terminal bars, exothermic welding, and machine screw-type fasteners that engage not less than two threads or that are secured with a nut.

New list item (6) under (A), Permitted Methods, should now clearly address the prohibition of sheet metal- and drywall-type screws. This text clearly states that "thread-forming machine screws that engage not less than two threads in the enclosure" are permitted. In application this would permit a screw type that cuts threads into an existing hole and a self-drilling/tapping type of screw (provided two threads are engaged).

Thread forming machine screws that engage not less than two threads in an enclosure are now recognized as suitable in 250.8.

250.20(D)
SEPARATELY DERIVED SYSTEMS

Chapter 2 Wiring and Protection

Article 250 Grounding and Bonding

Part II System Grounding

250.20 Alternating-Current Systems to Be Grounded

(D) Separately Derived Systems

Change Type: Revision

Change Summary:

- 250.20(D) now includes prescriptive text to determine if an alternate source is a separately derived source.

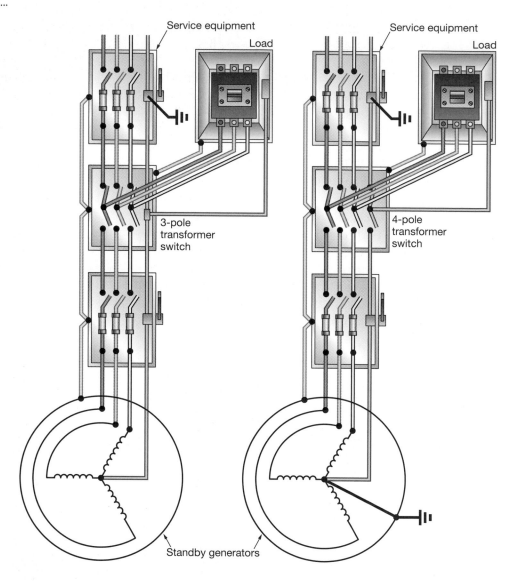

Standby generators

All sources of power other than utility service must be individually considered to determine if the power source is a separately derived system.

Revised 2008 NEC Text:

250.20 ALTERNATING-CURRENT SYSTEMS TO BE GROUNDED
(D) SEPARATELY DERIVED SYSTEMS. Separately derived systems, as covered in 250.20(A) or (B), shall be grounded as specified in 250.30(A). Where an alternate source such as an on site generator is provided with transfer equipment that includes a grounded conductor that is not solidly interconnected to the service-supplied grounded conductor, the alternate source (derived system) shall be grounded in accordance with 250.30(A).

(continued)

FPN No. 1: An alternate ac power source such as an on-site generator is not a separately derived system if the ~~neutral~~ <u>grounded conductor</u> is solidly interconnected to a service-supplied system ~~neutral~~ <u>grounded conductor.</u> <u>An example of such situations is where alternate source transfer equipment does not include a switching action in the grounded conductor and allows it to remain solidly connected to the service-supplied grounded conductor when the alternate source is operational and supplying the load served.</u>

FPN No. 2: For systems that are not separately derived and are not required to be grounded as specified in 250.30, see 445.13 for minimum size of conductors that must carry fault current.

Change Significance:

This revision to 250.20(D) provides positive text referencing the installation of alternate sources such as an onsite generator. This new text states that an alternate source that does not maintain a solid connection to a service-supplied grounded conductor shall be grounded in accordance with 250.30(A).

The information provided in FPN 1 has been expanded to explain when an alternate source such as an onsite generator is not a separately derived system. Where an onsite generator is installed and is connected to premises wiring through transfer switches, the generator is not separately derived unless the grounded conductor is switched and the connection to the service-supplied grounded conductor is opened.

A transformer represents the most common type of a separately derived system.

250.22
CIRCUITS NOT TO BE GROUNDED

Chapter 2 Wiring and Protection

Article 250 Grounding and Bonding

Part II System Grounding

250.22 Circuits Not to Be Grounded

Change Type: Revision

Change Summary:

- 250.22, Circuits Not to Be Grounded, has been revised with a new list item (5) requiring that secondary circuits of lighting systems for lighting fixtures installed below the normal water level of a swimming pool not be grounded.

Revised 2008 NEC Text:

250.22 CIRCUITS NOT TO BE GROUNDED.   The following circuits shall not be grounded:

(1) Circuits for electric cranes operating over combustible fibers in Class III locations, as provided in 503.155

(2) Circuits in health care facilities as provided in 517.61 and 517.160

(3) Circuits for equipment within electrolytic cell working zone as provided in Article 668

(4) Secondary circuits of lighting systems as provided in 411.5(A)

(5) Secondary circuits of lighting systems as provided in 680.23(A)(2)

Change Significance:

This section provides a list of circuits not permitted to be grounded. A new list item (5) has been added to require that secondary circuits of lighting systems for lighting fixtures installed below the normal water level of a swimming pool not be grounded. Although this requirement exists in Article 680, it is included in 250.22 (along with references to other chapters of the NEC) where circuits are not permitted to be grounded.

Where a transformer is used to supply an underwater lighting fixture, 680.23 requires that the transformer be of an isolated winding type with an ungrounded secondary and provided with a grounded metal barrier between the primary and secondary windings (and that it shall be listed as a swimming pool and spa transformer).

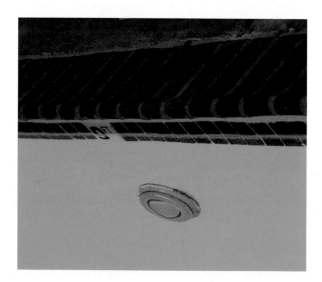

Conductors supplying underwater luminaries supplied from an isolated winding-type secondary are not permitted to be grounded.

Change Summary:

- The requirements for sizing main and system bonding jumpers in 250.28(D) have been revised to address three scenarios.
  (1) Sizing of a single main or system bonding jumper
  (2) Sizing of more than one main bonding jumper
  (3) Sizing of more than one system bonding jumper

The connection between the grounded conductor and the equipment grounding conductor at the service is called the main bonding jumper.

250.28
MAIN BONDING JUMPER AND SYSTEM BONDING JUMPER

Chapter 2 Wiring and Protection

Article 250 Grounding and Bonding

Part II System Grounding

250.28 Main Bonding Jumper and System Bonding Jumper

(D) Size

(1) General

(2) Main Bonding Jumper for Service With More Than One Enclosure

(3) Separately Derived System With More Than One Enclosure

Change Type: Revision

Revised 2008 NEC Text:

250.28 MAIN BONDING JUMPER AND SYSTEM BONDING JUMPER
(D) SIZE. Main bonding jumpers and system bonding jumpers shall ~~not~~ be <u>sized in accordance with 250.28(D)(1) through (D)(3)</u>.

(1) GENERAL. <u>Main bonding jumpers and system bonding jumpers shall not be</u> smaller than the sizes shown in Table 250.66. Where the supply conductors are larger than 1100 kcmil copper or 1750 kcmil aluminum, the bonding jumper shall have an area that is not less than $12\frac{1}{2}$ percent of the area of the largest phase conductor except that, where the phase conductors and the bonding jumper are of different materials (copper or aluminum), the minimum size of the bonding jumper shall be based on the assumed use of phase conductors of the same material as the bonding jumper and with an ampacity equivalent to that of the installed phase conductors.

(continued)

(2) MAIN BONDING JUMPER FOR SERVICE WITH MORE THAN ONE ENCLOSURE.   Where a service consists of more than a single enclosure as permitted in 230.71(A), the main bonding jumper for each enclosure shall be sized in accordance with 250.28(D)(1) based on the largest ungrounded service conductor serving that enclosure.

(3) SEPARATELY DERIVED SYSTEM WITH MORE THAN ONE ENCLOSURE.   Where a separately derived system supplies more than a single enclosure, the system bonding jumper for each enclosure shall be sized in accordance with 250.28(D)(1) based on the largest ungrounded feeder conductor serving that enclosure, or a single system bonding jumper shall be installed at the source and sized in accordance with 250.28(D)(1) based on the equivalent size of the largest supply conductor determined by the largest sum of the areas of the corresponding conductors of each set.

Change Significance:

This revision of 250.28(D) provides clear installation requirements where more than one main or system bonding jumper is installed. This section has been separated into three new second-level subdivisions as follows in order to separate requirements:

- Sizing of a single main or system bonding jumper
- Sizing of more than one main bonding jumper where more than one service disconnecting means is installed as permitted in 230.71(A)
- Sizing of more than one system bonding jumper, providing installation requirements for a single-system bonding jumper at the source or individual system bonding jumpers

The additional text helps to clarify sizing requirements for main bonding jumpers and system bonding jumpers where the service disconnecting means or the first system overcurrent device for separately derived systems consists of more than a single enclosure. The result is user-friendly text that provides the code user with clear installation requirements for minimum sizes of main and system bonding jumpers where multiple enclosures are used for either situation.

Notes:

_____

_____

_____

_____

_____

Change Summary:

- 250.30(A) has been revised to provide prescriptive requirements for terminating a grounded conductor on the load side of the "point of grounding" of separately derived systems.

Revised 2008 NEC Text:

250.30 GROUNDING SEPARATELY DERIVED ALTERNATING-CURRENT SYSTEMS

(A) GROUNDED SYSTEMS. A separately derived ac system that is grounded shall comply with 250.30(A)(1) through (A)(8). <u>Except as otherwise permitted in this article,</u> A̶ <u>a</u> g̶r̶o̶u̶n̶d̶i̶n̶g̶ ̶c̶o̶n̶n̶e̶c̶t̶i̶o̶n̶ <u>grounded conductor</u> shall not be <u>connected</u> m̶a̶d̶e̶ to <u>normally non-current carrying metal parts of equipment, to equipment grounding conductors, or be reconnected to ground</u> a̶n̶y̶ ̶g̶r̶o̶u̶n̶d̶e̶d̶ ̶c̶i̶r̶c̶u̶i̶t̶ ̶c̶o̶n̶d̶u̶c̶t̶o̶r̶ on the load side of the point of grounding of t̶h̶e̶ <u>a</u> separately derived system<u>.</u> e̶x̶c̶e̶p̶t̶ ̶a̶s̶ ̶o̶t̶h̶e̶r̶w̶i̶s̶e̶ ̶p̶e̶r̶m̶i̶t̶t̶e̶d̶ ̶i̶n̶ ̶t̶h̶i̶s̶ ̶a̶r̶t̶i̶c̶l̶e̶.̶

Change Significance:

This revision is part of the global effort within the NEC to provide clearer and more prescriptive text for all grounding and bonding requirements. The changes made in 250.30(A) clarify the present requirement in more prescriptive language.

The general rule of 250.30(A) clearly requires that on the load side of the point of grounding of a separately derived system a grounded conductor is not permitted to be connected to:

- Normally non current-carrying metal parts of equipment
- Equipment grounding conductors
- Ground, the earth

This code change adds more specific restrictions for grounded conductor connections in separately derived systems on the load side of the point of grounding of a separately derived system.

250.30(A)
GROUNDED
SYSTEMS

Chapter 2 Wiring and Protection

Article 250 Grounding and Bonding

Part II System Grounding

250.30 Grounding Separately Derived Alternating-Current Systems

(A) Grounded Systems

Change Type: Revision

Transformers are the most common type of separately derived system.

250.30(A)(4)
GROUNDING ELECTRODE CONDUCTOR, MULTIPLE SEPARATELY DERIVED SYSTEMS

Chapter 2 Wiring and Protection

Article 250 Grounding and Bonding

Part II System Grounding

250.30 Grounding Separately Derived Alternating-Current Systems

(A) Grounded Systems

(4) Grounding Electrode Conductor, Multiple Separately Derived Systems

Change Type: Revision

Change Summary:
- A new last sentence to 250.30(A)(4) now clearly requires that the grounding electrode connection be made at the same point the system bonding jumper is installed for separately derived systems installed with a common grounding electrode conductor.

Revised 2008 NEC Text:
250.30 GROUNDING SEPARATELY DERIVED ALTERNATING-CURRENT SYSTEMS
(A) GROUNDED SYSTEMS
(4) GROUNDING ELECTRODE CONDUCTOR, MULTIPLE SEPARATELY DERIVED SYSTEMS. Where more than one separately derived system is installed, it shall be permissible to connect a tap from each separately derived system to a common grounding electrode conductor. Each tap conductor shall connect the grounded conductor of the separately derived system to the common grounding electrode conductor. The grounding electrode conductors and taps shall comply with 250.30(A)(4)(a) through (A)(4)(c). <u>This connection shall be made at the same point on the separately derived system where the system bonding jumper is installed.</u>
(No change in remainder of 250.30(A)(4))

Change Significance:
250.30(A)(4) provides specific requirements for the installation of a common grounding electrode conductor and grounding electrode taps for more than one separately derived system.

This revision now clearly requires that the termination of the grounding electrode conductor or the grounding electrode conductor tap be located at the same point the system bonding jumper is installed.

The location of this termination is now clarified for multiple separately derived systems with the same text located in 250.30(A)(3) for single separately derived systems.

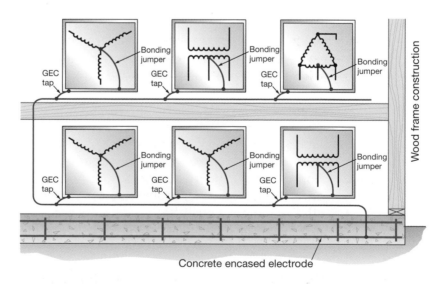

A common grounding electrode conductor is permitted to ground multiple separately derived systems.

Change Summary:

- The 2005 NEC text in 250.32(B)(2) has been revised into an exception to 250.32(B). The permission to use a grounded conductor as an equipment grounding conductor in a feeder or branch circuit to a building or structure has been removed. This method is now permissible in "existing premises wiring systems" only.

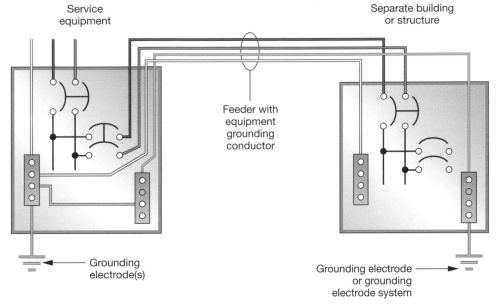

Service equipment

Separate building or structure

Feeder with equipment grounding conductor

Grounding electrode(s)

Grounding electrode or grounding electrode system

Buildings and/or structures supplied by a new feeder or branch circuit require an equipment grounding conductor.

250.32
BUILDINGS OR STRUCTURES SUPPLIED BY A FEEDER(S) OR BRANCH CIRCUIT(S)

Chapter 2 Wiring and Protection

Article 250 Grounding and Bonding

Part II System Grounding

250.32 Buildings or Structures Supplied by a Feeder(s) or Branch Circuit(s)

(B) Grounded Systems

Exception

Change Type: Revision

Revised 2008 NEC Text:

250.32 BUILDINGS OR STRUCTURES SUPPLIED BY A FEEDER(S) OR BRANCH CIRCUIT(S)
(B) GROUNDED SYSTEMS.   For a grounded system at the separate building or structure, ~~the connection to the grounding electrode and grounding or bonding of equipment, structures, or frames required to be grounded or bonded shall comply with either 250.32(B)(1) or (B)(2)~~ .

~~(1) EQUIPMENT GROUNDING CONDUCTOR.~~   an equipment grounding conductor as described in 250.118 shall be run with the supply conductors and <u>be</u> connected to the building or structure disconnecting means and to the grounding electrode(s). The equipment grounding conductor shall be used for grounding or bonding of equipment, structures, or frames required to be grounded or bonded. The equipment grounding conductor shall be sized in accordance with 250.122. Any installed grounded conductor shall not be connected to the equipment grounding conductor or to the grounding electrode(s).

~~(2) GROUNDED CONDUCTOR.~~   (Text is deleted and rolled into the new exception)

(continued)

Exception: For existing premises wiring systems only, the grounded conductor run with the supply to the building or structure shall be permitted to be connected to the building or structure disconnecting means and to the grounding electrode(s) and shall be used for grounding or bonding of equipment, structures, or frames required to be grounded or bonded where all the requirements of (1), (2), and (3) are met:

(1) *An equipment grounding conductor is not run with the supply to the building or structure~,~.*

(2) *There are no continuous metallic paths bonded to the grounding system in each building or structure involved~, and~.*

(3) *Ground-fault protection of equipment has not been installed on the supply side of the feeder(s)~,~.*

Where the grounded conductor is used for grounding in accordance with the provision of this exception, ~T~ the size of the grounded conductor shall not be smaller than the larger of either of the following:

(1) *That required by 220.61*

(2) *That required by 250.122*

Change Significance:

In the 2005 NEC, a feeder or branch circuit to a separate building or structure could utilize an equipment grounding conductor in accordance with 250.32(B)(1)—or the grounded conductor could serve a dual role if the requirements of 250.32(B)(2) were met.

The permission to use the grounded conductor of a feeder or branch circuit in a dual-role capacity (as grounded and equipment grounding conductor) to a building or structure has been removed. This method is now recognized in a new exception to 250.32(B)(1) as permissible in "existing premises wiring systems" only. This revision is similar to how the NEC addresses the grounding of frames of existing dryers and ranges as provided in the exception to 250.140.

Notes:

_____

_____

_____

_____

_____

_____

_____

Change Summary:
- A new Section 250.35 has been added to require an effective ground-fault path from a permanently installed generator to the first disconnecting means.

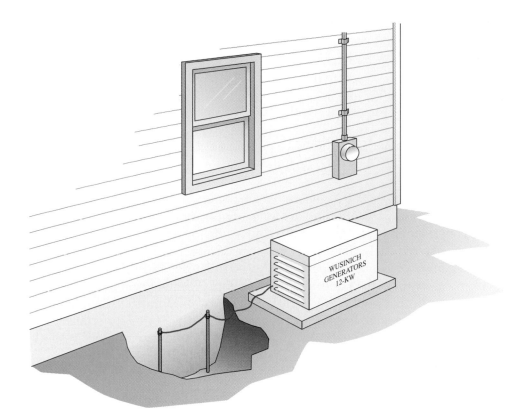

A permanently installed generator which does not have a direct connection to the grounded conductor of the (service supplied) normal system is a separately derived system and must be grounded.

250.35
PERMANENTLY INSTALLED GENERATORS

Chapter 2 Wiring and Protection

Article 250 Grounding and Bonding

Part II System Grounding

250.35 Permanently Installed Generators

(A) Separately Derived System

(B) Nonseparately Derived System

(1) Supply Side of Generator Overcurrent Device

(2) Load Side of Generator Overcurrent Device

Change Type: New

Revised 2008 NEC Text:
250.35 PERMANENTLY INSTALLED GENERATORS. A conductor that provides an effective ground-fault current path shall be installed with the supply conductors from a permanently installed generator(s) to the first disconnecting mean(s) in accordance with (A) or (B).

(A) SEPARATELY DERIVED SYSTEM. Where the generator is installed as a separately derived system, the requirements in 250.30 shall apply.

(B) NONSEPARATELY DERIVED SYSTEM. Where the generator is not installed as a separately derived system, an equipment bonding jumper shall be installed between the generator equipment grounding terminal and the equipment grounding terminal or bus of the enclosure of supplied disconnecting mean(s) in accordance with (B)(1) or (B)(2).

(continued)

(1) SUPPLY SIDE OF GENERATOR OVERCURRENT DEVICE.   The equipment bonding jumper on the supply side of each generator overcurrent device shall be sized in accordance with 250.102(C) based on the size of the conductors supplied by the generator.

(2) LOAD SIDE OF GENERATOR OVERCURRENT DEVICE.   The equipment grounding conductor on the load side of each generator overcurrent device shall be sized in accordance with 250.102(D) based on the rating of the overcurrent device supplied.

Change Significance:

This new section is necessary to address the installation of "permanently" installed generators. This new text will now require that all permanently installed generators are provided with equipment bonding jumpers in accordance with 250.102(C) or (D). The following example illustrates the need for this requirement.

Single-phase 120-/240-V generators are commonly installed at dwellings as optional standby systems and are located outside the building. If the generator is permanently installed, 250.34 does not apply. Many transfer switch suppliers use two pole devices. A connection made in this manner means that the generator is not a separately derived system, and thus 250.30 does not apply. Section 250.32 does not seem to apply either, because the generator is not "supplied"—it *is* the supply. The service grounding electrode system grounds the system because the generator neutral is solidly connected to the service neutral. A specific requirement for providing an effective fault current path from the generator to the first disconnect is necessary.

Permanently installed generators are common in rural areas where power outages may be of prolonged duration during winter storms.

Change Summary:
- The need for continuity of power is no longer a prerequisite for the installation of a high-impedance grounded neutral system.

Revised 2008 NEC Text:
250.36 HIGH-IMPEDANCE GROUNDED NEUTRAL SYSTEMS. High-impedance grounded neutral systems in which a grounding impedance, usually a resistor, limits the ground-fault current to a low value shall be permitted for 3-phase ac systems of 480 volts to 1000 volts where all the following conditions are met:

(1) The conditions of maintenance and supervision ensure that only qualified persons service the installation.

~~(2) Continuity of power is required.~~

~~(3)~~(**2**) Ground detectors are installed on the system.

~~(4)~~(**3**) Line-to-neutral loads are not served.

High-impedance grounded neutral systems shall comply with the provisions of 250.36(A) through (G).

Change Significance:
The installation of high-impedance grounded systems in accordance with 250.36 no longer requires that the designer convince the "authority having jurisdiction" that the system is installed due to the need for continuous service. This is a design issue. The need for continuity of power is subjective and better left on the design table.

In some installations, a high-impedance grounded neutral system may be desirable for increased personnel safety. The provision for "continuity of power" should never impact the safety of persons.

250.36
HIGH-IMPEDANCE GROUNDED NEUTRAL SYSTEMS

Chapter 2 Wiring and Protection

Article 250 Grounding and Bonding

Part II System Grounding

250.36 High-Impedance Grounded Neutral Systems

Change Type: Revision

High-impedance grounded neutral systems must have ground detectors installed.

250.52(A)(2)
METAL FRAME OF THE BUILDING OR STRUCTURE

Chapter 2 Wiring and Protection

Article 250 Grounding and Bonding

Part III Grounding Electrode System and Grounding Electrode Conductor

250.52 Grounding Electrodes

(A) Electrodes Permitted for Grounding

(2) Metal Frame of the Building or Structure

Change Type: Revision

Change Summary:
- The metal frame of a building or structure is not considered a grounding electrode where its connection to earth depends on a connection to an underground metal water pipe.

Revised 2008 NEC Text:
250.52 GROUNDING ELECTRODES
(A) ELECTRODES PERMITTED FOR GROUNDING

(2) METAL FRAME OF THE BUILDING OR STRUCTURE. The metal frame of the building or structure <u>that is connected to the earth by any of the following methods:</u> ~~where any of the following methods are used to make an earth connection:~~

(1) 3.0 m (10 ft) or more of a single structural metal member in direct contact with the earth or encased in concrete that is in direct contact with the earth

(2) ~~The~~ <u>Connecting the</u> structural metal frame ~~is~~ <u>to the reinforcing bars of a concrete-encased electrode as provided in 250.52(A)(3) or ground ring as provided in 250.52(A)(4)</u> ~~bonded to one or more of the grounding electrodes as defined in 250.52(A)(1), (A)(3), or (A)(4)~~

(3) ~~The~~ <u>Bonding the</u> structural metal frame ~~is bonded~~ to one or more of the grounding electrodes as defined in 250.52(A)(5) or (A)(6) that comply with 250.56~~, or~~

(4) <u>O</u>ther approved means of establishing a connection the earth

Change Significance:
250.52(A) provides specific requirements for grounding electrodes. To qualify as a grounding electrode, the individual requirements of 250.52 for each type of electrode must be met.

The metal frame of a building or structure is commonly recognized as a grounding electrode. However, the building steel must be in direct contact with the earth to alone qualify as an electrode. For example, building steel that sits on and is bolted to an existing concrete slab would not qualify as an electrode.

This revision now qualifies building steel as an electrode only where (1) 10 feet or more of a single structural metal member in direct contact with the earth or encased in concrete that is in direct contact with the earth exists; (2) the steel is connected to a concrete-encased electrode or a ground ring; (3) the steel is bonded to rod, pipe, or plate electrodes; or (4) another approved method is used. Bonding an underground metal water pipe to building steel does not qualify the building steel as an electrode.

Notes:

_____

_____

_____

The metal frame of a building or structure may qualify as a grounding electrode alone or by being connected to another electrode per 250.52(A)(2).

Change Summary:
- Where multiple concrete-encased electrodes exist separately in a building or structure, only one is required to be bonded to the grounding electrode system.

Revised 2008 NEC Text:
250.52 GROUNDING ELECTRODES
(A) ELECTRODES PERMITTED FOR GROUNDING
(3) CONCRETE-ENCASED ELECTRODE.   An electrode encased by at least 50 mm (2 in.) of concrete, located <u>horizontally</u> ~~within and~~ near the bottom <u>or vertically, and within that portion</u> of a concrete foundation or footing that is in direct contact with the earth, consisting of at least 6.0 m (20 ft) of one or more bare or zinc galvanized or other electrically conductive coated steel reinforcing bars or rods of not less than 13 mm (½ in.) in diameter, or consisting of at least 6.0 m (20 ft) of bare copper conductor not smaller than 4 AWG. Reinforcing bars shall be permitted to be bonded together by the usual steel tie wires or other effective means. <u>Where multiple concrete-encased electrodes are present at a building or structure, it shall be permissible to bond only one into the grounding electrode system.</u>

Change Significance:
The terms *horizontally* and *vertically* have been added to the first sentence in an effort to provide additional clarity. Horizontal rebar and vertical rebar (at least 20 feet) are now clearly acceptable as concrete-encased electrodes as long as they are located "within that portion" of concrete in direct contact with the earth.

A new last sentence addresses multiple concrete-encased electrodes on a single site or building. In many cases, a building or structure is structurally designed with more than one concrete-encased electrode. For example, a building may be elevated to allow parking on the first floor. The building may sit on a dozen poured concrete columns, each of which sits on a foundation constituting a concrete-encased electrode as outlined in 250.52(A)(3). The building design does not provide a conductive path between all 12 concrete-encased electrodes.

This revision now allows the installer to bond only one of the concrete-encased electrodes to the grounding electrode system. There is no prohibition to bonding them all, but this is not a requirement.

Notes:

_____

_____

_____

_____

250.52(A)(3)
CONCRETE-ENCASED ELECTRODE

Chapter 2 Wiring and Protection

Article 250 Grounding and Bonding

Part III Grounding Electrode System and Grounding Electrode Conductor

250.52 Grounding Electrodes

(A) Electrodes Permitted for Grounding

(3) Concrete-Encased Electrode

Change Type: Revision

A caisson-type concrete footing containing vertically installed reinforcing steel rebar is recognized as a concrete-encased electrode.

250.52(A)
(5) ROD AND PIPE ELECTRODES AND (6) OTHER LISTED ELECTRODES

Chapter 2 Wiring and Protection

Article 250 Grounding and Bonding

Part III Grounding Electrode System and Grounding Electrode Conductor

250.52 Grounding Electrodes

(A) Electrodes Permitted for Grounding

(5) Rod and Pipe Electrodes

(6) Other Listed Electrodes

Change Type: Revision and New

Change Summary:
- The requirements for rod and pipe electrodes have been revised for clarity.
- A new list item (6) has been included to recognize "other" listed electrodes, which would include chemical- or electrolytic-type grounding electrodes.

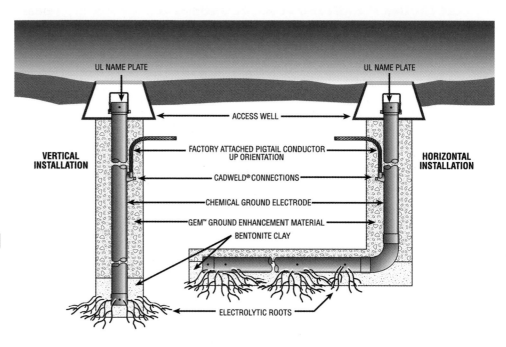

Chemical- or electrolytic-type grounding electrodes are now recognized in 250.52(A)(6). Courtesy of ERICO.

Revised 2008 NEC Text:
250.52 GROUNDING ELECTRODES
(A) ELECTRODES PERMITTED FOR GROUNDING

(5) ROD AND PIPE ELECTRODES. Rod and pipe electrodes shall not be less than ~~2.5~~ 2.44 m (8 ft) in length and shall consist of the following materials.

  (a) Grounding electrodes of pipe or conduit shall not be smaller than metric designator 21 (trade size ¾) and, where of ~~iron or~~ steel, shall have the outer surface galvanized or otherwise metal-coated for corrosion protection.

  (b) Grounding electrodes ~~of rods~~ of stainless steel, copper or zinc coated ~~iron or~~ steel shall be at least 15.87 mm (⅝ in.) in diameter,~~,~~ unless ~~Stainless steel rods less than 16 mm (⅝ in.) in diameter, nonferrous rods, or their equivalent shall be~~ listed and ~~shall~~ not ~~be~~ less than ~~13~~ 12.70 mm (½ in.) in diameter.

(6) OTHER LISTED ELECTRODES. Other listed grounding electrodes shall be permitted.

(continued)

Change Significance:

The requirements of 250.52(A)(5) have been revised to eliminate "iron" pipe or conduit and recognize only steel pipe or conduit that is galvanized or protected from corrosion. The requirements for rods are revised to recognize stainless rods or steel rods that are copper or zinc coated and $\frac{5}{8}$ inch in diameter. The text is further revised to permit any type of rod to be smaller than $\frac{5}{8}$ inch but not smaller than $\frac{1}{2}$ inch, provided the rod is listed.

A new list item (6) has been added to recognize "other listed electrodes" such as a chemical or electrolytic ground rod enhanced by the use of chemicals or salts to lower soil resistance. These enhanced grounding electrodes are listed and are used in many applications, from areas where soil resistance is high to specialized electronic applications where the design of the system seeks to enhance the grounding electrode system. The addition of this new list item to recognize other listed electrodes now clearly permits the use of these devices as a grounding electrode.

Notes:

_____

_____

_____

_____

_____

_____

_____

_____

_____

_____

250.54
AUXILIARY GROUNDING ELECTRODES

Chapter 2 Wiring and Protection

Article 250 Grounding and Bonding

Part III Grounding Electrode System and Grounding Electrode Conductor

250.54 Auxiliary Grounding Electrodes

Change Type: Revision

Change Summary:
- Where a grounding electrode is not required by the NEC but is installed for design purposes, it is no longer considered a "supplemental" grounding electrode. The term *auxiliary grounding electrode* clarifies that this electrode is not required and is installed for design purposes.

Revised 2008 NEC Text:
250.54 ~~SUPPLEMENTARY~~ <u>AUXILIARY</u> GROUNDING ELECTRODES. ~~Supplementary~~ <u>One or more</u> grounding electrodes shall be permitted to be connected to the equipment grounding conductors specified in 250.118 and shall not be required to comply with the electrode bonding requirements of 250.50 or 250.53(C) or the resistance requirements of 250.56, but the earth shall not be used as an effective ground-fault current path as specified in 250.4(A)(5) and 250.4(B)(4).

Change Significance:
The primary purpose of grounding electrodes in the NEC are to ground electrical systems for all of the reasons set forth in 250.4. However, grounding electrodes are often designed and installed in electrical installations for purposes other than meeting the minimum requirements of the NEC. For example, an electrical installation of pole lights in a parking lot may be designed with a ground rod driven at each pole light with a 6 AWG copper bonded to each pole base. This is not a requirement of the NEC.

The term *supplemental* leads the code user to believe that a "supplemental electrode" is compensating for some type of deficiency. Where electrodes are installed for design purposes, they are "auxiliary." This revision clarifies the requirements of 250.54 by replacing the term *supplemental* with the term *auxiliary.*

An electrode installed at a pole-mounted luminaire location would be considered an auxiliary grounding electrode.

Change Summary:
- 250.56 has been revised to prohibit the metal frame of a building/structure or a concrete-encased electrode from augmenting a rod, pipe, or plate electrode that does not have a resistance to ground of 25 ohms or less.

Revised 2008 NEC Text:
250.56 RESISTANCE OF ROD, PIPE, AND PLATE ELECTRODES. A single electrode consisting of a rod, pipe, or plate that does not have a resistance to ground of 25 ohms or less shall be augmented by one additional electrode of any of the types specified by 250.52(A)(4)(2) through (A)(8)(7). Where multiple rod, pipe, or plate electrodes are installed to meet the requirements of this section, they shall not be less than 1.8 m (6 ft) apart.

FPN: The paralleling efficiency of rods longer than 2.5 m (8 ft) is improved by spacing greater than 1.8 m (6 ft).

Change Significance:
This section has been debated in many forums for many NEC cycles. This requirement mandates that a single electrode consisting of a rod, pipe, or plate that does not have a resistance to ground of 25 ohms or less shall be augmented by one additional electrode. 250.52(A)(2) clearly permits the metal frame of a building or structure to "stand alone" as a grounding electrode. 250.52(A)(3) clearly permits a concrete-encased electrode to "stand alone" as a grounding electrode.

Using these two electrodes to "augment" a made electrode was incorrect. This section will now permit a rod, pipe, or plate electrode to be augmented by "(4) a ground ring, (5) rod or pipe electrode, (6) other listed electrodes such as chemical or electrolytic, (7) plate electrode, or (8) other local metal underground systems or structures such as a well."

250.56
RESISTANCE OF ROD, PIPE, AND PLATE ELECTRODES

Chapter 2 Wiring and Protection

Article 250 Grounding and Bonding

Part III Grounding Electrode System and Grounding Electrode Conductor

250.56 Resistance of Rod, Pipe, and Plate Electrodes

Change Type Revision

Rod-type grounding electrodes must be installed with at least 8 feet of the rod in contact with the soil.

250.64

GROUNDING ELECTRODE CONDUCTOR INSTALLATION

Change Type: Revision

Change Summary:

- New text is added to 250.64, 66, and 68 to clarify that those rules do not apply to grounding electrode conductor installations that may be installed for other reasons—such as auxiliary (supplemental) grounding conductors installed for electromagnetic compatibility, lightning protection, establishing ground planes for antennas, and similar reasons.

A cell tower may have grounding electrodes installed as a design consideration.

Revised 2008 NEC Text:

250.64 GROUNDING ELECTRODE CONDUCTOR INSTALLATION. Grounding electrode conductors <u>at the service, at each building or structure where</u>

(continued)

supplied by a feeder(s) or branch circuit(s), or at a separately derived system shall be installed as specified in 250.64(A) through (F).

250.66 SIZE OF ALTERNATING-CURRENT GROUNDING ELECTRODE CONDUCTOR.   The size of the grounding electrode conductor at the service, at each building or structure where supplied by a feeder(s) or branch circuit(s), or at a separately derived system of a grounded or ungrounded ac system shall not be less than given in Table 250.66, except as permitted in 250.66(A) through (C).

250.68 GROUNDING ELECTRODE CONDUCTOR AND BONDING JUMPER CONNECTION TO GROUNDING ELECTRODES.   The connections of grounding electrode conductor at the service, at each building or structure where supplied by a feeder(s) or branch circuit(s), or at a separately derived system and associated bonding jumper(s) shall be made as specified 250.68(A) and (B).

Change Significance:

The revised definition of "grounding electrode" in Article 100 reads as follows:

GROUNDING ELECTRODE CONDUCTOR.   A conductor used to connect the system grounded conductor or the equipment to a grounding electrode or to a point on the grounding electrode system.

The revision of this definition will now qualify a conductor from a remotely located antenna (equipment) to a ground rod (grounding electrode) as a grounding electrode conductor. This would now impose all of the rules for the installation, sizing, and connection of this conductor to the rules in 250.64, 66, and 68. These rules for grounding electrode conductors are intended to be applied where systems or equipment are grounded at the service, at each building or structure where supplied by a feeder(s) or branch circuit(s), or at a separately derived system. Where grounding electrodes are installed for other reasons, the rules in 250.64, 66, and 68 will not apply.

Notes:

_____

_____

_____

_____

_____

_____

250.64(D)
SERVICE WITH MULTIPLE DISCONNECTING MEANS ENCLOSURES

Chapter 2 Wiring and Protection

Article 250 Grounding and Bonding

Part III Grounding Electrode System and Grounding Electrode Conductor

250.64 Grounding Electrode Conductor Installation

(D) Service With Multiple Disconnecting Means Enclosures

(1) Grounding Electrode Conductor Taps

(2) Individual Grounding Electrode Conductors

(3) Common Location

Change Type: Revision

Change Summary:
- 250.64(D) has been revised into three second-level subdivisions for clarity. The requirements for the installation of the grounding electrode conductor in a service with multiple disconnecting means enclosures are now logically separated as follows:
  - **(1)** Grounding Electrode Conductor Taps
  - **(2)** Individual Grounding Electrode Conductors
  - **(3)** Common Location

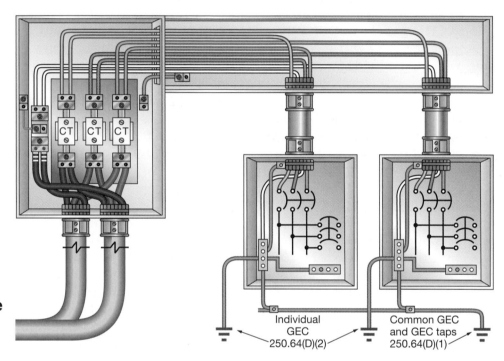

A common grounding electrode conductor and taps or individual grounding electrode conductors are permitted.

Revised 2008 NEC Text:
250.64 GROUNDING ELECTRODE CONDUCTOR INSTALLATION
(D) SERVICE WITH MULTIPLE DISCONNECTING MEANS ENCLOSURES.
Where a service consists of more than a single enclosure as permitted in 230.71(A), <u>grounding electrode connections shall be made in accordance with (1), (2) or (3).</u>

(1) GROUNDING ELECTRODE CONDUCTOR TAPS. Where the service is installed as permitted by 230.40, Exception No. 2, a common grounding electrode conductor and grounding electrode conductor taps shall be installed. The common grounding electrode conductor shall be sized in accordance with 250.66, based on the sum of the circular mil area of the largest ungrounded service entrance conductor(s). Where the service-entrance conductors connect directly to a service drop or service lateral, the common grounding electrode conductor shall be sized in accordance with Table

(continued)

250.66, Note 1. A tap conductor shall extend to the inside of each service disconnecting means enclosure. The grounding electrode conductor taps shall be sized in accordance with 250.66 for the largest conductor serving the individual enclosure. The tap conductors shall be connected to the common grounding electrode conductor <u>by exothermic welding or with connectors listed as grounding and bonding equipment</u> in such a manner that the common grounding electrode conductor remains without a splice or joint.

(2) INDIVIDUAL GROUNDING ELECTRODE CONDUCTORS. <u>A grounding electrode conductor shall be connected between the grounded conductor in each service equipment disconnecting means enclosure and the grounding electrode system. Each grounding electrode conductor shall be sized in accordance with 250.66 based on the service-entrance conductor(s) supplying the individual service disconnecting means.</u>

(3) COMMON LOCATION. <u>A grounding electrode conductor shall be connected to the grounded service conductor(s) in a wireway or other accessible enclosure on the supply side of the service disconnecting means. The connection shall be made with exothermic welding or a connector listed as grounding and bonding equipment. The grounding electrode conductor shall be sized in accordance with 250.66 based on the service-entrance conductor(s) at the common location where the connection is made.</u>

Change Significance:

The revision in 250.64(D)(1) cited previously has been reformatted from previous text with only the "new" requirement underlined. This revision now recognizes two methods of connecting the grounding electrode in services with multiple disconnecting means enclosures.

Previous code text recognized only grounding electrode conductor taps and did not recognize the common practice of installing individual grounding electrode conductors to each service enclosure. These methods are now clearly illustrated in 250.64(D)(1) and (D)(2).

The termination of grounding electrode conductor taps to the common grounding electrode conductor is now clarified in 250.64(D)(1). Only the exothermic welding process or connectors listed as grounding and bonding equipment are permitted. 250.64(D)(3) has been added to clearly require that the grounding electrode connection for services with multiple disconnecting means enclosures be made at an accessible location.

Notes:

250.64(F)

INSTALLATION TO ELECTRODE(S)

Chapter 2 Wiring and Protection

Article 250 Grounding and Bonding

Part III Grounding Electrode System and Grounding Electrode Conductor

250.64 Grounding Electrode Conductor Installation

(F) Installation to Electrode(s)

Change Type: Revision

Bonding jumpers from grounding electrodes are permitted to terminate on a common busbar.

Change Summary:

*   250.64(F), Installation to Electrode(s), is separated into three list items for clarity. Requirements for terminating bonding jumpers and grounding electrode conductors from grounding electrodes to copper and aluminum busbar are logically relocated here from 250.64(C).

Revised 2008 NEC Text:

250.64 GROUNDING ELECTRODE CONDUCTOR INSTALLATION (F) INSTALLATION TO ELECTRODE(S). ~~A~~ <u>Grounding electrode conductor(s) and bonding jumpers interconnecting grounding electrodes</u> shall be ~~permitted~~ <u>installed in accordance with (1), (2) or (3).</u> The grounding electrode conductor shall be sized for the largest grounding electrode conductor required among all the electrodes connected to it.

<u>(1) The grounding electrode conductor shall be permitted</u> to be run to any convenient grounding electrode available in the grounding electrode system <u>where the other electrode(s), if any, are connected by bonding jumpers per 250.53(C).</u>~~, or~~

<u>(2) Grounding electrode conductor(s) shall be permitted</u> to one or more grounding electrode(s) individually, <u>or to the aluminum or copper busbar as permitted in 250.64(C).</u>

<u>(3) Bonding jumper(s) from grounding electrode(s) shall be permitted to be connected to an aluminum or copper busbar not less than 6 mm × 50 mm ($\frac{1}{4}$ in. × 2 in.). The busbar shall be securely fastened and shall be installed in an accessible location. Connections shall be made by a listed connector or by the exothermic welding process. The grounding electrode conductor shall be permitted to be run to the busbar. Where aluminum busbars are used, the installation shall comply with 250.64(A).</u>

Change Significance:

250.64(F) has been separated into list items for clarity. Provisions from 250.64(C)(3) have been incorporated into 250.64(F)(3) because they are more appropriate in this first-level subdivision, which deals with connections to electrodes.

This first-level subdivision addresses the installation of grounding electrode conductors to the electrodes and bonding jumpers interconnecting grounding electrodes. List item (1) addresses a GEC permitted to be run to any grounding electrode, provided all of the electrodes are bonded. List item (2) addresses a single or multiple grounding electrode conductors run individually to different grounding electrodes.

List item (3) has been added to 250.64(F). These requirements were previously located in 250.64(C), which addresses the requirement for grounding electrode conductors to be "continuous." The relocation of these requirements from 250.64(C) to (F) is user friendly and places this text in a more logical format. This new list item addresses the requirements for connecting bonding jumpers and grounding electrode conductors from grounding electrodes to busbar.

Change Summary:

- This revision clarifies that in general "all mechanical elements" (including hardware) used to terminate a grounding electrode conductor must be accessible. The exception for fireproofed metal has been revised to include all mechanical elements.

Revised 2008 NEC Text:

250.68 GROUNDING ELECTRODE CONDUCTOR AND BONDING JUMPER CONNECTION TO GROUNDING ELECTRODES

(A) ACCESSIBILITY. ~~The connection of~~ All mechanical elements used to terminate a grounding electrode conductor or bonding jumper to a grounding electrode shall be accessible.

Exception No. 1: An encased or buried connection to a concrete-encased, driven, or buried grounding electrode shall not be required to be accessible.

Exception No. 2: ~~An e~~ Exothermic or irreversible compression connections used at terminations, together with the mechanical means used to attach such terminations to fire-proofed structural metal whether or not the mechanical means is reversible, shall not be required to be accessible.

Change Significance:

This revision provides needed clarity for the general requirement of "accessibility" in 250.68 for the connection of a grounding electrode conductor or bonding jumper to a grounding electrode. Previous code text required that the conductor connection be accessible. The revised text now requires that "all mechanical elements" involved in the termination (including bolts, nuts, and screws) be accessible.

Exception No. 2 has been revised to allow all mechanical means of termination (as well as the conductor termination) to be covered with fireproofing material.

250.68

GROUNDING ELECTRODE CONDUCTOR AND BONDING JUMPER CONNECTION TO GROUNDING ELECTRODES

Chapter 2 Wiring and Protection

Article 250 Grounding and Bonding

Part III Grounding Electrode System and Grounding Electrode Conductor

250.68 Grounding Electrode Conductor and Bonding Jumper Connection to Grounding Electrodes

(A) Accessibility

Exception No. 2

Change Type: Revision

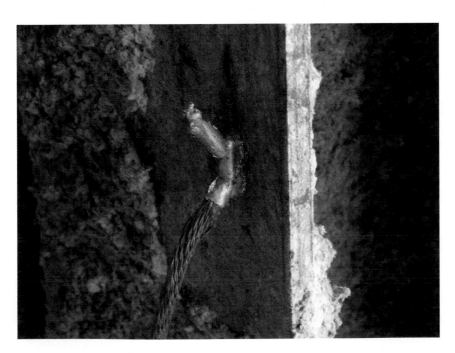

Grounding electrode conductor connections to building steel are permitted to be covered with fireproofing.

250.94

BONDING FOR OTHER SYSTEMS

Chapter 2 Wiring and Protection

Article 250 Grounding and Bonding

Part V Bonding

250.94 Bonding for Other Systems

Change Type: Revision

Change Summary:

* 250.94 has been revised to require an intersystem bonding termination. This device will be located and installed as now required in this section to create a dedicated and well-defined location for terminating the grounding conductors required in Articles 770, 800, 810, 820, and 830.

Requirements for intersystem bonding require accessible means for connecting at least three intersystem bonding jumpers.

Revised 2008 NEC Text:

250.94 BONDING FOR OTHER SYSTEMS.   An intersystem bonding termination for connecting intersystem bonding and grounding conductors required for other systems shall be provided external to enclosures at the service equipment and at the disconnecting means for any additional buildings or structures. The intersystem bonding termination shall be accessible for connection and inspection. The intersystem bonding termination shall have the capacity for connection of not less than three intersystem bonding conductors. The intersystem bonding termination device shall not interfere with opening a service or metering equipment enclosure. The intersystem bonding termination shall be one of the following:

(1) A set of terminals securely mounted to the meter enclosure and electrically connected to the meter enclosure. The terminals shall be listed as grounding and bonding equipment.

(2) A bonding bar near the service equipment enclosure, meter enclosure or raceway for service conductors. The bonding bar shall be connected with a minimum 6 AWG copper conductor to an equipment grounding conductor(s)

(continued)

in the service equipment enclosure, meter enclosure or exposed nonflexible metallic raceway.

(3) A bonding bar near the grounding electrode conductor. The bonding bar shall be connected to the grounding electrode conductor with a minimum 6 AWG copper conductor.

Exception: In existing buildings or structures where any of the intersystem bonding and grounding conductors required by 770.93, 800.100(B), 810.21(F), 820.100(B), 830.100(B) exist, installation of the intersystem bonding termination is not required.

An accessible means external to enclosures for connecting intersystem bonding and grounding electrode conductors shall be <u>permitted</u> ~~provided~~ at the service equipment and at the disconnecting means for any additional buildings or structures by at least one of the following means:

(1) *Exposed nonflexible metallic raceways*

(2) *Exposed grounding electrode conductor*

(3) *Approved means for the external connection of a copper or other corrosion-resistant bonding or grounding conductor to the grounded raceway or equipment*

FPN No. 1: A 6 AWG copper conductor with one end bonded to the grounded nonflexible metallic raceway or equipment and with 150 mm (6 in.) or more of the other end made accessible on the outside wall is an example of the approved means covered in 250.94, Exception, item (3).

FPN No. 2: (No change)

Change Significance:

This is one of several changes in this cycle that are supplemented by a new definition of "intersystem bonding termination" in Article 100 to improve the requirements related to intersystem bonding and grounding of communication systems. The intent of these changes is to create a dedicated and well-defined location for terminating the grounding conductors required in Articles 770, 800, 810, 820, and 830.

Attachment of the grounding conductors required by 770.93, 800.100(B), 810.21(F), 820.100(B), and 830.100(B) to the intersystem bonding termination provided external to enclosures at the service equipment and at the disconnecting means for any additional buildings or structures accomplishes "intersystem bonding."

The general rule in the revised requirement will locate a set of terminals at the meter enclosure or a bonding bar at the service equipment, meter enclosure, or raceway for service conductors. A bonding bar may also be located near the grounding electrode conductor. Each intersystem bonding termination shall have the capacity for connection of not less than three intersystem bonding conductors. This will allow for intersystem bonding between the power system and communication systems, satellite systems, and CATV systems.

The new exception exempts existing buildings and structures from the requirement to install an intersystem bonding termination. However, this exception includes the previous code text of 250.94 and requires an accessible means external to enclosures for connecting intersystem bonding and grounding electrode conductors.

250.104 (A)(2)
BUILDINGS OF MULTIPLE OCCUPANCY

Chapter 2 Wiring and Protection

Article 250 Grounding and Bonding

Part V Bonding

250.104 Bonding of Piping Systems and Exposed Structural Steel

(A) Metal Water Piping

(2) Buildings of Multiple Occupancy

Change Type: Revision

Change Summary:
- The size of the bonding jumper required in 250.104(A)(2) has been clarified by specifying that the code user apply the rating of the overcurrent protective device supplying the occupancy as indicated in Table 250.122.

Revised 2008 NEC Text:
250.104 BONDING OF PIPING SYSTEMS AND EXPOSED STRUCTURAL STEEL
(A) METAL WATER PIPING
(2) BUILDINGS OF MULTIPLE OCCUPANCY.   In buildings of multiple occupancy where the metal water piping system(s) installed in or attached to a building or structure for the individual occupancies is metallically isolated from all other occupancies by use of nonmetallic water piping, the metal water piping system(s) for each occupancy shall be permitted to be bonded to the equipment grounding terminal of the panelboard or switchboard enclosure (other than service equipment) supplying that occupancy. The bonding jumper shall be sized in accordance with Table 250.122, <u>based on the rating of the overcurrent protective device for the circuit supplying the occupancy.</u>

Change Significance:
The text added to the last sentence of this section provides clarity and usability by specifying that the overcurrent device supplying the occupancy is used to determine the size of the bonding jumper. Where multiple occupancies exist in a building and metal water piping systems are isolated from each other, the size of the required bonding jumper is based on the rating of the overcurrent protective device supplying the occupancy as indicated in Table 250.122.

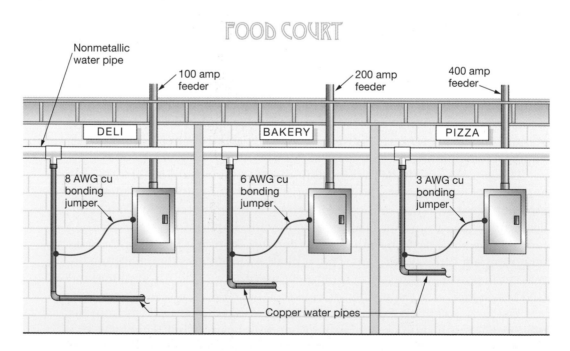

Metal water piping that does not qualify as a grounding electrode due to plastic pipe used for the incoming domestic water supply must be bonded per 250.104(A).

Change Summary:
- All Class 1 non power-limited circuits operating at 50 volts or more are now required to be grounded.

Revised 2008 NEC Text:
250.112 FASTENED IN PLACE OR CONNECTED BY PERMANENT WIRING METHODS (FIXED)—SPECIFIC

(I) ~~POWER-LIMITED~~ **REMOTE-CONTROL, SIGNALING, AND FIRE ALARM CIRCUITS.** <u>Equipment supplied by Class 1 circuits shall be grounded unless operating at less than 50 volts.</u> Equipment supplied by Class 1 power-limited circuits, ~~and Class 1,~~ <u>by</u> Class 2, and Class 3 remote-control and signaling circuits, and by fire alarm circuits, shall be grounded where system grounding is required by Part II or Part VIII of this article.

Change Significance:
The first sentence has been revised to clarify grounding requirements for Class 1 circuits. This revision provides prescriptive text requiring that all Class 1 circuits operating at 50 volts or more be grounded.

The second sentence remains unchanged and addresses Class 1 power-limited circuits, Class 2 and Class 3 remote-control and signaling circuits, and fire alarm circuits as follows: "Equipment supplied by Class 1 power-limited circuits, Class 2, and Class 3 remote-control and signaling circuits, and by fire alarm circuits, shall be grounded where system grounding is required by Part II or Part VIII of this article."

250.112(I)
REMOTE-CONTROL, SIGNALING, AND FIRE ALARM CIRCUITS

Chapter 2 Wiring and Protection

Article 250 Grounding and Bonding

Part VI Equipment Grounding and Equipment Grounding Conductors

250.112 Fastened in Place or Connected by Permanent Wiring Methods (Fixed)—Specific

(I) Remote-Control, Signaling, and Fire Alarm Circuits

Change Type: Revision

Class 1 circuits of 50 volts or more must be grounded.

250.119

IDENTIFICATION OF EQUIPMENT GROUNDING CONDUCTORS

Chapter 2 Wiring and Protection

Article 250 Grounding and Bonding

Part VI Equipment Grounding and Equipment Grounding Conductors

250.119 Identification of Equipment Grounding Conductors

(A) Conductors Larger Than 6 AWG

Change Type: Revision

An exception is added to permit low-voltage thermostat wiring to use the color green for other than equipment grounding purposes.

Change Summary:

- A new exception is added to permit conductors with green insulation used in power-limited Class 2 or Class 3 circuits operating at less than 50 volts to be used for other than equipment grounding purposes.
- The identification of equipment grounding conductors larger than 6 AWG is clarified by specifying that the marking occur at the termination of the conductor.

Revised 2008 NEC Text:

250.119 IDENTIFICATION OF EQUIPMENT GROUNDING CONDUCTORS (No change to parent text.)

Exception: Power limited, Class 2, or Class 3 circuit cables containing only circuits operating at less than 50 volts shall be permitted to use a conductor with green insulation for other than equipment grounding purposes.

(A) CONDUCTORS LARGER THAN 6 AWG. Equipment grounding conductors larger than 6 AWG shall comply with 250.119(A)(1) and (A)(2).

(1) An insulated or covered conductor larger than 6 AWG shall be permitted, at the time of installation, to be permanently identified as an equipment grounding conductor at each end and at every point where the conductor is accessible.

Exception: Conductors larger than 6 AWG shall not be required to be marked in conduit bodies that contain no splices or unused hubs.

(2) Identification shall encircle the conductor and shall be accomplished by one of the following:

    a. Stripping the insulation or covering from the entire exposed length

    b. Coloring the ~~exposed~~ insulation or covering green <u>at the termination</u>

    c. Marking the ~~exposed~~ insulation or covering with green tape or green adhesive labels <u>at the termination</u>

Change Significance:

A new exception has been added to permit conductors with green insulation used in power-limited Class 2 or Class 3 circuits operating at less than 50 volts to be used for other than equipment grounding purposes. There are many low-voltage applications (such as thermostat wiring) that use the color green for other than equipment grounding purposes.

The second revision shown in 250.119(A)(2) provides clear guidance to code users by clarifying that the identification of the equipment grounding conductor is only required at the termination. The common practice of a few wraps of green tape is all that is required to comply with this requirement. This revision now correlates with the marking provisions for grounded conductors larger than 6 AWG by specifying that the identification occur at terminations.

Notes:

_____

_____

Change Summary:
- 250.122(C) has been modified to recognize the use of a single equipment grounding conductor with multiple circuits in a cable tray.

Revised 2008 NEC Text:
250.122 SIZE OF EQUIPMENT GROUNDING CONDUCTORS
(C) MULTIPLE CIRCUITS.   Where a single equipment grounding conductor is run with multiple circuits in the same raceway, ~~or~~ cable <u>or cable tray</u>, it shall be sized for the largest overcurrent device protecting conductors in the raceway, ~~or~~ cable <u>or cable tray. Equipment grounding conductors installed in cable trays shall meet the minimum requirements of 392.3(B)(1)(c).</u>

Change Significance:
This revision now clearly permits the long-standing practice of using a common equipment grounding conductor as permitted in 250.122(C) in a cable tray. A new last sentence requires that where an equipment grounding conductor is installed with multiple circuits in a cable tray the requirements of 392.3(B)(1)(c) are met. This requires that the equipment grounding conductor installed with multiple circuits in a cable tray be insulated, covered, or bare and not smaller than 4 AWG.

250.122(C)
MULTIPLE CIRCUITS

Chapter 2 Wiring and Protection

Article 250 Grounding and Bonding

Part VI Equipment Grounding and Equipment Grounding Conductors

250.122 Size of Equipment Grounding Conductors

(C) Multiple Circuits

Change Type: Revision

A single-equipment grounding conductor is permitted for multiple circuits in a cable tray.

250.122(D)
MOTOR CIRCUITS

Chapter 2 Wiring and Protection

Article 250 Grounding and Bonding

Part VI Equipment Grounding and Equipment Grounding Conductors

250.122 Size of Equipment Grounding Conductors

(D) Motor Circuits

Change Type: Revision

Change Summary:

- 250.122(D) has been completely revised to provide specific guidance to determine adequate sizes of equipment grounding conductors for motor installations.

Revised 2008 NEC Text:

250.122 SIZE OF EQUIPMENT GROUNDING CONDUCTORS
(D) MOTOR CIRCUITS.   Equipment grounding conductors for motor circuits shall be sized in accordance with (D)(1) or (D)(2).

(1) General. The equipment grounding conductor size shall not be smaller than determined by 250.122(A) based on the rating of the branch-circuit short-circuit and ground-fault protective device.

(2) Instantaneous-Trip Circuit Breaker and Motor Short-Circuit Protector. Where the overcurrent device is an instantaneous-trip circuit breaker or a motor short-circuit protector, the equipment grounding conductor shall be sized not smaller than that given by 250.122(A) using the maximum permitted rating of a dual element time-delay fuse selected for branch-circuit short-circuit and ground-fault protection in accordance with 430.52(C)(1), Exception No. 1.

Change Significance:

The previous text in the 2005 NEC permitted a motor installation to size the equipment grounding conductor according to the rating of the motor overload protective device. This meant that the nameplate current of a motor (almost always lower than table current values) was used to determine the overload rating in accordance with 430.32, which in most cases is 125% of the nameplate current rating. This value was applied in 250.122(A) to determine the size of the EGC. The branch-circuit, short-circuit, ground-fault protective device for the motor was sized using table current (the values in 430.52 and 430.52 Exception No. 1), which allows the next higher standard rated device to be applied.

The net result of this method could result in equipment grounding conductors smaller than the circuit conductors and undersized according to Table 250.122. This new text requires an EGC that is (1) based on the BC, SC, or GF device and 250.122(A) or (2) based on applying 430.52 Exception No. 1 with dual-element time-delay fuses and 250.122(A).

Motors are provided with overcurrent protection through the use of overload and of branch-circuit short-circuit ground-fault protective devices.

Change Summary:
* The permission to run multiconductor cables in parallel with undersized equipment grounding conductors where ground fault protection is applied has been removed.

Revised 2008 NEC Text:
250.122 SIZE OF EQUIPMENT GROUNDING CONDUCTORS
(F) CONDUCTORS IN PARALLEL. Where conductors are run in parallel in multiple raceways or cables as permitted in 310.4, the equipment grounding conductors, where used, shall be run in parallel in each raceway or cable. ~~One of the methods in 250.122(F)(1) or (F)(2) shall be used to ensure the equipment grounding conductors are protected.~~

~~**(1) BASED ON RATING OF OVERCURRENT PROTECTIVE DEVICE.**~~ Each parallel equipment grounding conductor shall be sized on the basis of the ampere rating of the overcurrent device protecting the circuit conductors in the raceway or cable in accordance with Table 250.122.

~~**(2) GROUND FAULT PROTECTION OF EQUIPMENT INSTALLED.** Where ground fault protection of equipment is installed, each parallel equipment grounding conductor in a multiconductor cable shall be permitted to be sized in accordance with Table 250.122 on the basis of the trip rating of the ground fault protection where the following conditions are met:~~

~~(1) Conditions of maintenance and supervision ensure that only qualified persons will service the installation.~~

~~(2) The ground fault protection equipment is set to trip at not more than the ampacity of a single ungrounded conductor of one of the cables in parallel.~~

~~(3) The ground fault protection is listed for the purpose of protecting the equipment grounding conductor.~~

Change Significance:
No products have ever been listed or introduced for this GFPE application.

The general rule in 250.122(F) requires that where EGCs are run in parallel the EGCs must be sized in accordance with Table 250.122 based on the overcurrent protective device protecting the parallel conductors.

250.122(F)(2) was added in the NEC in an effort to allow the use of multiconductor cable assemblies in parallel. The reason a typical cable assembly (not custom made) cannot be applied in a parallel feeder is due to the size of the equipment grounding conductors in the cable assembly. For example, a three-phase, three-wire, 480-volt/800-amp parallel feeder is to be installed and we are considering two raceways or two cable assemblies. We will use 500-kcmil THW copper conductors. The 75°C ampere rating for a 500-kcmil copper is 380. Thus, 2 x 380 = 760.

We are permitted in 240.4(B) to protect these conductors at 800 amps. First level subdivision 250.122(F) requires EGCs sized at 800 amps. The EGC size, at 800 amps, is 1/0 AWG copper. Considering a typical cable assembly with 500-kcmil copper conductors, the EGC will be sized on the maximum overcurrent protective device rating for the 500-kcmil copper. The standard EGC would be 3 AWG, making the cable unusable in a parallel feeder.

250.122(F)
CONDUCTORS IN PARALLEL

Chapter 2 Wiring and Protection

Article 250 Grounding and Bonding

Part VI Equipment Grounding and Equipment Grounding Conductors

250.122 Size of Equipment Grounding Conductors

(F) Conductors in Parallel

Change Type: Revision

Parallel conductor installations require that where an equipment grounding conductor is installed it shall be of full size according to the rating of the OCPD.

250.146(A)
SURFACE MOUNTED BOX

Chapter 2 Wiring and Protection

Article 250 Grounding and Bonding

Part VII Methods of Equipment Grounding

250.146 Connecting Receptacle Grounding Terminal to Box

(A) Surface Mounted Box

Change Type: Revision

Change Summary:
- A listed exposed cover with a device securely attached may serve as the grounding and bonding means, provided the cover is listed, the device is permanently secured or locked to the cover, and the cover is provided with flat corners for a solid connection to the box.

Revised 2008 NEC Text:
250.146 CONNECTING RECEPTACLE GROUNDING TERMINAL TO BOX (A) SURFACE MOUNTED BOX. Where the box is mounted on the surface, direct metal-to-metal contact between the device yoke and the box or a contact yoke or device that complies with 250.146(B) shall be permitted to ground the receptacle to the box. At least one of the insulating washers shall be removed from receptacles that do not have a contact yoke or device that complies with 250.146(B) to ensure direct metal-to-metal contact. This provision shall not apply to cover-mounted receptacles unless the box and cover combination are listed as providing satisfactory ground continuity between the box and the receptacle. <u>A listed exposed work cover shall be permitted to be the grounding and bonding means when (1) the device is attached to the cover with at least two fasteners that are permanent (such as a rivet) or have a thread locking or screw locking means and (2) when the cover mounting holes are located on a flat non-raised portion of the cover.</u>

Change Significance:
This revision will now allow an exposed cover to serve as the grounding and bonding means for a device where the following requirements are met:
- The exposed work cover shall be listed.
- The device is attached to the cover with at least two fasteners that are permanent or have a thread or screw locking means.
- The cover mounting holes shall be located on a flat/non-raised portion of the cover.
- With the device securely attached to the exposed cover, the screws through the crushed corner of the exposed work cover complete the grounding and bonding connection. The flat or crushed corners allow for more contact area, providing a solid grounding and bonding connection where the cover is secured to the box.

The additional fastening screws have not been installed in this installation.

250.166
SIZE OF THE DIRECT-CURRENT GROUNDING ELECTRODE CONDUCTOR

Chapter 2 Wiring and Protection

Article 250 Grounding and Bonding

Part VIII Direct-Current Systems

250.166 Size of the Direct-Current Grounding Electrode Conductor

Change Type: Revision

Change Summary:
- The requirements for sizing grounding electrode conductors for DC systems have been clarified. The provisions of 250.166(A) or (B) shall apply unless an electrode of the types listed in 250.166(C), (D), or (E) are installed.

Revised 2008 NEC Text:
250.166 SIZE OF THE DIRECT-CURRENT GROUNDING ELECTRODE CONDUCTOR. The size of the grounding electrode conductor for a dc system shall be as specified in 250.166(A) <u>and (B), except as permitted in 250.166(C)</u> through (E).

(A) Not Smaller Than the Neutral Conductor

(B) Not Smaller Than the Largest Conductor

(C) Connected to Rod, Pipe, or Plate Electrodes

(D) Connected to a Concrete-Encased Electrode

(E) Connected to a Ground Ring

Change Significance:
Previous code text contained conflicting requirements. This change provides clarity with user-friendly text for applying the requirements for GECs in DC systems. This revision clarifies that in general all DC systems have a grounding electrode conductor sized in accordance with 250.166(A) or (B).

250.166(A) addresses DC systems consisting of a three-wire balancer set or balancer windings, requiring in general that the GEC be not smaller than the neutral. 250.166(B) addresses DC systems other than those in 250.166(A), requiring in general that the GEC be not smaller than the largest conductor supplied by the system. This revision further clarifies that GECs in DC systems run to electrodes of the rod, pipe, plate, ground ring, or concrete-encased electrode types be installed in accordance with 250.166(C), (D), and (E) as follows:

(C) Connected to Rod, Pipe, or Plate Electrodes. Shall not be required to be larger than 6 AWG copper wire or 4 AWG aluminum wire.

(D) Connected to a Concrete-Encased Electrode. Shall not be required to be larger than 4 AWG copper wire.

(E) Connected to a Ground Ring. The portion of the grounding electrode conductor that is the sole connection to the grounding electrode shall not be required to be larger than the conductor used for the ground ring.

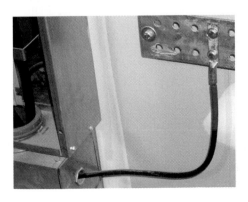

DC systems require a grounding electrode conductor sized according to 250.166.

250.168

DIRECT-CURRENT SYSTEM BONDING JUMPER

Chapter 2 Wiring and Protection

Article 250 Grounding and Bonding

Part VII Direct-Current Systems

250.168 Direct-Current System Bonding Jumper

Change Type: Revision

Change Summary:

- 250.168 has been modified to require a "system bonding jumper" where a DC system is to be grounded.

Revised 2008 NEC Text:

250.168 DIRECT-CURRENT SYSTEM BONDING JUMPER. For direct-current systems that are to be grounded, an unspliced bonding jumper shall be used to connect the equipment grounding conductor(s) to the grounded conductor at the source or the first system disconnecting means where the system is grounded. For dc systems, The size of the bonding jumper shall not be smaller than the system grounding electrode conductor specified in 250.166 and shall comply with the provisions of 250.28(A), (B), and (C).

Change Significance:

Previous code text in 250.168 specified how to size a DC bonding jumper. However, there were no other specific requirements to mandate that the DC bonding jumper be installed. There were no specific provisions for the installer or enforcement authority to apply.

250.24(B) requires that the provisions of 250.28 apply to the main bonding jumper in service-supplied AC systems. The provisions of 250.30(A)(1) for a system bonding jumper apply only to an AC system as detailed in 250.30(A).

This revision now corrects that deficiency by requiring a system bonding jumper where a DC system is to be grounded. This requirement continues to size the conductor according to 250.166 and additionally requires that the bonding jumper material, construction, and attachment comply with 250.28(A), (B), and (C).

Solar photovoltaic systems and other DC systems require a system bonding jumper when the system is grounded.

Change Summary:

* Article 280 has been significantly revised limiting application to surge arrestors over 1 kV along with editorial and style manual corrections.

Article 280
SURGE ARRESTERS, OVER 1 KV

Chapter 2 Wiring and Protection

Article 280 Surge Arresters, Over 1 kV

Part I General

280.1 Scope

(A) Rating

(1) Soildly Grounded Systems

(2) Impedance or Ungrounded System

(B) Silicon Carbide Types

280.5 Listing

Change Type: Revision

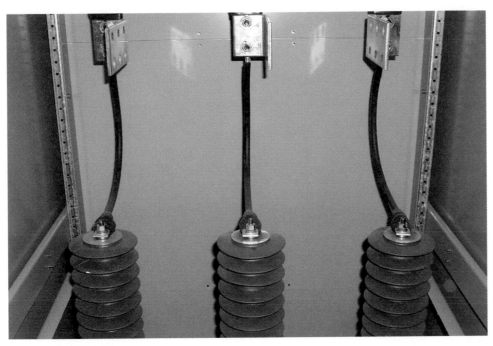

All surge arrestors must be listed.

Revised 2008 NEC Text:

ARTICLE 280 SURGE ARRESTERS, OVER 1 KV

I. GENERAL

280.1 SCOPE. This article covers general requirements, installation requirements, and connection requirements for surge arresters installed on premises wiring systems <u>over 1 kV.</u>

~~**280.2 DEFINITION.**~~ (Relocated to Article 100)

<u>**280.2 USES NOT PERMITTED.** A surge arrester shall not be installed where the rating of the surge arrester is less than the maximum continuous phase-to-ground power frequency voltage available at the point of application.</u>

280.4 SURGE ARRESTOR SELECTION. <u>The surge arresters shall comply with 280.4(A) and (B).</u>

(A) RATING. The rating of <u>a</u> ~~the~~ surge arrester shall be equal to or greater than the maximum continuous ~~phase-to-ground power frequency~~ <u>operating</u> voltage available at the point of application.

<u>**(1) SOLIDLY GROUNDED SYSTEMS.** The maximum continuous operating voltage shall be the phase-to-ground voltage of the system.</u>

(continued)

(2) IMPEDANCE OR UNGROUNDED SYSTEM.   The maximum continuous operating voltage shall be the phase-to-phase voltage of the system.

(B) SILICON CARBIDE TYPES.   The rating of a silicon carbide-type surge arrester shall be not less than 125 percent of the rating specified in 280.4(A).

280.5 LISTING.   A surge arrester shall be a listed device.

Change Significance:

Article 280 has been significantly revised. The following are a few of the significant changes in this revision:

* The title as well as the scope of Article 280 have been revised to clarify that this article applies only to surge arresters over 1 kV. Article 285 now covers both surge arresters and SPDs or TVSSs rated at 1 kV or less. Transient voltage surge suppressors (TVSSs) are also known as SPDs.

* The technology of low-voltage surge arresters and TVSSs or SPDs is now basically the same, and UL has combined them under the same standard (UL 1449). The difference is the location of the device: the line side (type 1) or load side (type 2) of the service disconnect overcurrent protection.

* The definition of "surge arrester" in 280.2 has been relocated to Article 100 because this term is used in more than one article. Existing sections within Article 280 have been editorially modified for clarity, usability, and conformance to the NEC manual of style.

* The selection of surge arrestors has been modified in 280.4 to be technically correct and for clarity. 280.4(A) now addresses ratings with two second-level subdivisions: (1) Solidly Grounded Systems and (2) Impedance or Ungrounded Systems.

* A new requirement for listing has been added in 280.5 mandating that all surge arresters be listed. Requirements in 280.4(A) for short-circuit current markings have been deleted.

* The new requirement for the surge arrestor to be a listed device will include these markings.

Notes:

_____

_____

_____

_____

_____

_____

Change Summary:

- Article 285 has been significantly revised and has been given the title "Surge Protective Devices (SPDs) 1 kV or Less"—expanding application up to 1 kV.

Article 285
SURGE PROTECTIVE DEVICES (SPDS) 1 KV OR LESS

Chapter 2 Wiring and Protection

Article 285 Surge Protective Devices (SPDs) 1 kV or Less

Part I General

285.1 Scope

Part III Connecting SPDs (Surge Arrestor or TVSS)

285.23 Type 1 SPDs (Surge Arrestor)

285.24 Type 2 SPDs (TVSS)

285.25 Type 3 SPDs (TVSS)

Change Type: Revision

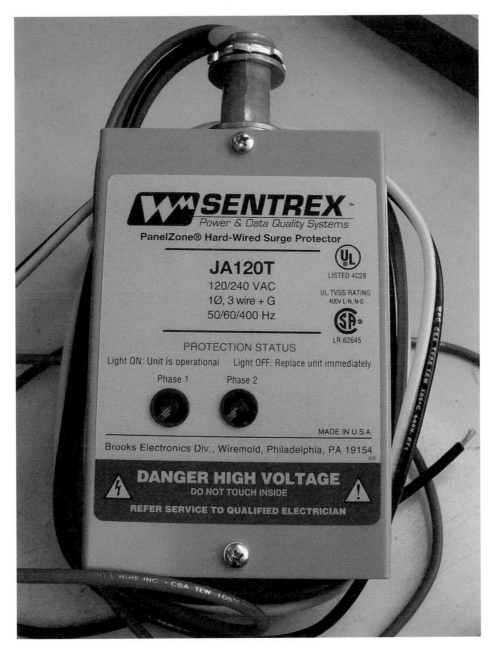

The term *transient voltage surge suppressor* (TVSS) has been replaced with the term *surge protective device* (SPD).

Revised 2008 NEC Text:

ARTICLE 285 <u>SURGE PROTECTIVE DEVICES (SPDS) 1 KV OR LESS</u> ~~TRANSIENT VOLTAGE SURGE SUPPRESSORS: TVSSS~~

(continued)

I. GENERAL
285.1 SCOPE. This article covers general requirements, installation requirements, and connection requirements for <u>SPDs (surge arrestors and</u> transient voltage surge suppressors (TVSSs)) permanently installed on premises wiring systems <u>1 kV or less</u>.

<u>FPN No 1: Surge arresters less than 1 kV are also known as Type 1 SPDs</u>
<u>FPN No. 2: Transient Voltage Surge Suppressors (TVSSs) are also known as Type 2 and Type 3 SPDs.</u>

~~**285.2 DEFINITION.**~~   (Revised and relocated to Article 100)

III. CONNECTING SPDs (SURGE ARRESTOR OR TVSS).

~~**TRANSIENT VOLTAGE SURGE SUPPRESSORS**~~

285.23 TYPE 1 SPDS (SURGE ARRESTOR).   (Supply side of service OCPD)

285.24 TYPE 2 SPDS (TVSS).   (Load side of service OCPD)

285.25 TYPE 3 SPDS (TVSS).   (Load side of branch circuit OCPD)

Change Significance:
Article 285 has been significantly revised. The following are a few of the significant changes in this revision.
*   The title and scope of Article 285 have been revised to clarify that this article applies only to surge protective devices up to 1 kV.
*   The term *transient voltage surge suppressor* (TVSS) and *surge arrestor* will continue to be used for clarity. These terms will eventually be discontinued and the term *SPD with Types 1, 2, and 3* will be used.
*   Two new FPNs have been added to 285.1 to provide information for the code user.
*   FPN No. 1 explains that surge arresters less than 1 kV are also known as type 1 SPDs.
*   FPN No. 2 explains that TVSSs are also known as type 2 and type 3 SPDs.
*   Article 285 now covers both surge arresters and SPDs or TVSSs rated at 1 kV or less. The technology of both low-voltage surge arresters and TVSSs or SPDs is now basically the same, and UL has combined them under the same standard (UL 1449). The difference is the location of the device: the line side (type 1) or load side (type 2) of the service disconnect overcurrent protection or the load side (type 3) of the branch-circuit overcurrent device.
*   The definition of "Transient Voltage Surge Suppressors (TVSSs)" in 285.2 has been revised and relocated to Article 100 titled "Surge Protective Devices (SPDs)."
*   New requirements have been added to Part III of Article 285 with specific installation requirements for types 1 through 3 SPDs.

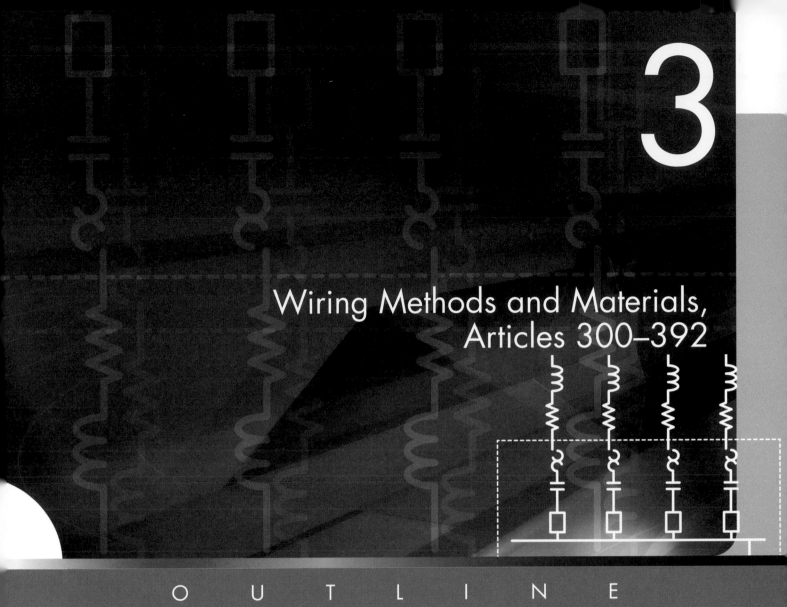

3

Wiring Methods and Materials, Articles 300–392

OUTLINE

Change Summary:

- The installation of all raceways and cable assemblies under metal corrugated sheet roof decking must be installed at least 1-½ inches from the decking to prevent damage from screws used to repair the roof.

Revised 2008 NEC Text:

300.4 PROTECTION AGAINST PHYSICAL DAMAGE
(E) CABLES AND RACEWAYS INSTALLED UNDER ROOF DECKING. A cable- or raceway-type wiring method, installed in exposed or concealed locations under metal-corrugated sheet roof decking, shall be installed and supported so the nearest outside surface of the cable or raceway is not less than 38 mm (1½ in.) from the nearest surface of the roof decking.

FPN: Roof decking material is often repaired or replaced after the initial raceway or cabling and roofing installation and may be penetrated by the screws or other mechanical devices designed to provide "hold down" strength of the water-proof membrane or roof insulating material.

Exception: Rigid metal conduit and intermediate metal conduit shall not be required to comply with 300.4(E).

Change Significance:

This new first-level subdivision will now prohibit the common practice of installing raceways and cable assemblies in direct contact with metal corrugated sheet roof decking. It is important to note that this applies only to metal decking that constitutes the "roof."

An informational fine print note (FPN) is included to illustrate the problem. The waterproofing and insulating material above conventional metal corrugated sheeting has a life span and will continuously need repair or replacement. Roofing material manufacturers' installation standards mandate a self-tapping deck screw to penetrate the underside of the corrugated steel decking by at least 1 inch and up to 1-¼ inches. Minimum penetration provides "hold-down" strength for the insulating material above, but becomes a hazard to anything within the lower penetration proximity. Therefore, raceways located in close proximity to the underside of the decking are exposed to severe physical damage.

This is happening every day during re-roofing projects that are only repairing or replacing the waterproofing material on the surface of the roof of an occupied building. It is not the first installation that is a concern because typically the exterior insulation and waterproofing material of the flat roof are already in place before the electrician begins to install the new wiring methods. An exception exists that permits rigid metal conduit (RMC) and intermediate metal conduit (IMC) to be installed in contact with the roof decking due to their physical properties.

300.4(E)
CABLES AND RACEWAYS INSTALLED UNDER ROOF DECKING

Chapter 3 Wiring Methods and Materials

Article 300 Wiring Methods

Part I General Requirements

300.4 Protection Against Physical Damage

(E) Cables and Raceways Installed Under Roof Decking

Change Type: New

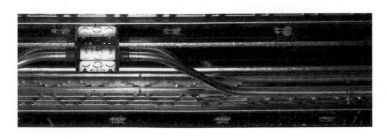

All wiring methods except RMC and IMC must be kept a minimum of 1½-inches from metal roof decking.

300.5(B), (C)
WET LOCATIONS AND UNDERGROUND CABLES UNDER BUILDINGS

Chapter 3 Wiring Methods and Materials

Article 300 Wiring Methods

Part I General Requirements

300.5 Underground Installations

(B) Wet Locations

(C) Underground Cables Under Buildings

Change Type: Revision

Change Summary:

- 300.5(B) is renamed "Wet Locations" and new text is provided to clarify that the interior of all raceways and enclosures installed underground are "wet locations."
- Clarification is provided in 300.5(C) requiring all underground cable installed under a building to be installed in a raceway.

Revised 2008 NEC Text:
300.5 UNDERGROUND INSTALLATIONS
(B) WET LOCATIONS ~~LISTING~~.   <u>The interior of enclosures or raceways installed underground shall be considered to be a wet location.</u> Insulated conductors and cables installed in <u>these</u> enclosures or raceways in underground installations shall be listed for use in wet locations <u>and shall comply with 310.8(C). Any connections or splices in an underground installation shall be approved for wet locations.</u>

(C) UNDERGROUND CABLES UNDER BUILDINGS.   Underground cable installed under a building shall be in a raceway ~~that is extended beyond the outside walls of the building~~.

Change Significance:
The title of 300.5(B) has been appropriately changed to "Wet Locations." The previous text simply required that all cables and insulated conductors installed in underground enclosures or raceways be listed for use in wet locations. There is confusion in the field with respect to the classification of the "interior" of a raceway installed underground. This revision clarifies that the interior of raceways and enclosures installed underground are indeed "wet locations." Compliance with 310.8(C) now requires that cables and conductors installed in these locations meet one of the following: (1) Be moisture-impervious metal-sheathed, (2) be types MTW, RHW, RHW-2, TW, THW, THW-2, THHW, THWN, THWN-2, XHHW, XHHW-2, or ZW, or (3) be of a type listed for use in wet locations.

300.5(C) allows underground cable or direct buried cables to be installed under a concrete slab under a building but requires a raceway where the cable is "under the building." The previous text was not clear. The deleted text inferred that a raceway was only required if the underground cable or direct buried cables "extended beyond the outside walls of the building."

Conductors installed in wet locations are required by 310.8(C) to be types MTW, RHW, RHW-2, TW, THW, THW-2, THHW, THWN, THWN-2, XHHW, XHHW-2, ZW, or be listed for wet locations.

Change Summary:

- The requirements for the protection of conductors as they emerge from grade have been clarified and limited to "conductors and cables."
- Conductors and cables installed as specified in column 5 for irrigation control and landscape lighting (which are limited to not more than 30 volts) do not require physical protection.

Revised 2008 NEC Text:

300.5 UNDERGROUND INSTALLATIONS
(D) PROTECTION FROM DAMAGE
(1) EMERGING FROM GRADE.   Direct-buried conductors and <u>cables</u> ~~enclosures~~ emerging from grade <u>and specified in columns 1 and 4 of Table 300.5</u> shall be protected by enclosures or raceways extending from the minimum cover distance below grade required by 300.5(A) to a point at least 2.5 m (8 ft) above finished grade. In no case shall the protection be required to exceed 450 mm (18 in.) below finished grade.

Change Significance:

The requirements of 300.5(D)(1) for the protection of conductors emerging from grade have been clarified to apply only to "conductors and cables." The previous text actually required the protection of enclosures by another enclosure or raceway, which was not intended by the technical committee.

The requirement for physical protection of conductors and cables has been further clarified with qualifying text limiting the scope of this requirement to columns 1 and 4 of Table 300.5. It is important to note that circuits identified in column 5 of Table 300.5(D) are now exempt from this physical protection requirement. These circuits are limited to not more than 30 volts and are no longer required to be physically protected as they emerge from grade.

Conductors and cables requiring protection in accordance with 300.5(D)(1) as they emerge from grade are as follows.

- **Column 1:** Direct Burial Cables or Conductors
- **Column 4:** Residential Branch Circuits Rated 120 Volts or Less with GFCI Protection and Maximum Overcurrent Protection of 20 Amperes

Conductors and cables *not* requiring protection in accordance with 300.5(D)(1) as they emerge from grade are as follows.

- **Column 5:** Circuits for Control of Irrigation and Landscape Lighting Limited to Not More Than 30 Volts and Installed with Type UF or in Other Identified Cable or Raceway

Note that columns 2 and 3 of 300.5(D)(1) address raceway installations.

300.5(D)
PROTECTION FROM DAMAGE

Chapter 3 Wiring Methods and Materials

Article 300 Wiring Methods

Part I General Requirements

300.5 Underground Installations

(D) Protection From Damage

(1) Emerging From Grade

Change Type: Revision

In general, all conductors emerging from grade must be protected by an enclosure or raceway at least 8 feet above finished grade.

300.6(B)
ALUMINUM METAL EQUIPMENT

Chapter 3 Wiring Methods and Materials

Article 300 Wiring Methods

Part I General Requirements

300.6 Protection Against Corrosion and Deterioration

(B) Aluminum Metal Equipment

Change Type: Revision

Change Summary:
- The requirements of 300.6(B) for supplementary corrosion protection for nonferrous metals in concrete, or direct contact with earth, are now clarified and limited to aluminum. Red brass is permitted and does not require supplementary corrosion protection.

Revised 2008 NEC Text:
300.6 PROTECTION AGAINST CORROSION AND DETERIORATION (B) ALUMINUM ~~NON-FERROUS~~ METAL EQUIPMENT. Aluminum ~~Non-ferrous~~ raceways, cable trays, cablebus, auxiliary gutters, cable armor, boxes, cable sheathing, cabinets, elbows, couplings, nipples, fittings, supports, and support hardware embedded or encased in concrete or in direct contact with the earth shall be provided with supplementary corrosion protection.

Change Significance:
This revision clarifies that the requirements of 300.6(B) apply only to aluminum materials. Prior to the revision of 300.6 in the 2002 NEC, the requirement for ferrous and nonferrous materials installed in concrete or in direct contact with the earth was that (1) the material being used was judged suitable for the condition or (2) approved corrosion protection was provided.

This clarification is necessary to permit the use of red brass conduits. Both aluminum and red brass are listed per UL 6A and are considered nonferrous metals. The *UL Electrical Construction Equipment Directory* includes limitations and special conditions of use for listed products. Under the product category "Conduit, Rigid Nonferrous Metallic (DYWV)," aluminum conduit is required to have supplementary corrosion protection when installed in concrete or in soil. UL does not require supplementary protection for red brass conduit. Red brass conduit has been used for many years as the preferred method for wiring swimming pool lighting and has held up in these harsh corrosive environments without problems.

Aluminum conductors or raceways exposed to moisture will corrode.

Change Summary:

- The FPN to 300.7(B) now includes a multiplier for the code user to apply when determining the expansion of aluminum raceways exposed to different temperatures. This multiplier is applied to the values in Table 352.44.
- The FPN also references the code user to Table 355.44 for expansion characteristics of reinforced thermosetting resin conduit (RTRC).

Expansion characteristics for 100 feet of conduit installed in an area with a 70°F change in temperature from winter to summer*	
Rigid Nonmetallic Conduit: Type RNMC (Dimension taken directly from Table 352.44(A))	2.84 or 2 7/8 inches
Rigid Metal Conduit: Type RMC (Steel) Table dimension of 2.84 multiplied by .2	.568 or 9/16 inch
Rigid Metal Conduit: Type RMC (Aluminum) Table dimension of 2.84 multiplied by .4	1.136 or 3/16 inch

* Note that as each dimension is converted to a fraction in this example, it is rounded up to the next larger sixteenth of an inch.

All raceways will expand and contract differently depending on the change in temperature and the material involved.

300.7(B)
EXPANSION FITTINGS

Chapter 3 Wiring Methods and Materials

Article 300 Wiring Methods

Part I General Requirements

300.7 Raceways Exposed to Different Temperatures

(B) Expansion Fittings

Change Type: Revision

Revised 2008 NEC Text:

300.7 RACEWAYS EXPOSED TO DIFFERENT TEMPERATURES

(B) EXPANSION FITTINGS. Raceways shall be provided with expansion fittings where necessary to compensate for thermal expansion and contraction.

FPN: Tables 352.44(A) and 355.44 provides the expansion information for polyvinyl chloride (PVC) and for reinforced thermosetting resin conduit (RTRC) respectively. A nominal number for steel conduit can be determined by multiplying the expansion length in this table 352.44 by 0.20. The coefficient of expansion for steel electrical metallic tubing, intermediate metal conduit, and rigid conduit is 11.70_10^{-56} (0.0000117 mm per mm of conduit for each °C in temperature change) [6.50_10^{-56} (0.0000065 in. per inch of conduit for each °F in temperature change)].

A nominal number for aluminum conduit and aluminum electrical metallic tubing can be determined by multiplying the expansion length in Table 352.44(A) by 0.40. The coefficient of expansion for aluminum electrical metallic tubing and aluminum rigid metal conduit is 2.34×10^{-5} (0.0000234 mm per mm of conduit for each °C in temperature change) [1.30×10^{-5} (0.000013) in. per inch of conduit for each °F in temperature change].

Change Significance:

A new second paragraph has been added to include aluminum to complete the family of metals commonly used. This new text includes a multiplier for

(continued)

determining the expansion of aluminum raceways exposed to different temperatures. This multiplier is applied to the values in Table 352.44(A).

This revision also changes the exponential value of the coefficient of expansion associated with steel raceways. It does not change the resultant calculations associated with determining the length of expansion or contraction change per unit of temperature change. This change is made only to correlate this FPN with the coefficient in Table 352.44(A).

It is imperative that installations of raceways exposed to different temperatures determine the amount of expansion and the need for expansion joints. Without expansion joints, raceways will buckle and distort and will damage equipment and enclosures as they expand and contract.

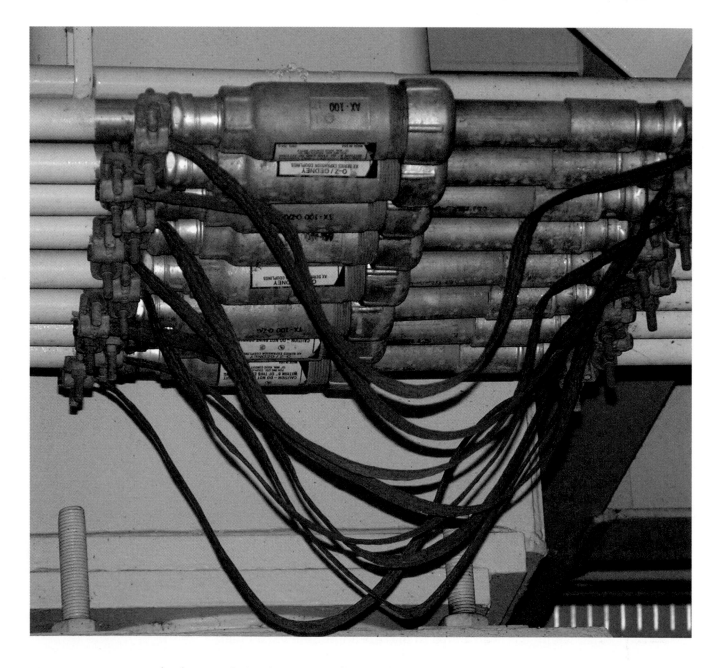

Raceways are required to be provided with expansion fittings where necessary to compensate for thermal expansion.

Change Summary:
- All above-grade installations of raceways in "wet locations" now recognize the interior of the raceway as a "wet location."

Revised 2008 NEC Text:
300.9 RACEWAYS IN WET LOCATIONS ABOVE GRADE.   Where raceways are installed in wet locations above grade, the interior of these raceways shall be considered to be a wet location. Insulated conductors and cables installed in raceways in wet locations above grade shall comply with 310.8(C).

Change Significance:
This new section clarifies that the interior of any raceway installed above grade in a wet location is a "wet location." Multiple proposals and comments were submitted on this issue seeking clarity for both above-ground and underground installations. Similar text now exists in 300.5(B) to clearly require that all conductors and cables installed underground be listed for wet locations. There is clearly a misunderstanding in the field that a raceway installed above grade in a wet location may contain conductors not listed for use in a wet location. This new section has settled that issue. This requirement now refers the code user to 310.8(C), as follows, for requirements for all conductors and cables installed above grade in wet locations.

310.8 LOCATIONS
(C) WET LOCATIONS.   Insulated conductors and cables used in wet locations shall comply with one of the following:
(1) Be moisture-impervious metal-sheathed;
(2) Be types MTW, RHW, RHW-2, TW, THW, THW-2, THHW, THWN, THWN-2, XHHW, XHHW-2, ZW;
(3) Be of a type listed for use in wet locations.

300.9
RACEWAYS IN WET LOCATIONS ABOVE GRADE

Chapter 3 Wiring Methods and Materials

Article 300 Wiring Methods

Part I General Requirements

300.9 Raceways in Wet Locations Above Grade

Change Type: New

Conductors installed in raceways in wet locations must be rated with the suffix W.

300.11(A)
SECURED IN PLACE

Chapter 3 Wiring Methods and Materials

Article 300 Wiring Methods

Part I General Requirements

300.11 Securing and Supporting

(A) Secured in Place

(2) Non–Fire-Rated Assemblies

Change Type: Revision

Change Summary:
- Where an independent means of secure support such as a support wire is installed in a non–fired-rated assembly to support wiring methods and materials listed in 300.11(A), the support wire is permitted to be attached to the assembly.

Revised 2008 NEC Text:
300.11 SECURING AND SUPPORTING
(A) SECURED IN PLACE
(2) NON–FIRE-RATED ASSEMBLIES.   Wiring located within the cavity of a non–fire-rated floor–ceiling or roof–ceiling assembly shall not be secured to, or supported by, the ceiling assembly, including the ceiling support wires. An independent means of secure support shall be provided <u>and shall be permitted to be attached to the assembly</u>.

Exception: The ceiling support system shall be permitted to support branch-circuit wiring and associated equipment where installed in accordance with the ceiling system manufacturer's instructions.

Change Significance:
This revision provides correlation with methods permitted in 300.11(A)(1) for fire-rated assemblies. For example, the addition of this text permits an independent support wire installed in a non–fire-rated ceiling assembly supporting a cable assembly to be attached to the lower end or to the ceiling. The intent behind this permission is to provide a method of stabilizing the installation. Independent support wires not attached to the assembly will swing freely in the ceiling space and create safety concerns for installer/maintainers who access the ceiling space in the future.

Support wires for wiring in a suspended ceiling are permitted to be attached to the ceiling.

Change Summary:

- A new exception to 300.12 clarifies that where raceways and cables are installed into the bottom of open-bottom equipment (such as switchboards, motor control centers, and transformers), it is not required to mechanically secure the conduits to the equipment. Electrical continuity is required in 300.10.

Revised 2008 NEC Text:

300.12 MECHANICAL CONTINUITY—RACEWAYS AND CABLES.   Metal or nonmetallic raceways, cable armors, and cable sheaths shall be continuous between cabinets, boxes, fittings, or other enclosures or outlets.

Exception <u>No. 1</u>: Short sections of raceways used to provide support or protection of cable assemblies from physical damage shall not be required to be mechanically continuous.

<u>*Exception No. 2: Raceways and cables installed into the bottom of open bottom equipment, such as switchboards, motor control centers, and floor or pad-mounted transformers, shall not be required to be mechanically secured to the equipment.*</u>

Change Significance:

The general requirement of 300.12 requires that all raceways, cable armors, and cable sheaths be continuous or mechanically secured to the box, cabinet, or enclosure they enter. Where conduits or cables are installed into the bottom of open-bottom equipment (such as switchboards, motor control centers, and transformers), it is not possible to mechanically secure the conduits to the equipment. New Exception No. 2 now permits the installation of such raceways and cables without them being mechanically secure to the equipment.

Raceways and cable sheaths entering equipment in this manner are still required to be bonded by 300.10 to provide electrical continuity. Raceways, including their end fittings, entering equipment in this manner must also not rise more than 3 inches above the bottom of the enclosure (as required in 408.5).

300.12
MECHANICAL CONTINUITY— RACEWAYS AND CABLES

Chapter 3 Wiring Methods and Materials

Article 300 Wiring Methods

Part I General Requirements

300.12 Mechanical Continuity—Raceways and Cables

Exceptions

Change Type: Revision and New

Mechanical continuity is not required for a raceway entering the bottom of an open-bottom enclosure.

300.19(B)
FIRE-RATED CABLES AND CONDUCTORS

Chapter 3 Wiring Methods and Materials

Article 300 Wiring Methods

Part I General Requirements

300.19 Supporting Conductors in Vertical Raceways

(B) Fire-Rated Cables and Conductors

Change Type: New

Change Summary:
- 300.19 now specifically addresses the installation of "fire-rated cables and conductors." This new requirement mandates that these conductors be installed in accordance with manufacturers' instructions.

Revised 2008 NEC Text:
300.19 SUPPORTING CONDUCTORS IN VERTICAL RACEWAYS (B) FIRE-RATED CABLES AND CONDUCTORS. <u>Support methods and spacing intervals for fire-rated cables and conductors shall comply with any restrictions provided in the listing of the electrical circuit protective system used and in no case shall exceed the values in Table 300.19(A).</u>
(Existing (B) becomes (C))

Change Significance:
This new first-level subdivision specifically addresses the support methods and spacing intervals for fire-rated cables and conductors. Testing has shown that in some cases fire-rated conductors in a vertical raceway weakened when exposed to fire conditions. This is currently being addressed in the appropriate UL standard. The support requirements for these conductors and cables may be more conservative than that presently required in Table 300.19(A).

This new requirement may seem to be redundant due to the fact that 110.3(B) already requires that the installation of all listed and labeled equipment be in accordance with the listing and labeling. The technical committee, however, felt that it was necessary to clarify and add specific requirements that apply only to electrical circuit protective systems.

300.19 provides maximum spacing intervals for conductors in vertical raceways, support requirements for fire-rated cables/conductors, and permitted support methods.

Change Summary:
- 300.20 has been clarified by requiring that *ferrous* metal enclosures or *ferrous* metal raceways have conductors grouped to avoid heating through induction.

Revised 2008 NEC Text:

300.20 INDUCED CURRENTS IN <u>FERROUS</u> METAL ENCLOSURES OR <u>FERROUS</u> METAL RACEWAYS

(A) CONDUCTORS GROUPED TOGETHER.   Where conductors carrying alternating current are installed in <u>ferrous</u> metal enclosures or <u>ferrous</u> metal raceways, they shall be arranged so as to avoid heating the surrounding <u>ferrous</u> metal by induction. To accomplish this, all phase conductors and, where used, the grounded conductor and all equipment grounding conductors shall be grouped together.

Exception No. 1: Equipment grounding conductors for certain existing installations shall be permitted to be installed separate from their associated circuit conductors where run in accordance with the provisions of 250.130(C).

Exception No. 2: A single conductor shall be permitted to be installed in a ferromagnetic enclosure and used for skin-effect heating in accordance with the provisions of 426.42 and 427.47.

Change Significance:
The addition of the qualifying term *ferrous* excludes aluminum or red brass from the grouping requirements of 300.20(A). Aluminum and red brass are nonferrous and where used will not be subject to the inductive heating that will occur in all ferrous metals.

300.20(A)
CONDUCTORS GROUPED TOGETHER

Chapter 3 Wiring Methods and Materials

Article 300 Wiring Methods

Part I General Requirements

300.20 Induced Currents in Ferrous Metal Enclosures or Ferrous Metal Raceways

(A) Conductors Grouped Together

Change Type: Revision

Inductive heating would occur if these MI cables were terminated in a ferrous metal.

300.22(C)
OTHER SPACES USED FOR ENVIRONMENTAL AIR

Chapter 3 Wiring Methods and Materials

Article 300 Wiring Methods

Part I General Requirements

300.22(C) Other Spaces Used for Environmental Air

(1) Wiring Methods

Change Type: Revision

Change Summary:

• 300.22(C) addresses permitted wiring methods in "other spaces" and now recognizes that raceways such as electrical nonmetallic tubing (ENT) or optical fiber raceways may be installed in recognized raceways in these spaces.

Revised 2008 NEC Text:

300.22(C) OTHER SPACES USED FOR ENVIRONMENTAL AIR (1) WIRING METHODS.   The wiring methods for such other space shall be limited to totally enclosed, nonventilated, insulated busway having no provisions for plug-in connections, Type MI cable, Type MC cable without an overall nonmetallic covering, Type AC cable, or other factory-assembled multiconductor control or power cable that is specifically listed for the use, or listed prefabricated cable assemblies of metallic manufactured wiring systems without nonmetallic sheath. Other types of cables, ~~and~~ conductors, <u>and raceways</u> shall be <u>permitted to be</u> installed in electrical metallic tubing, flexible metallic tubing, intermediate metal conduit, rigid metal conduit without an overall nonmetallic covering, flexible metal conduit, or, where accessible, surface metal raceway or metal wireway with metal covers or solid bottom metal cable tray with solid metal covers.

Change Significance:

It is a common practice for raceways such as ENT, Article 352, and Optical Fiber/Communications raceways found in Articles 725, 770, 800, and 820 to be pulled into metal conduits in 300.22(C) spaces. This common practice allows cable to be removed and replaced without interrupting other services. The metal raceways are terminated or end in equipment outside the 300.22(C) space. These installations commonly refer to the raceways installed inside another raceway (such as EMT) as innerduct. This practice allows, for example, the removal, replacement, or addition of fiber optic cable without interfering with or damaging other conductors.

Raceways such as ENT may be installed in EMT or other permitted raceways in a 300.22(C) space.

Change Summary:

- A new Note 3 to Table 300.50 allows a reduction in burial depth for wiring methods other than RMC or IMC in industrial establishments. Where conditions of maintenance and supervision ensure that qualified persons will service the installation, the burial depth may be reduced by 6 inches for each 2 inches of concrete installed.

Revised 2008 NEC Text:

NEW NOTE 3 TO TABLE 300.50 MINIMUM COVER REQUIREMENTS

Note 3. In industrial establishments, where conditions of maintenance and supervision ensure that qualified persons will service the installation, the minimum cover requirements, for other than rigid metal conduit and intermediate metal conduit, shall be permitted to be reduced 150 mm (6-inches) for each 50 mm (2-inches) of concrete or equivalent placed entirely within the trench over the underground installation.

Change Significance:

This revision provides a reduction in burial depth for industrial establishments only. These requirements are located in Part II of Article 300 and apply only to installations greater than 600 volts nominal.

Where an underground installation consists of wiring methods other than RMC or IMC, the permitted reduction in depth may be applied in industrial establishments only.

Note 3 now permits an industrial establishment where conditions of maintenance and supervision ensure that qualified persons will service the installation to reduce the depth requirements of Table 300.50. For each 2 inches of concrete or equivalent placed entirely within the trench over the underground installation, the depth required in Table 300.50 may be reduced by 6 inches.

For example, an installation of rigid nonmetallic conduit (RNMC) containing conductors operating at 33 kV is required by column 2 of Table 300.50 to be installed at least 24 inches deep. The burial depth may be reduced to 12 inches if 4 inches of concrete are placed entirely within the trench over the underground installation.

Table 300.50
MINIMUM COVER REQUIREMENTS

Chapter 3 Wiring Methods and Materials

Article 300 Wiring Methods

Part II Requirements for Over 600 Volts, Nominal

300.50 Underground Installations

(A) General

Table Note 3

Change Type: New

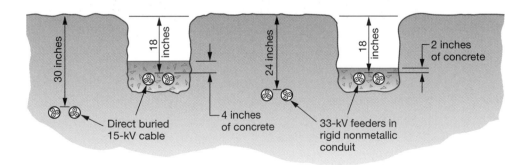

Minimum cover requirements may be modified where concrete is placed over the installation.

310.4

CONDUCTORS IN PARALLEL

Chapter 3 Wiring Methods and Materials

Article 310 Conductors for General Wiring

310.4 Conductors in Parallel

(A) General

(B) Conductor Characteristics

(C) Separate Cables or Raceways

(D) Ampacity Adjustment

(E) Equipment Grounding Conductors

Change Type: Revision

Change Summary:
* 310.4 has been editorially revised for usability and clarity.

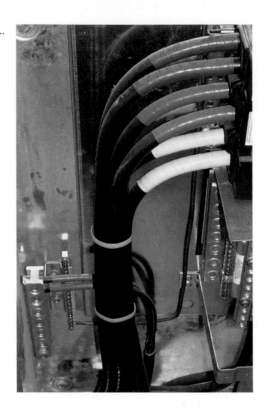

In general, only conductors 1/0 AWG and larger may be installed in parallel.

Revised 2008 NEC Text:

310.4 CONDUCTORS IN PARALLEL

(A) GENERAL. Aluminum, copper-clad aluminum, or copper conductors of size 1/0 AWG and larger, comprising each phase, polarity, neutral, or grounded circuit conductor, shall be permitted to be connected in parallel (electrically joined at both ends).

Exception No. 1: Conductors in sizes smaller than 1/0 AWG shall be permitted to be run in parallel to supply control power to indicating instruments, contactors, relays, solenoids, and similar control devices, <u>or for frequencies of 360 Hz and higher,</u> provided all of the following apply:

(a) They are contained within the same raceway or cable.

(b) The ampacity of each individual conductor is sufficient to carry the entire load current shared by the parallel conductors.

(c) The overcurrent protection is such that the ampacity of each individual conductor will not be exceeded if one or more of the parallel conductors become inadvertently disconnected.

Exception No. 2: Under engineering supervision, grounded neutral conductors in sizes 2 AWG and larger shall be permitted to be run in parallel for existing installations.

FPN <u>to Exception No. 2</u>: Exception No. 2 can be used to alleviate overheating of neutral conductors in existing installations due to high content of triplen harmonic currents.

(continued)

(B) CONDUCTOR CHARACTERISTICS. The paralleled conductors in each phase, polarity, neutral, grounded circuit conductor, or equipment grounding conductor shall comply with all of the following:

(1) Be the same length

(2) Have the same conductor material

(3) Be the same size in circular mil area

(4) Have the same insulation type

(5) Be terminated in the same manner

(C) SEPARATE CABLES OR RACEWAYS. Where run in separate cables or raceways, the cables or raceways <u>with conductors shall have the same number of conductors and</u> shall have the same electrical characteristics. Conductors of one phase, polarity, neutral, ~~or~~ grounded circuit conductor, <u>or equipment grounding conductor</u> shall not be required to have the same physical characteristics as those of another phase, polarity, neutral, ~~or~~ grounded circuit conductor, <u>or equipment grounding conductor</u> to achieve balance.

(D) AMPACITY ADJUSTMENT. Conductors installed in parallel shall comply with the provisions of 310.15(B)(2)(a).

(E) EQUIPMENT GROUNDING CONDUCTORS. Where <u>parallel</u> equipment grounding conductors are used, ~~with conductors in parallel~~ they shall ~~comply with the requirements of this section except that they shall~~ be sized in accordance with 250.122. <u>Sectioned equipment grounding conductors smaller than 1/0 AWG shall be permitted in multiconductor cables in accordance with 310.13, provided the combined circular mil area in each cable complies with 250.122.</u>

Change Significance:

The intent of the technical committee in this change is to reorganize 310.4, increasing clarity and usability without making substantive changes. The requirements are logically separated into five first-level subdivisions.

- 310.4(A) is titled "General" and contains the basic permission for parallel connections. The previous exception recognizing 620.12(A)(1) is deleted because it is redundant. The previous Exception No. 2 is deleted and incorporated into Exception No. 1.

- 310.4(B) is titled "Conductor Characteristics" and contains the previous requirement.

- 310.4(C) is titled "Separate Cables or Raceways" and contains the previous requirement.

- 310.4(D) is titled "Ampacity Adjustment" and contains the previous requirement.

- 310.4(E) is titled "Equipment Grounding Conductors" and contains the previous requirement to size the EGC according to 250.122. A new last sentence permits EGCs smaller than 1/0 AWG where they are "sectioned" within a multiconductor cable and the circular mil area meets the requirements of 250.122.

310.15(B)(2)

ADJUSTMENT FACTORS

(a) MORE THAN THREE CURRENT-CARRYING CONDUCTORS IN A RACEWAY OR CABLE

Chapter 3 Wiring Methods and Materials

Article 310 Conductors for General Wiring

310.15 Ampacities for Conductors Rated 0–2000 Volts

(B) Tables

(2) Adjustment Factors

(a) More Than Three Current-Carrying Conductors in a Raceway or Cable

Change Type: Revision

Change Summary:

- 310.15(B)(2)(a) is revised to clarify that single conductors and multiconductor cables installed without maintaining spacing for a "continuous" length of 24 inches must be derated.

- The terms *stacked* and *bundled* are deleted and replaced with the language "installed without maintaining spacing" for clarity.

Revised 2008 NEC Text:

310.15 AMPACITIES FOR CONDUCTORS RATED 0–2000 VOLTS
(B) TABLES
(2) ADJUSTMENT FACTORS.
(a) More Than Three Current-Carrying Conductors in a Raceway or Cable. Where the number of current-carrying conductors in a raceway or cable exceeds three, or where single conductors or multiconductor cables are <u>installed without maintaining</u> spacing ~~stacked or bundled~~ <u>for a continuous length</u> longer than 600 mm (24 in.) ~~without maintaining spacing~~ and are not installed in raceways, the allowable ampacity of each conductor shall be reduced as shown in Table 310.15(B)(2)(a). Each current-carrying conductor of a paralleled set of conductors shall be counted as a current-carrying conductor.

Change Significance:

This revision and the technical committee statement clearly points out that where single conductors or multiconductor cables are installed the adjustment factors are required to be applied only where the lack of spacing occurs in a "continuous" length. For example, if multiple MC cables pass through sleeves in five separate locations and do not maintain spacing for 18 inches in each location the adjustment factors of 310.15(B)(2)(a) do not apply. Once a conductor or cable is no longer bundled or stacked (i.e., "installed without maintaining spacing"), the air around each conductor will lower the conductor operating temperature before it gets to the next bundle and there is no need to reduce the ampacity.

It is important to note that the terms *stacked* and *bundled* are deleted and replaced with the language "installed without maintaining spacing" for clarity. Physically "stacking" or "bundling" cables is not required to have a lack of "spacing."

Physically stacking or bundling is not required to have a lack of "spacing."

Change Summary:

- Conductors or cables installed in conduits exposed to direct sunlight on or above rooftops must now include the temperature adder as listed in Table 310.15(B)(2)(c) to determine ambient temperature.

310.15(B)(2)(c) adjustments assuming 100°F ambient temperature

Above 36 inches to bottom of conduit, no change, 100°F ambient

24 inches to bottom of conduit
100°F + 25°F = 125°F ambient

11 inches to bottom of conduit
100°F + 30°F = 130°F ambient

3 3/8 inches to bottom of conduit
100°F + 40°F = 140°F ambient

3/8 inches to bottom of conduit
100°F + 60°F = 160°F ambient

Temperatures on a rooftop exposed to direct sunlight are significantly hotter than other outdoor applications exposed to sunlight.

310.15(B)(2)
ADJUSTMENT FACTORS
(C) CONDUITS EXPOSED TO SUNLIGHT ON ROOFTOPS

Chapter 3 Wiring Methods and Materials

Article 310 Conductors for General Wiring

310.15 Ampacities for Conductors Rated 0–2000 Volts

(B) Tables

(2) Adjustment Factors

(c) Conduits Exposed to Sunlight on Rooftops

Change Type: New

Revised 2008 NEC Text:

310.15 AMPACITIES FOR CONDUCTORS RATED 0–2000 VOLTS
(B) TABLES
(2) ADJUSTMENT FACTORS.
(c) Conduits Exposed to Sunlight on Rooftops. Where conductors or cables are installed in conduits exposed to direct sunlight on or above rooftops, the adjustments shown in Table 310-15(B)(2)(c) shall be added to the outdoor temperature to determine the applicable ambient temperature for application of the correction factors in Tables 310.16 and 310.18.

FPN: One source for the average ambient temperatures in various locations is the ASHRAE Handbook–*Fundamentals.*

Table 310.15(B)(2)(c) Ambient Temperature Adjustment for Conduits Exposed to Sunlight On or Above Rooftops

Distance Above Roof to Bottom of Conduit	Temperature Adder C°	F°
0 thru 13 mm (1/2 in.)	33	60
Above 13 mm (½ in.)—90 mm (3-½ in.)	22	40
Above 90 mm (3-½ in.)—300 mm (12 in.)	17	30
Above 300 mm (12 in.)—900 mm (36 in.)	14	25

FPN to Table 310.15(B)(2)(c): The temperature adders in Table 310.15(B)(2)(c) are based on the results of averaging the ambient temperatures. The highest sustained ambient temperature in the location should be considered to minimize the risk of damage to the insulation by overheating.

Change Significance:

A new third-level subdivision, 310.15(B)(2)(c), now requires that corrections be made to ambient temperature for installations of conductors or cables in conduits on or above rooftops. This revision requires that the ambient temperature be increased prior to selection of a correction factor in Table 310.16 or Table 310.18.

In the 2005 cycle, an informational FPN was added to 310.10 to inform the code user that conductors installed in conduit in close proximity to rooftops experience large increases in temperature above the ambient. This FPN is no longer necessary and is deleted because it has become a requirement in 310.15(B)(2(c).

A study has been performed to provide data to support this revision. The study is titled "Effect of Rooftop Exposure on Ambient Temperatures Inside Conduits, November 2005." A new FPN No. 1 provides useful information for finding average ambient temperatures in various locations using an American Society of Heating, Refrigerating, and Air-Conditioning (ASHRAE) handbook. The new FPN alerts the code user that the temperature adders are based on the results of averaging ambient temperatures. The highest sustained average should be considered to prevent damage to conductors.

Notes:

_____

_____

_____

_____

_____

_____

_____

_____

_____

Change Summary:

- The minimum depth of boxes for outlets, devices, and utilization equipment in 314.24 is completely revised in a user-friendly format with prescriptive requirements.

314.24 Minimum Depth of Boxes for Outlets, Devices, and Utilization Equipment

(A) Outlet Boxes Without Enclosed Devices or Utilization Equipment
(B) Outlet and Device Boxes with Enclosed Devices
(C) Utilization Equipment
    (1) Large Equipment
    (2) Conductors Larger than 4 AWG
    (3) Conductors 8, 6, or 4 AWG
    (4) Conductors 12 or 10 AWG
    (5) Conductors 14 AWG and Smaller

314.24 now provides the code user with comprehensive prescriptive requirements.

Revised 2008 NEC Text:

314.24 MINIMUM DEPTH OF BOXES FOR OUTLETS, DEVICES AND UTILIZATION EQUIPMENT. Outlet and device boxes shall have sufficient depth to allow equipment installed within them to be mounted properly and with sufficient clearance to prevent damage to conductors within the box.

(A) OUTLET BOXES WITHOUT ENCLOSED DEVICES OR UTILIZATION EQUIPMENT. No box shall have an internal depth of less than 12.7 mm (½ in.)

(B) OUTLET AND DEVICE BOXES WITH ENCLOSED DEVICES. Boxes intended to enclose flush devices shall have an internal depth of not less than 23.8 mm (¹⁵/₁₆ in.)

(C) UTILIZATION EQUIPMENT. Outlet and device boxes that enclose utilization equipment shall have a minimum internal depth that accommodates the rearward projection of the equipment and the size of the conductors that supply the equipment. The internal depth shall include, where used, that of any extension boxes, plaster rings, or raised covers. The internal depth shall comply with all applicable provisions of (C)(1) through (C)(5).

(1) LARGE EQUIPMENT. Boxes that enclose utilization equipment that projects more than 48 mm (1⅞ in.) rearward from the mounting plane of the box shall have a depth that is not less than the depth of the equipment plus 6 mm (¼ in.).

(continued)

314.24
MINIMUM DEPTH OF BOXES FOR OUTLETS, DEVICES, AND UTILIZATION EQUIPMENT

Chapter 3 Wiring Methods and Materials

Article 314 Outlet, Device, Pull, and Junction Boxes; Conduit Bodies; Fittings; and Handhole Enclosures

Part II Installation

314.24 Minimum Depth of Boxes for Outlets, Devices, and Utilization Equipment

(A) Outlet Boxes Without Enclosed Devices or Utilization Equipment

(B) Outlet and Device Boxes With Enclosed Devices

(C) Utilization Equipment

(1) Large Equipment

(2) Conductors Larger Than 4 AWG

(3) Conductors 8, 6, or 4 AWG

(4) Conductors 12 or 10 AWG

(5) Conductors 14 AWG and Smaller

Exception

Change Type: Revision and New

(2) CONDUCTORS LARGER THAN 4 AWG. Boxes that enclose utilization equipment supplied by conductors larger than 4 AWG shall be identified for their specific function.

(3) CONDUCTORS 8, 6, OR 4 AWG. Boxes that enclose utilization equipment supplied by 8, 6, or 4 AWG conductors shall have an internal depth that is not less than 52.4 mm (2$\frac{1}{16}$ in.).

(4) CONDUCTORS 12 OR 10 AWG. Boxes that enclose utilization equipment supplied by 12 or 10 AWG conductors shall have an internal depth that is not less than 30.2 mm (1$\frac{3}{16}$ in.). Where the equipment projects rearward from the mounting plane of the box by more than 25 mm (1 in.), the box shall have a depth not less than that of the equipment plus 6 mm ($\frac{1}{4}$ in.).

(5) CONDUCTORS 14 AWG AND SMALLER. Boxes that enclose utilization equipment supplied by 14 AWG or smaller conductors shall have a depth that is not less than 23.8 mm ($\frac{15}{16}$ in.).

    Exception to (1) through (5): Utilization equipment that is listed to be installed with specified boxes shall be permitted.

Change Significance:

This revision provides the code user with a comprehensive list of requirements for the depth of boxes. Previous code text addressed only "outlet boxes," providing a minimum depth of $\frac{1}{2}$ inch overall—with a $\frac{15}{16}$-inch requirement for boxes enclosing flush devices.

The general rule of 314.24 now requires that outlet and device boxes shall have sufficient depth to allow equipment installed within them to be mounted properly and with sufficient clearance to prevent damage to conductors within the box. This general requirement is then prescriptively addressed in three new first-level subdivisions as follows:

- **(A) Outlet Boxes Without Enclosed Devices or Utilization Equipment.** This contains the previous basic rule that no box shall have an internal depth of less $\frac{1}{2}$ inch.
- **(B) Outlet and Device Boxes With Enclosed Devices.** This contains the previous rule that boxes intended to enclose flush devices shall have an internal depth of not less than $\frac{15}{16}$ inch.
- **(C) Utilization Equipment.** These requirements specifically address equipment that will utilize electric energy. Five second-level subdivisions address all types of configurations, and an exception is added for utilization equipment listed for use with specific boxes.

Notes:

_____

_____

_____

Change Summary:

- 314.27 has been revised to clarify that boxes used for ceiling- and wall-mounted lighting outlets be designed for the purpose and support a minimum of 50 pounds.
- Lighting fixtures weighing more than 50 pounds must be independently supported, or the box must be listed and marked with the maximum weight.

Revised 2008 NEC Text:

314.27 OUTLET BOXES

(A) BOXES AT LUMINAIRE ~~(LIGHTING FIXTURE)~~ OUTLETS. Boxes used at luminaire ~~(lighting fixture)~~ or lampholder outlets <u>in a ceiling</u> shall be designed for the purpose <u>and shall be required to support a luminaire weighing a minimum of 23 kg (50 lb.). Boxes used at luminaire or lampholder outlets in a wall shall be designed for the purpose and shall be marked on the interior of the box to indicate the maximum weight of the luminaire that is permitted to be supported by the box in the wall, if other than 23 kg (50 lb).</u> At every outlet used exclusively for lighting, the box shall be designed or installed so that a luminaire ~~(lighting fixture)~~ may be attached.

Exception: A wall-mounted luminaire ~~(fixture)~~ weighing not more than 3 kg (6 lb) shall be permitted to be supported on other boxes or plaster rings that are secured to other boxes, provided the luminaire ~~(fixture)~~ or its supporting yoke is secured to the box with no fewer than two No. 6 or larger screws.

(B) MAXIMUM LUMINAIRE ~~(FIXTURE)~~ WEIGHT. Outlet boxes or fittings <u>designed for the support of luminaires and</u> installed as required by 314.23 shall be permitted to support luminaires ~~(lighting fixtures)~~ weighing 23 kg (50 lb) or less. A luminaire ~~(lighting fixture)~~ that weighs more than 23 kg (50 lb) shall be supported independently of the outlet box unless the outlet box is listed <u>and marked</u> for the <u>maximum</u> weight to be supported.

Change Significance:

This revision clarifies requirements for wall- and ceiling-mounted lighting fixtures. Ceiling-mounted boxes for lighting fixtures must be designed for the purpose and capable of supporting a minimum of 50 pounds.

Wall-mounted boxes for lighting fixtures must be designed for the purpose, capable of supporting a minimum of 50 pounds, and be marked indicating that the box can handle more than 50 pounds or that it can handle less than 50 pounds. Lighting fixtures greater than 50 pounds must be independently supported unless the box is listed and marked for the maximum weight supported.

Notes:

_____

_____

314.27
OUTLET BOXES

Chapter 3 Wiring Methods and Materials

Article 314 Outlet, Device, Pull, and Junction Boxes; Conduit Bodies; Fittings; and Handhole Enclosures

Part II Installation

314.27 Outlet Boxes

(A) Boxes at Luminaire Outlets

(B) Maximum Luminaire Weight

Change Type: Revision

Boxes for fixtures greater than 50 pounds must be listed and marked with the maximum weight permitted.

314.27(E)
UTILIZATION EQUIPMENT

Chapter 3 Wiring Methods and Materials

Article 314 Outlet, Device, Pull, and Junction Boxes; Conduit Bodies; Fittings; and Handhole Enclosures

Part II Installation

314.27 Outlet Boxes

(E) Utilization Equipment

Change Type: New

Change Summary:

- A new first-level subdivision 314.27(E) is added to address box support requirements for utilization equipment other than ceiling-suspended fans.

Revised 2008 NEC Text:

314.27 OUTLET BOXES

(E) UTILIZATION EQUIPMENT. Boxes used for the support of utilization equipment other than ceiling-suspended (paddle) fans shall meet the requirements of 314.27(A) and (B) for the support of a luminaire that is the same size and weight.

Exception: Utilization equipment weighing not more than 3 kg (6 lb) shall be permitted to be supported on other boxes or plaster rings that are secured to other boxes, provided the equipment or its supporting yoke is secured to the box with no fewer than two No. 6 or larger screws.

Change Significance:

This new text provides installation requirements for boxes supporting utilization equipment other than ceiling-suspended fans. This will apply only to equipment that *utilizes* electrical energy. Where outlet boxes support fire alarm or security devices, for example, they are required in general to meet the requirements of 314.27 (A) and (B).

The exception permits utilization equipment weighing not more than 6 pounds to be supported by boxes installed outside the requirements of 314.27 (A) and (B), provided the device or equipment is supported by at least two No. 6 or larger screws.

314.27(E) provides requirements for utilization equipment and allows equipment weighing not more than 6-pounds to be supported with two number 6 screws.

Change Summary:

- The requirements of 314.28 have been clarified to not apply to a GEC.
- The term *splices* has been added to the title of 314.28(2) to clarify that these requirements may be used for a straight-through box where the box is used only to *splice in* and not to *pull through* conductors.

Revised 2008 NEC Text:

314.28 PULL AND JUNCTION BOXES AND CONDUIT BODIES

(A) MINIMUM SIZE. For raceways containing conductors of 4 AWG or larger <u>that are required to be insulated</u>, and for cables containing conductors of 4 AWG or larger, the minimum dimensions of pull or junction boxes installed in a raceway or cable run shall comply with (A)(1) through (A)(3). Where an enclosure dimension is to be calculated based on the diameter of entering raceways, the diameter shall be the metric designator (trade size) expressed in the units of measurement employed.

(1) Straight Pulls. (No change to existing text)

(2) Angle or U Pulls <u>or Splices.</u> (No change to existing text)

Change Significance:

This revision in 314.28(A) clarifies that the sizing requirements in 314.28(A)(1), (2), and (3) apply only to conductors of 4 AWG or larger that are "required to be insulated." Where a grounding electrode conductor is installed in a raceway, it is not required to be insulated and the provisions of 314.28 will not apply.

The requirements for sizing boxes for a straight pull are clearly defined: the box must be eight times the trade size of the raceway. The addition of the term *splices* in 314.28(A)(2) will now clearly inform the user of this code that conduits installed in a straight-through fashion for splices and not for pulling through the box may be installed in accordance with 314.28(A)(2).

314.28(A)
MINIMUM SIZE

Chapter 3 Wiring Methods and Materials

Article 314 Outlet, Device, Pull, and Junction Boxes; Conduit Bodies; Fittings; and Handhole Enclosures

Part II Installation

314.28 Pull and Junction Boxes and Conduit Bodies

(A) Minimum Size

(1) Straight Pulls

(2) Angle or U Pulls or Splices

Change Type: Revision

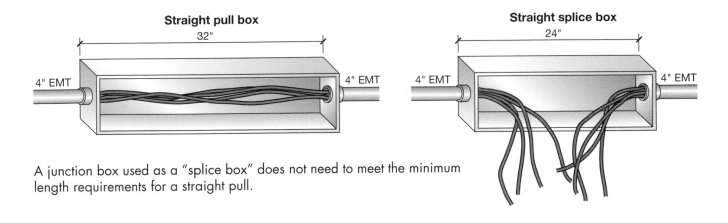

A junction box used as a "splice box" does not need to meet the minimum length requirements for a straight pull.

Notes:

314.30
HANDHOLE ENCLOSURES

Chapter 3 Wiring Methods and Materials

Article 314 Outlet, Device, Pull, and Junction Boxes; Conduit Bodies; Fittings; and Handhole Enclosures

Part II Installation

314.30 Handhole Enclosures

(C) Enclosed Wiring

Change Type: Revision

Change Summary:

- Handhole enclosures are now required to be identified for use in underground systems. All handhole enclosures are installed in the ground. The requirement for all conductors and any splices or terminations, if present, to be listed as suitable for wet locations is clarified to apply to all handhole enclosures—not just those without a bottom.

Revised 2008 NEC Text:

314.30 HANDHOLE ENCLOSURES.   Handhole enclosures shall be designed and installed to withstand all loads likely to be imposed <u>on them. They shall be identified for use in underground systems.</u>

FPN: See ANSI/SCTE 77-2002, *Specification for Underground Enclosure Integrity,* for additional information on deliberate and nondeliberate traffic loading that can be expected to bear on underground enclosures.

(C) <u>ENCLOSED WIRING.</u> ~~HANDHOLE ENCLOSURES WITHOUT BOTTOMS.~~ ~~Where handhole enclosures without bottoms are installed, a~~ <u>All</u> enclosed conductors and any splices or terminations, if present, shall be listed as suitable for wet locations.

Change Significance:

Handhole enclosures are now clearly required to be identified for use in underground systems. Equipment recognized as suitable for the specific purpose, function, use, environment, or application is said to be "identified" for the use. See the Article 100 definition of "identified." This revision will prohibit an installer from building a custom handhole enclosure from wood or metal, which will deteriorate rapidly underground.

All handhole enclosures are installed in the ground, and all are subject to moisture and water. 314.30(C) has been revised to require all conductors and any splices or terminations, if present, in handhole enclosures to be listed as suitable for wet locations. This change clarifies that conductors, splices, terminations, and the like shall be listed for wet locations in all handhole enclosures—not just those without a bottom.

All wiring in a handhole must be listed for use in wet locations, not just handholes without a bottom.

Change Summary:

* New text in 320.10 for permitted use of type AC cable recognizes that this wiring method is identified for use as branch circuits and feeders.

Revised 2008 NEC Text:

320.10 USES PERMITTED.   Type AC cable shall be permitted as follows:

(1) <u>For feeders and branch circuits</u> ~~I~~in both exposed and concealed work

(2) In cable trays

(3) In dry locations

(4) Embedded in plaster finish on brick or other masonry, except in damp or wet locations

(5) To be run or fished in the air voids of masonry block or tile walls where such walls are not exposed or subject to excessive moisture or dampness

FPN: The "Uses Permitted" is not an all-inclusive list.

Change Significance:

This revision now clearly recognizes that type AC cable is permitted to be used as a branch circuit or feeder. Type AC cable has been used in applications as feeders and branch circuits for many years. However, this revision provides the code user with clear and unambiguous text for proper application of the NEC. The common numbering format in Chapter 3 for raceways and cable assemblies provides a user-friendly format to quickly identify uses permitted and not permitted in Section XXX.10 for uses permitted and Section XXX.12 for uses not permitted. Cable assemblies in Chapter 3 are presently recognized for use as follows:

Branch Circuits Only	Branch Circuit or Feeder	Service, Branch Circuit, or Feeder	Not Defined
322 Type: FC	320 Type: AC	326 Type: IGS	328 Type: MV
324 Type: FCC	334 Type: NM [See exclusion as service in 334.12(3)]	330 Type: MC	336: Type TC
	340 Type: UF [See exclusion as service in 334.12(1)]	332 Type: MI	
		338 Type: SE & USE	

Notes:

_____

_____

_____

320.10

USES PERMITTED

Chapter 3 Wiring Methods and Materials

Article 320 Armored Cable, Type AC

Part II Installation

320.10 Uses Permitted

Change Type: Revision

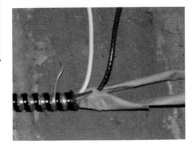

The metal sheath of type AC cable is recognized in 250.118 as an equipment grounding conductor.

330.10

USES PERMITTED

Chapter 3 Wiring Methods and Materials

Article 330 Metal-Clad Cable Type MC

Part II Installation

330.10 Uses Permitted

(A) General Uses

(B) Specific Uses

(4) Installed Outside Buildings or as Aerial Cable

Change Type: Revision

Change Summary:
The installation of type MC cable in wet locations now clearly requires:
- A metallic covering impervious to moisture, or
- A lead sheath or moisture-impervious jacket, or
- Conductors listed for wet location with a *corrosion-resistant jacket provided over the metallic sheath*

Revised 2008 NEC Text:
330.10 USES PERMITTED

(A) GENERAL USES.   Type MC cable shall be permitted as follows:

(11) In wet locations where any of the following conditions are met:

   a. The metallic covering is impervious to moisture.

   b. A lead sheath or moisture-impervious jacket is provided under the metal covering.

   c. The insulated conductors under the metallic covering are listed for use in wet locations <u>and a corrosion-resistant jacket is provided over the metallic sheath.</u>

   (Note: No change to the remaining 11 list items in 334.10(A))

(B) SPECIFIC USES.   Type MC cable shall be permitted to be installed in compliance with Parts II and III of Article 725 and 770.133 as applicable and in accordance with 330.10(B)(1) through (B)(4).

(4) INSTALLED OUTSIDE OF BUILDINGS <u>OR STRUCTURES</u> OR AS AERIAL CABLE.   Type MC cable installed outside of buildings <u>or structures</u> or as aerial cable shall comply with 225.10, 396.10, and 396.12.

Change Significance:
The text in 330.10(A)(11) list items (a) and (b) clearly requires a metallic covering or jacket "impervious" to moisture. The previous text of 300.10(A)(11)(c) permitted an interlocking type MC cable, which is not impervious to moisture, to be installed in a wet location—provided the insulated conductors under the metallic covering were listed for use in a wet location.

   The requirements of 330.10(A)(11)(a) and (11)(b) prohibit moisture/water from entering the cable core of MC cable. The requirement of 330.10(A)(11)(c) has been revised to require an overall moisture-impervious jacket over the metal covering to prevent water from entering the cable core, where it could migrate to conductor terminations. This type of MC cable with a protective jacket is readily available for installation in wet locations.

   The title of second-level subdivision 330.10(B)(4) has been editorially revised to clearly permit installation of type MC cable outside "structures" where the provisions of Articles 330, 225.10, 396.10, and 396.12 are met.

Notes:

_____

_____

A corrosion-resistant jacket is required on MC cable used in a wet location.

334.12
USES NOT PERMITTED

Chapter 3 Wiring Methods and Materials

Article 334 Nonmetallic-Sheathed Cable: Types NM, NMC, and NMS

Part II Installation

334.12 Uses Not Permitted

(A) Types NM, NMC, and NMS

(B) Types NM and NMS

Change Type: Revision and New

Change Summary:
- A new exception in 334.12(A)(1) will permit the installation of types NM, NMC, and NMS cables in types I and II construction when installed in raceway.

Revised 2008 NEC Text:
334.12 USES NOT PERMITTED

(A) TYPES NM, NMC, AND NMS.   Types NM, NMC, and NMS cables shall not be permitted as follows:

(1) In any dwelling or structure not specifically permitted in 334.10(1), (2), and (3) <u>Exception: Type NM, NMC, and NMS cable shall be permitted in Types I and II construction when installed within raceways permitted to be installed in Types I and II construction.</u>

(Note: No change to list items (2) through (10))

(B) TYPES NM AND NMS.   Types NM and NMS cables shall not be used under the following conditions or in the following locations:

(1) Where exposed to corrosive fumes or vapors

(2) Where embedded in masonry, concrete, adobe, fill, or plaster

(3) In a shallow chase in masonry, concrete, or adobe and covered with plaster, adobe, or similar finish

(4) <u>In wet or damp locations</u> ~~Where exposed or subject to excessive moisture or dampness~~

Change Significance:
The addition of the exception in 334.12(1) will permit the installation of types NM, NMC, and NMS cables in types I and II construction when installed in a permitted raceway. This revision is driven by the presence of listed type NM cables that consist of power, communications, and signaling conductors under a common jacket. This exception will allow the use of such a cable as well as any type NM, NMC, or NMS cable in a building of type I and II construction where installed in permitted raceways.

The revised text in 334.12 (B)(4) provides clarity and usability by prohibiting the use of type NM and NMS in a "wet" or "damp" location. The previous text—which read "Where exposed or subject to excessive moisture or dampness"—is subjective. The revised text is easy to read, apply, and enforce. The terms *wet location* and *damp location* are defined in Article 100.

Types NM, NMC, and NMS are not permitted in "wet or damp locations."

334.15
EXPOSED WORK

Chapter 3 Wiring Methods and Materials

Article 334 Nonmetallic-Sheathed Cable: Types NM, NMC, and NMS

Part II Installation

334.15 Exposed Work

(A) To Follow Surface

(B) Protection from Physical Damage

(C) In Unfinished Basements and Crawl Spaces

Change Type: Revision

Change Summary:

- The provisions of 334.15 for exposed installations of types NM, NMC, and NMS have been revised to:
  - Include crawl spaces in 334.15(C)
  - Require that cable in shallow chase/groove comply with 300.4(E)
  - Provide clear requirements for the protection of type NM, NMC, or NMS cables entering conduit or tubing

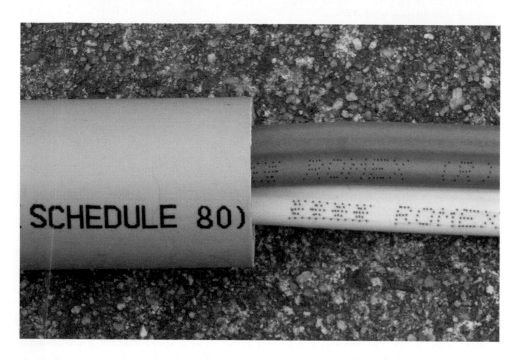

The jacket of type NM cable (commonly referred to as romex) is not designed to protect the conductors from physical damage. The provisions of 300.4 and 334.15 mandate additional protection of type NM where subject to damage.

Revised 2008 NEC Text:

334.15 EXPOSED WORK. In exposed work, except as provided in 300.11(A), cable shall be installed as specified in 334.15(A) through (C).

(A) TO FOLLOW SURFACE. Cable shall closely follow the surface of the building finish or of running boards.

(B) PROTECTION FROM PHYSICAL DAMAGE. Cable shall be protected from physical damage where necessary by rigid metal conduit, intermediate metal conduit, electrical metallic tubing, Schedule 80 PVC ~~rigid non-metallic~~ conduit, or other approved means. Where passing through a floor, the cable shall be enclosed in rigid metal conduit, intermediate metal conduit, electrical metallic tubing, Schedule 80 PVC ~~rigid nonmetallic~~ conduit, or other approved means extending at least 150 mm (6 in.) above the floor.
~~Where~~ Type NMC cable ~~is~~ installed in shallow chases or grooves in masonry, concrete, or adobe, ~~the cable~~ shall be protected ~~against nails or screws~~

(continued)

by a steel plate at least 1.59 mm (¹⁄₁₆-in.) thick <u>in accordance with the requirements in 300.4(E)</u> and covered with plaster, adobe, or similar finish.

(C) IN UNFINISHED BASEMENTS <u>AND CRAWL SPACES</u>.   Where cable is run at angles with joists in unfinished basements <u>and crawl spaces</u>, it shall be permissible to secure cables not smaller than two 6 AWG or three 8 AWG conductors directly to the lower edges of the joists. Smaller cables shall be run either through bored holes in joists or on running boards. NM cable <u>installed</u> ~~used~~ on <u>the</u> ~~a~~ wall of an unfinished basement shall be permitted to be installed in a listed conduit or tubing <u>or shall be protected in accordance with 300.4.</u> Conduit or tubing shall <u>be provided with</u> ~~utilize~~ a ~~nonmetallic~~ <u>suitable insulating bushing</u> or adapter at the point the cable enters the raceway. <u>The NM cable sheath shall extend through the conduit or tubing and into the outlet or device box not less than 6 mm (¹⁄₄ in.). The cable shall be secured within 300 mm (12 in.) of the point where the cable enters the conduit or tubing.</u> Metal conduit, tubing, and metal outlet boxes shall be ~~grounded~~ <u>connected to an equipment grounding conductor</u>.

Change Significance:

The first paragraph of first-level subdivision 334.15(B) has been revised by deleting the term *rigid nonmetallic.* The remaining text "Schedule 80 PVC conduit" requires rigid nonmetallic conduit in accordance with Article 352.

The second paragraph of first-level subdivision 334.15(B) has been revised to include shallow "grooves" and to remove prescriptive text requiring a steel plate. New text requires that cables installed in shallow chases or grooves be installed in accordance with the requirements in 300.4(E). This revision continues to require a steel plate, but allows cables in chases or grooves to be protected by a raceway in 300.4(E) Exception No. 1 or a listed steel plate smaller than 1/16 inch as permitted in 300.4(E) Exception No. 2.

The title and text of 334.15(C) have been revised to include crawl spaces. This revision will now allow the installation methods permitted in an unfinished basement to be used in crawl spaces. In addition, the requirements for protection of NM cable on the wall of an unfinished basement have been clarified. Protection may now be achieved through installation in a listed conduit or tubing or protected in accordance with 300.4. Where conduit or tubing is used for protection, the installation is now required as follows:

* The conduit or tubing shall be provided with a suitable insulating bushing or adapter at the point the cable enters the raceway, and

* The NM cable sheath shall extend through the conduit or tubing and into the outlet or device box not less than 6 mm (1/4 inch), and

* The cable shall be secured within 300 mm (12 inches) of the point where the cable enters the conduit or tubing

These requirements for the protection of type NM cable mirror the existing requirements in exceptions to 312.5(C) and 314.17(C).

The term *grounded* in the last sentence has been deleted and for clarity and usability has been replaced with new text requiring connection to an equipment grounding conductor. Where metal conduit/tubing and metal boxes are used, they shall be "connected to an equipment grounding conductor."

334.80

AMPACITY

Chapter 3 Wiring Methods and Materials

Article 334 Nonmetallic-Sheathed Cable: Types NM, NMC, and NMS

Part II Installation

334.80 Ampacity

Change Type: Revision

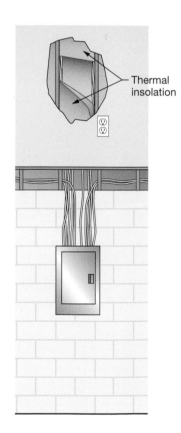

Thermal insolation

Type NM cable installed in thermal insulation may be subject to the derating requirements of 310.15(B)(2(a).

Change Summary:

- Where two or more type NM cables pass through wood framing and the opening is for any reason to be fire-stopped or filled, such cables must be derated whether or not they are bundled—and the provisions of 310.15(A)(2), Exception, shall not apply.

- A new last paragraph requires the derating of ampacity where more than two NM cables are installed in contact with thermal insulation without maintaining spacing between cables.

Revised 2008 NEC Text:

334.80 AMPACITY.   The ampacity of Types NM, NMC, and NMS cable shall be determined in accordance with 310.15. The ampacity shall be in accordance with the 60°C (140°F) conductor temperature rating. The 90°C (194°F) rating shall be permitted to be used for ampacity derating purposes, provided the final derated ampacity does not exceed that for a 60°C (140°F) rated conductor. The ampacity of Types NM, NMC, and NMS cable installed in cable tray shall be determined in accordance with 392.11.

Where more than two NM cables, containing two or more current-carrying conductors, ~~are bundled together and pass~~ are installed, without maintaining spacing between the cables, through the same opening in wood framing that is to be fire- or draft-stopped using thermal insulation, caulk, or sealing foam, the allowable ampacity of each conductor shall be adjusted in accordance with Table 310.15(B)(2)(a) and the provisions of 310.15(A)(2), Exception, shall not apply.

Where more than two NM cables containing two or more current-carrying conductors are installed in contact with thermal insulation without maintaining spacing between cables, the allowable ampacity of each conductor shall be adjusted in accordance with Table 310.15(B)(2)(a).

Change Significance:

The second paragraph of 334.80 has been revised to clarify that cables (two or more, each with two or more current-carrying conductors) need only pass through the same opening in wood framing that is to be fire- or draft-stopped using thermal insulation, caulk, or sealing foam to require derating in accordance with 310.15(B)(2)(a). Previous text addressed cables that were "bundled together" and passed through the same opening in wood framing that is to be fire- or draft-stopped using thermal insulation or sealing foam.

The term *caulk* has been added to the list of materials that are commonly used to fire-stop or draft-stop. Any packing or material used to achieve fire-stopping or draft-stopping requires derating the cables where two or more cables pass through wood framing. Almost all applications of type NM cables will have two or more current-carrying conductors in each cable.

A new last paragraph has been added to require the derating of ampacity of type NM cables where more than two cables (each with two or more current-carrying conductors) are installed in contact with thermal insulation, without maintaining spacing between cables. For example, this will require derating of cables stapled on top of each other in outside walls to be insulated or where cables are run through the same openings in insulated ceilings.

Change Summary:

- A new exception in 336.10(7) will now permit power and control tray cable marked TC-ER to be run without continuous support to utilization equipment, or between cable trays for not more than 6 feet. The cable shall not be exposed to physical damage and shall be mechanically supported where it leaves or enters the cable tray.

336.10
USES PERMITTED

Chapter 3 Wiring Methods and Materials

Article 336 Power and Control Tray Cable, Type TC

Part II Installation

336.10 Uses Permitted

Exception

Change Type: New

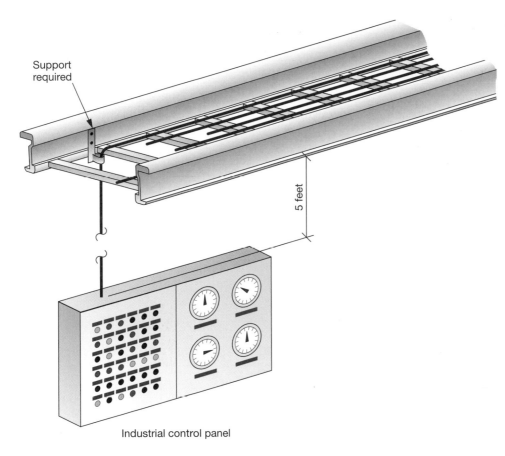

Support required

5 feet

Industrial control panel

Type TC-ER is permitted to travel not more than 6 feet without continuous support.

Revised 2008 NEC Text:
336.10 USES PERMITTED

Type TC cable shall be permitted to be used as follows:

(1) through **(6)** & **(8)** (No change)

(7) In industrial establishments where the conditions of maintenance and supervision ensure that only qualified persons service the installation, and where the cable is continuously supported and protected against physical damage using mechanical protection, such as struts, angles, or channels, Type TC tray cable that complies with the crush and impact requirements of Type MC cable and is identified for such use with the marking Type TC–ER shall be permitted between a cable tray and the utilization equipment or device. The cable shall be secured at intervals not exceeding 1.8 m

(continued)

(6 ft). Equipment grounding for the utilization equipment shall be provided by an equipment grounding conductor within the cable. In cables containing conductors sized 6 AWG or smaller, the equipment grounding conductor shall be provided within the cable or, at the time of installation, one or more insulated conductors shall be permanently identified as an equipment grounding conductor in accordance with 250.119(B).

Exception: Where not subject to physical damage, Type TC-ER shall be permitted to transition between cable trays and between cable trays and utilization equipment or devices for a distance not to exceed 1.8 m (6 ft) without continuous support. The cable shall be mechanically supported where exiting the cable tray to ensure that the minimum bending radius is not exceeded.

Change Significance:

Type TC cable marked with the suffix ER is listed to meet the crush and impact requirements of type MC cable. The suffix ER designates "exposed run." List item (7) in 336.10 for uses permitted is limited to industrial establishments where the conditions of maintenance and supervision ensure that only qualified persons service the installation.

This new exception will permit unsupported lengths of TC-ER cable from cable tray to utilization equipment or to another tray only in those industrial establishments identified in 336.10(7). The length of unsupported cable is limited to 6 feet. To apply this exception, however, "The cable shall be <u>mechanically supported</u> where exiting the cable tray to ensure that the minimum bending radius is not exceeded." "Mechanically supported" means that the cable must be held in place by mechanical means such as a cable strap bolted to the wall of the cable tray.

Notes:

_____

_____

_____

_____

_____

_____

_____

Change Summary:
- Article 338 now includes prescriptive text on "uses not permitted" for type SE cable in 338.12(A) and type USE cable in 338.12(B).

Revised 2008 NEC Text:
338.12 USES NOT PERMITTED
(A) SERVICE-ENTRANCE CABLE. Service-Entrance Cable (SE) shall not be used under the following conditions or in the following locations.
(1) Where subject to physical damage unless protected in accordance with 230.50(A)
(2) Underground with or without a raceway
(3) For exterior branch circuits and feeder wiring unless the installation complies with the provisions of Part I of Article 225 and is supported in accordance with 334.30 or is used as messenger supported wiring as permitted in Part II of Article 396

(B) UNDERGROUND SERVICE-ENTRANCE CABLE. Underground Service-Entrance Cable (USE) shall not be used under the following conditions or in the following locations:
(1) For interior wiring
(2) For above ground installations except where USE cable emerges from the ground and is terminated in an enclosure at an outdoor location and the cable is protected in accordance with 300.5(D)
(3) As aerial cable unless it is a multiconductor cable identified for use above ground and installed as messenger supported wiring in accordance with 225.10 and Part II of Article 396

Change Significance:
Article 338 did not address "uses not permitted" in previous editions of the NEC. This revision conforms Article 338 for SE and USE cable to the parallel numbering format of other cable assembly articles. For example, "uses permitted" is always addressed in XXX.10 and "uses not permitted" in XXX.12. The addition of prescriptive text outlining "uses not permitted" is user friendly and adds clarity. Without specifically identifying "uses not permitted," the user of the code had to look at what is specifically permitted to make a determination on prohibited use. 338.10 for "uses permitted" is editorially modified to correspond with the new 338.12.

338.12(A) now clearly prohibits type SE cable from being installed (1) where exposed to physical damage, (2) underground with or without a raceway, and (3) as exterior branch circuits and feeder wiring unless installed in accordance with Part I of Article 225 and supported in accordance with 334.30 or where installed as messenger-supported wiring as permitted in Part II of Article 396.

338.12(B) now clearly prohibits USE cable from being installed (1) as interior wiring, (2) above ground unless terminated in an enclosure where emerging from ground at an outdoor location and the cable is protected in accordance with 300.5(D), and (3) as aerial cable unless it is a multiconductor cable identified for use above ground and installed as messenger-supported wiring in accordance with 225.10 and Part II of Article 396.

338.12
USES NOT PERMITTED

Chapter 3 Wiring Methods and Materials

Article 338 Service-Entrance Cable, Types SE and USE

Part II Installation

338.12 Uses Not Permitted

(A) Service-Entrance Cable

(B) Underground Service-Entrance Cable

Change Type: New

Type SE cable is not permitted underground with or without a raceway.

342.30(C)
UNSUPPORTED RACEWAYS (IMC)

344.30(C) Unsupported Raceways (RMC)

352.30(C) Unsupported Raceways (PVC)

358.30(C) Unsupported Raceways (EMT)

Chapter 3 Wiring Methods and Materials

Article 342 Intermediate Metal Conduit, Type IMC

Article 344 Rigid Metal Conduit, Type RMC

Article 352 Rigid Polyvinyl Chloride Conduit, Type PVC

Article 358 Electrical Metallic Tubing, Type EMT

Part II Installation

342.30 Securing and Supporting

344.30 Securing and Supporting

352.30 Securing and Supporting

358.30 Securing and Supporting

(C) Unsupported Raceways (in each section)

Change Type: New

Change Summary:

- Raceway types IMC, RMC, PVC, and EMT are now permitted to be installed in lengths up to 18 inches without support. This is permitted only where (1) oversized, concentric, or eccentric knockouts are not encountered; (2) the raceway is unbroken (no couplings); and (3) the raceway terminates in an outlet box, junction box, device box, cabinet, or other termination at each end of the raceway.

Revised 2008 NEC Text:

3XX.30 SECURING AND SUPPORTING   (342 IMC, 344 RMC, 352 PVC & 358 EMT)

(C) UNSUPPORTED RACEWAYS.   Where oversized, concentric or eccentric knockouts are not encountered, Type *(342* IMC, *344* RMC, *352* PVC *& 358* EMT)IMC shall be permitted to be unsupported where the raceway is not more than 450 mm (18 in.) and remains in unbroken lengths (without coupling). Such raceways shall terminate in an outlet box, junction box, device box, cabinet, or other termination at each end of the raceway.

Change Significance:

This new first-level subdivision will allow "unsupported raceways" for type IMC, RMC, PVC, and EMT provided that oversized, concentric, or eccentric knockouts are not encountered. Where the raceway is unsupported, it is limited to 18 inches in length and must be unbroken; no couplings are permitted. The raceway ends must terminate in an outlet box, junction box, device box, or cabinet.

This revision recognizes that installations of very short lengths of raceways (including IMC, RMC, PVC, and EMT terminated in enclosures) are secure and do not need to be supported. It is typical to see short, unsupported lengths of these raceways between junction boxes, cabinets, wireways, and equipment. These installations are secure and do not represent a safety hazard. These installations were code violations that have been overlooked for decades due to the fact that there was not a safety hazard and the integrity of the installation was not compromised. This revision has occurred in the XXX.30 section of Articles, 342 IMC, 344 RMC, 352 PVC, and 358 EMT.

Under certain conditions, 18-inch lengths of types IMC, RMC, PVC, and EMT are permitted to be unsupported.

Change Summary:

* 344.10 has been revised to clarify and differentiate permitted uses of steel RMC, aluminum RMC, and red brass RMC. The requirements for corrosion protection of all types of RMC have been clarified in accordance with the listing information.

Article 344 provides requirements for the installation and construction of steel, aluminum, and red brass rigid conduit.

Revised 2008 NEC Text:

344.10 USES PERMITTED

(A) ~~ALL~~ ATMOSPHERIC CONDITIONS AND OCCUPANCIES.

(1) GALVANIZED STEEL AND STAINLESS STEEL RMC. ~~Use of~~ Galvanized steel and stainless steel RMC shall be permitted under all atmospheric conditions and occupancies.

(2) RED BRASS RMC. Red brass RMC shall be permitted to be installed for direct burial and swimming pool applications.

(3) ALUMINUM RMC. Aluminum RMC shall be permitted to be installed where judged suitable for the environment. Rigid aluminum conduit encased in concrete or in direct contact with the earth shall be provided with approved supplementary corrosion protection.

(4) FERROUS RACEWAYS AND FITTINGS. Ferrous raceways and fittings protected from corrosion solely by enamel shall be permitted only indoors and in occupancies not subject to severe corrosive influences.

(continued)

344.10
USES PERMITTED

Chapter 3 Wiring Methods and Materials

Article 344 Rigid Metal Conduit, Type RMC

Part II Installation

344.10 Uses Permitted

(A) Atmospheric Conditions and Occupancies

(1) Galvanized Steel and Stainless Steel RMC

(2) Red Brass RMC

(3) Aluminum RMC

(4) Ferrous Raceways and Fittings

(B) Corrosive nvironments

(1) Galvanized Steel, Stainless Steel and Red Brass RMC, Elbows, Couplings, and Fittings

(2) Supplementary Protection of Aluminum RMC

(C) Cinder Fill

(D) Wet Locations

Exception

Change Type: Revision

(B) ~~CORROSION~~ <u>CORROSIVE</u> ENVIRONMENTS.
(1) <u>GALVANIZED STEEL, STAINLESS STEEL AND RED BRASS RMC,</u> ELBOWS, COUPLINGS, AND FITTINGS.   <u>Galvanized steel, stainless steel and red brass</u> RMC~,~ elbows, couplings, and fittings shall be permitted to be installed in concrete, in direct contact with the earth, or in areas subject to severe corrosive influences where protected by corrosion protection and judged suitable for the condition.

(2) <u>SUPPLEMENTARY PROTECTION OF ALUMINUM RMC.</u>   <u>Aluminum RMC shall be provided with approved supplementary corrosion protection where encased in concrete or in direct contact with the earth.</u>

(C) CINDER FILL.   <u>Galvanized steel, stainless steel and red brass</u> RMC shall be permitted to be installed in or under cinder fill where subject to permanent moisture where protected on all sides by a layer of noncinder concrete not less than 50 mm (2 in.) thick; where the conduit is not less than 450mm (18 in.) under the fill; or where protected by corrosion protection and judged suitable for the condition.

(D) WET LOCATIONS.   (No change)

Change Significance:

This revision provides clarity and usability for the code user when determining permitted uses for different types of RMC and the requirements for corrosion protection.

344.10(A) now clarifies that (1) galvanized and stainless steel RMC are permitted in all atmospheric conditions and occupancies, (2) red brass is permitted for direct burial and swimming pools, and (3) aluminum is permitted "where judged suitable for the environment" and supplementary corrosion protection is required where installed in concrete or in direct contact with earth.

Corrosion protection in corrosive environments is now clearly separated with requirements for steel in 344.10(B)(1) and aluminum in (B)(2). 344.10(C) for cinder fill has been revised to clarify that only galvanized steel, stainless steel, and red brass RMC are permitted. This clarifies that aluminum RMC may not be installed in cinder fill as detailed in 344.10(C).

Notes:

_____

_____

_____

_____

_____

Change Summary:

- The previous text in the parallel numbering format of XXX.12 for uses not permitted for raceways in hazardous locations was not "all inclusive" of permitted applications. This revision recognizes that other articles may specifically permit the raceway to be installed under given conditions.

Revised 2008 NEC Text:

3XX.12 USES NOT PERMITTED.   *FMC, PVC, HDPE, NUCC, RTRC, LFNC, EMT, ENT, ELLULAR Concrete Floor Raceways, Cellular Metal Floor Raceways, Nonmetallic Wireways, Multioutlet Assemblies and Surface Nonmetallic Raceways* shall not be used in the following: (x) In any hazardous (classified) location, except ~~other than~~ as permitted by other Articles in this Code ~~in 5XX.XX and 5XX.X~~.

Change Significance:

The use of a "laundry list" of specific permitted applications for raceways in hazardous locations is not user friendly. This revision now recognizes that any "other article" may permit a given raceway to be used in a specific application. This revision has occurred in 12 different articles in the .12 section for "uses not permitted."

Flexible metal conduit (FMC), for example, in the 2005 NEC was in general prohibited in hazardous locations except where permitted in 501.10(B) and 504.20. This would lead the code user to believe that these are the "only" permitted applications for FMC in a hazardous location. FMC is permitted, for example, to contain nonincendive field wiring in a Class III Division I location.

It is impossible to note every permitted application of all raceways in hazardous locations in the raceway articles. The code user must follow the arrangement of the NEC as detailed in 90.3, which mandates that the provisions of Chapters 1 through 4 be supplemented and/or modified by Chapters 5, 6, and 7.

3XX.12
USES NOT PERMITTED

Chapter 3 Wiring Methods and Materials

Articles 348, 352, 353, 354, 355, 356, 358, 362, 372, 374, 378, 380, and 388

Part II Installation

3XX.12 Uses Not Permitted

Change Type: Revision

Article 348	Flexible Metal Conduit: Type FMC
Article 352	Rigid Polyvinyl Chloride Conduit: Type PVC
Article 353	High Density Polyethylene Conduit: Type HDPE
Article 354	Nonmetallic Underground Conduit with Conductors: Type NUCC
Article 355	Reinforced Thermosetting Resin Conduit: Type RTRC
Article 356	Liquidtight Flexible Nonmetallic Conduit: Type LFNC
Article 358	Electrical Metallic Tubing: Type EMT
Article 362	Electrical Nonmetallic Tubing: Type ENT
Article 372	Cellular Concrete Floor Raceways
Article 374	Cellular Metal Floor Raceways
Article 378	Nonmetallic Wireways
Article 380	Multioutlet Assembly
Article 388	Surface Nonmetallic Raceways

The common numbering sequence of all circular raceways provides the code user with quick access to uses permitted in the 3XX.10 section and uses not permitted in the 3XX.12 section.

348.12
USES NOT PERMITTED

Chapter 3 Wiring Methods and Materials

Article 348 Flexible Metal Conduit: Type FMC

Part II Installation

348.12 Uses Not Permitted

Change Type: Revision

Change Summary:
• The use of FMC is now clearly prohibited in wet locations.

Revised 2008 NEC Text:
348.12 USES NOT PERMITTED
FMC shall not be used in the following:
(1) In wet locations ~~unless the conductors are approved for the specific conditions and the installation is such that liquid is not likely to enter raceways or enclosures to which the conduit is connected~~

Change Significance:
This revision clarifies that FMC is not permitted in any "wet location."

The previous code text permitted the use of FMC in a wet location where (1) the conductors were approved for the specific conditions and (2) the installation was such that liquid was not likely to enter raceways or enclosures to which the conduit is connected. FMC by design readily allows water to enter the raceway. This allows water to accumulate in the FMC and moisture to migrate into the connected equipment and enclosures. Liquid tight flexible metal conduit (LFMC) is readily available and is listed for use in wet locations.

Flexible metal conduit (FMC) is not permitted in wet locations.

Notes:

_____

_____

Change Summary:
* The grounding and bonding requirements for both FMC and LFMC are clarified by adding the language "after installation" with respect to flexibility.

Revised 2008 NEC Text:
348.60 GROUNDING AND BONDING &
350.60 GROUNDING AND BONDING
Where used to connect equipment where flexibility is required <u>after installation</u>, an equipment grounding conductor shall be installed.

Where flexibility is not required <u>after installation</u>, FMC *or LFMC* shall be permitted to be used as an equipment grounding conductor when installed in accordance with 250.118(5). *250.118(6) for LFMC.*

Where required or installed, equipment grounding conductors shall be installed in accordance with 250.134(B).

Where required or installed, equipment bonding jumpers shall be installed in accordance with 250.102.

Change Significance:
The addition of the clarifier "after installation" provides clear direction to the code user that an EGC shall be installed where FMC or LFMC is used due to vibration or movement of equipment. This revision reflects changes that occurred during the 2005 NEC cycle in 250.188, which provides the code user with permitted types of equipment grounding conductors. The requirements of 250.118 for FMC and LFMC where flexibility is required read as follows: "250.118(5)(d) for FMC and 250.118(6)(e) for LFMC." Where used to connect equipment where flexibility is necessary after installation, an equipment grounding conductor shall be installed.

Both FMC and LFMC are permitted to serve as an EGC under specific provisions listed in 250.118. However, where flexibility is required "after installation" for applications such as motors equipment continuously moving or vibrating shall have an EGC installed. For example, 2 × 4 lighting fixtures are installed in an acoustical lay-in tile ceiling. The installation consists of EMT with 5-foot FMC tails to each fixture. This installation does not require "flexibility after installation." In this situation, a flexible wiring method is used for "ease of installation" not for "flexibility after installation."

348.60 and 350.60
GROUNDING AND BONDING

Chapter 3 Wiring Methods and Materials

Article 348 Flexible Metal Conduit: Type FMC

Part II Installation

348.60 Grounding and Bonding

Article 350 Liquidtight Flexible Metal Conduit, Type LFMC

Part II Installation

350.60 Grounding and Bonding

Change Type: Revision

Flexibility "after installation" is required in this installation.

350.30
SECURING AND SUPPORTING

Chapter 3 Wiring Methods and Materials

Article 350 Liquidtight Flexible Metal Conduit, Type LFMC

Part II Installation

350.30 Securing and Supporting

(A) Securely Fastened

Change Type: Revision

Change Summary:
- Requirements for securely fastening LFMC have been revised to permit unsecured portions of the raceway where flexibility is required. Where LTFMC is "fished," the exception has been clarified as referring to the concealed space only.

**2.5 feet of LFMC to motor
(no support required)**

Pump Motor A-1

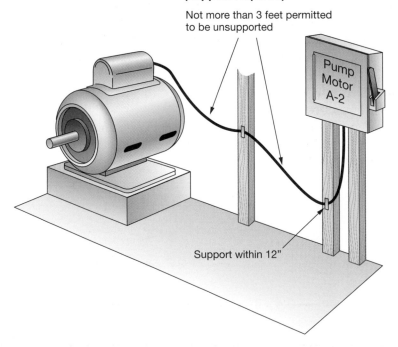

**5 feet of LFMC to motor
(support required)**

Not more than 3 feet permitted
to be unsupported

Pump Motor A-2

Support within 12"

LFMC support requirements allow limited lengths of unsupported raceway.

(continued)

Revised 2008 NEC Text:
350.30 SECURING AND SUPPORTING. LFMC shall be securely fastened in place and supported in accordance with 350.30(A) and (B).

(A) SECURELY FASTENED. LFMC shall be securely fastened in place by an approved means within 300 mm (12 in.) of each box, cabinet, conduit body, or other conduit termination and shall be supported and secured at intervals not to exceed 1.4 m (4½ ft).

Exception No. 1: Where LFMC is fished between access points through concealed spaces in finished buildings or structures and supporting is impractical.

Exception No. 2: ~~Lengths not exceeding 900 mm (3 ft) at terminals~~ *Where flexibility is necessary after installation, lengths shall not exceed the following:*

(1) 900 mm (3 ft) for metric designators 16 through 35 (trade sizes ½ through 1¼)

(2) 1200 mm (4 ft) for metric designators 41 through 53 (trade sizes 1½ through 2)

(3) 1500 mm (5 ft) for metric designators 63 (trade size 2½) and larger

Exception No. 3: *(No change)*
Exception No. 4: *(No change)*

Change Significance:
In the 2005 NEC cycle, Exception No. 2 to 348.30 was revised as seen in this change to permit unsecured lengths of FMC where flexibility is required. LFMC is now permitted to be installed without being secured within 12 inches of conduit termination where flexibility is required. The requirement for secure/support at intervals of 4-½ feet is applied individually according to the size of the LFMC and the new text in Exception No. 2.

The new text clearly outlines that this permission is limited. For example, a 2-½-foot piece of 1-inch LFMC from a motor disconnecting means to the motor would not need to be secured where flexibility is required. However, if this piece of 1-inch LFMC were 5 feet long, secure/support would be required within 12 inches of the motor disconnect and at least one point of additional support required to ensure that not more than 3 feet of 1-inch LFMC was unsecured.

Exception No. 1 has been revised to clarify that only the "concealed" portion of "fished" LFMC is permitted to be unsecured. A similar change has been made for FMC in 348.30 Exception No. 1.

Notes:

_____

_____

_____

_____

Article 352
RIGID POLYVINYL CHLORIDE CONDUIT: TYPE PVC

Chapter 3 Wiring Methods and Materials

Article 352 Rigid Polyvinyl Chloride Conduit, Type PVC

Part I General

352.1

352.2

Change Type: Revision

Change Summary:

- Article 352 has been renamed to address only one type of RNC: polyvinyl chloride conduit (PVC). Three separate articles for RNC now exist in the NEC to provide clearly defined installation and construction requirements for types PVC in 352, HDPE in 353, and RTRC in 355.

Revised 2008 NEC Text:

ARTICLE 352 RIGID ~~NONMETALLIC~~ <u>POLYVINYL CHLORIDE</u> CONDUIT: TYPE <u>PVC</u> ~~RNC~~

I. GENERAL

352.1 SCOPE.   This article covers the use, installation, and construction specifications for rigid ~~nonmetallic~~ <u>polyvinyl chloride</u> conduit (<u>PVC</u> ~~RNC~~) and associated fittings.

FPN: Refer to Article 353 for High Density Polyethylene Conduit: Type HDPE and Article 355 for Reinforced Thermosetting Resin Conduit: Type RTRC.

352.2 DEFINITION

RIGID ~~NONMETALLIC~~ <u>POLYVINYL CHLORIDE</u> CONDUIT (<u>PVC</u> ~~RNC~~).   A <u>rigid</u> nonmetallic ~~raceway~~ <u>conduit</u> (RNC) of circular cross section, with integral or associated couplings, connectors, and fittings for the installation of electrical conductors and cables.

	Rigid PVC	HDPE	RTRC
NEC article[a]	352	353	355
U.S. listing standards	UL 651	UL 651A and UL 651B[b]	UL 1684 and UL 1684A
Listed trade sizes	1/2 – 6	1 – 6	1/2 – 6
Types listed for aboveground use	Schedule 40 or 80	See 353.10(5)	Type AG or XW
Types listed for direct burial	Schedule 40 or 80	Yes	Type BG, AG, or XW
Types listed for physical damage	Schedule 80	No	Type XW
Can be used where concealed	Yes	No	Yes
Can be used for directional boring	If approved	Yes	If approved
Maximum support span[c]	See NEC table 352.30	N/A	See NEC table 355.30
Temperature rating	Maximum 50°C ambient	N/A	-40°C – 110°C continuous operating
Can be used inside buildings	Schedule 40 or 80	No	Type AG or XW
Expansion characteristics	3.38×10^{-5} in/in/°F	N/A	1.5×10^{-5} in/in/°F
	6.084×10^{-5} mm/mm/°C		2.7×10^{-5} mm/mm/°C
Weather resistant	Yes	No	Yes
Approximate weight	Schedule 40 – 232 lbs	Schedule 40 – 143 lbs	Type AG or BG – 82 lbs
(4-inch trade size per 100')	Schedule 80 – 308 lbs	Schedule 80 – 198 lbs	Type XW – 282 lbs
Low halogen	No	No	Yes

a   NEC Article references pertain to the 2008 National Electric Code.
b   UL 651A applies to discrete length HDPE and UL 651B to continuous length HDPE.
c   Unless otherwise listed, the maximum support span distances for both rigid PVC and RTRC are identical.

Three separate types of rigid nonmetallic conduit exist in the NEC: PVC in 352, HDPE in 353, and RTRC in 355.

Change Significance:

This revision of the title of this article, the scope, the definition, and other changes limit the application of Article 352 to only PVC. There are three separate types of rigid nonmetallic conduits recognized for use in the NEC.

(continued)

Prior to the 2005 edition of the NEC, all three types of RNC were applied under the provisions of Article 352. The materials, construction, and permitted applications of the three types of RNC are very different. The three types of RNC recognized for use in the NEC are as follows:

- **Article 352 Rigid Polyvinyl Chloride Conduit: Type PVC** (Renamed in 2008)
- **Article 353 High Density Polyethylene Conduit: Type HDPE Conduit** (New in 2005)
- **Article 355 Reinforced Thermosetting Resin Conduit: Type RTRC** (New in 2008)

A new Article 353 for HDPE-type RNC was added to the NEC in 2005. A new Article 355 for RTRC conduit is added in the 2008 NEC. These new articles and the renaming of Article 352 to address type PVC conduit provide clarity and usability. The separation of PVC, HDPE, and RTRC products into three separate articles covering the full range of RNC will benefit the code user by better defining the installation and construction specifications for each conduit type.

Article 352 is limited to one type of rigid nonmetallic conduit, Type PVC, polyvinyl chloride conduit.

352.10(F)

EXPOSED

Chapter 3 Wiring Methods and Materials

Article 352 Rigid Polyvinyl Chloride Conduit, Type PVC

Part II Installation

352.10 Uses Permitted

(F) Exposed

Change Type: Revision

Change Summary:
- The use of type PVC conduit where subject to physical damage has been clarified. The addition of a new FPN identifies Schedule 80 as being identified for use in areas of physical damage.

Revised 2008 NEC Text:
352.10 USES PERMITTED
(F) EXPOSED.   PVC conduit ~~RNC~~ shall be permitted for exposed work. ~~where not subject to physical damage if identified for such use.~~ PVC conduit used exposed in areas of physical damage shall be identified for the use.

 FPN: PVC Conduit, Type Schedule 80, is identified for areas of physical damage.

Change Significance:
The use of type PVC rigid nonmetallic conduit where exposed to physical damage is revised for clarification and usability. An informational FPN was added to help the code user to identify a known and readily available raceway, Schedule 80–type PVC for use in areas of physical damage. Schedule 80–type PVC rigid nonmetallic conduit is recognized in 300.5(D) for protecting conductors emerging from ground from damage, along with conduit types RMC and IMC.

Schedule 80 PVC is identified for use in areas subject to physical damage.

Change Summary:
- Three significant changes have occurred in Article 353 for type HDPE.
  - New FPN with scope of Article 353
  - Revised uses permitted above ground encased in 2 inches of concrete
  - New FPN describes methods of joining type HDPE

Revised 2008 NEC Text:
353.1 SCOPE.   This article covers the use, installation, and construction specification for high density polyethylene (HDPE) conduit and associated fittings.

FPN: Refer to Article 352 for Rigid Polyvinyl Chloride Conduit: Type PVC and Article 355 for Reinforced Thermosetting Resin Conduit: Type RTRC.

353.10 USES PERMITTED.
(1) through **(4)** (No change)

(5) Aboveground, except as prohibited in Section 353.12, where encased in not less than 50 mm (2 in.) of concrete.

353.48 JOINTS.   All joints between lengths of conduit and between conduit and couplings, fittings, and boxes, shall be made by an approved method.

FPN: HDPE Conduit can be joined using either heat fusion, electrofusion or mechanical fittings.

Change Significance:
Three significant changes have occurred in Article 353 for type HDPE. A new FPN has been added to the scope of Article 353. This informational FPN informs the code user that two other articles address different types of RNC. The three types of RNC are PVC in 352, HDPE in 353, and RTRC in 355.

A new list item (5) has been added to the uses permitted for HDPE in 353.10. HDPE is now permitted aboveground where encased in 2 inches of concrete except as prohibited in 353.12. A new FPN in 353.48 describes the three methods used to join lengths of HDPE conduit: heat fusion, electrofusion, and mechanical fittings.

Article 353
HIGH-DENSITY POLYETHYLENE CONDUIT: TYPE HDPE CONDUIT

Chapter 3 Wiring Methods and Materials

Article 353 High-Density Polyethylene Conduit, Type HDPE Conduit

Part I General

353.1 Scope

Part II Installation

353.10 Uses Permitted

353.48 Joints

Change Type: Revision and New

HDPE conduit may be joined using heat fusion, electrofusion, or mechanical fittings.

Article 355
REINFORCED THERMOSETTING RESIN CONDUIT: TYPE RTRC

Chapter 3 Wiring Methods and Materials

Article 355 Reinforced Thermosetting Resin Conduit, Type RTRC

Part I General

II. Installation

III. Construction Specifications

Entire Article

Change Type: New

RTRC, reinforced thermosetting resin conduit, is now recognized in Article 355.

Change Summary:

- A new Article 355 for RTRC has been added to the 2008 NEC. Three separate articles for RNC now exist in the NEC to provide clearly defined installation and construction requirements for types PVC in 352, HDPE in 353, and RTRC in 355.

Revised 2008 NEC Text:

ARTICLE 355 REINFORCED THERMOSETTING RESIN CONDUIT: TYPE RTRC. I. GENERAL

355.1 SCOPE.   This article covers the use, installation, and construction specification for reinforced thermosetting resin conduit (RTRC) and associated fittings.

FPN: Refer to Article 352 for Rigid Polyvinyl Chloride Conduit: Type PVC and Article 353 for High Density Polyethylene Conduit: Type HDPE.

355.2 DEFINITION

REINFORCED THERMOSETTING RESIN CONDUIT (RTRC).   A rigid nonmetallic conduit (RNC) of circular cross section, with integral or associated couplings, connectors, and fittings for the installation of electrical conductors and cables.

(Entire new Article 355 for RTRC in the parallel numbering format used for all raceway articles)

Change Significance:

This new article, in addition to Articles 352 and 353, comprehensively addresses the three types of rigid nonmetallic conduit permitted by the NEC. As seen previously in the scope and definition of this new Article 355, only type RTRC conduit is addressed. The FPN included with the scope of the article provides the code user with direction to use Article 352 for type PVC and 353 for type HDPE.

There are three separate types of rigid nonmetallic conduits recognized for use in the NEC. Prior to the 2005 edition of the NEC, all three types of RNC were applied under the provisions of Article 352. The materials, construction, and permitted applications of the three types of RNC are very different. The three types of RNC recognized for use in the NEC are as follows:

- **Article 352 Rigid Polyvinyl Chloride Conduit: Type PVC** (Renamed in 2008)
- **Article 353 High Density Polyethylene Conduit: Type HDPE Conduit** (New in 2005)
- **Article 355 Reinforced Thermosetting Resin Conduit: Type RTRC** (New in 2008)

This new article for RTRC and the renaming of Article 352 to address type PVC conduit provides clarity and usability. The separation of PVC, HDPE, and RTRC products into three separate articles covering the full range of RNC will benefit the code user by better defining the installation and construction specifications for each conduit type.

Change Summary:

- A new exception has been added to 362.30 for securing and supporting ENT, allowing ENT to be fished through concealed walls in unbroken lengths without support.

Revised 2008 NEC Text:

362.30 SECURING AND SUPPORTING.   ENT shall be installed as a complete system in accordance with 300.18 and shall be securely fastened in place and supported in accordance with 362.30(A) and (B).

(A) SECURELY FASTENED.   ENT shall be securely fastened at intervals not exceeding 900 mm (3 ft). In addition, ENT shall be securely fastened in place within 900 mm (3 ft) of each outlet box, device box, junction box, cabinet, or fitting where it terminates.

> *Exception No. 1:* (No change)
> *Exception No. 2:* (No change)
> <u>*Exception No. 3: For concealed work in finished buildings or prefinished wall panels where such securing is impracticable, unbroken lengths (without coupling) of ENT shall be permitted to be fished.*</u>

Change Significance:

This revision recognizes the common practice of using ENT in finished walls for protecting conductors and cables. This new exception is similar to one for EMT in 358.30(A).

ENT has the flexibility to be fished through concealed spaces. This new exception allows ENT to be fished in existing walls in the same manner as cable assemblies such as types NM, AC, and MC.

362.30
SECURING AND SUPPORTING

Chapter 3 Wiring Methods and Materials

Article 362 Electrical Nonmetallic Tubing, Type ENT

Part II Installation

362.30 Securing and Supporting

(A) Securely Fastened

Exception

Change Type: New

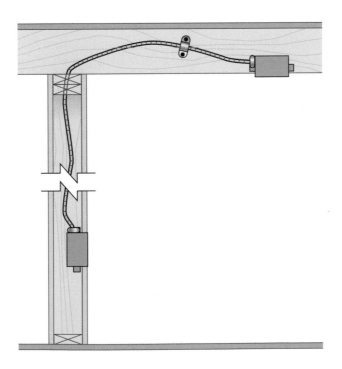

ENT where fished into existing walls does not need to be supported.

366.2
DEFINITIONS

Chapter 3 Wiring Methods and Materials

Article 366 Auxiliary Gutters

Part I General

366.2 Definitions

Change Type: Revision

Change Summary:

- The definition of "auxiliary gutter" has been revised for clarity and usability.
- An "auxiliary gutter" is used to "supplement wiring space" at equipment.

Revised 2008 NEC Text:

366.2 DEFINITIONS

METALLIC AUXILIARY GUTTERS.   A ~~S~~sheet metal enclosures <u>used to supplement wiring spaces at meter centers, distribution centers, switchboards, and similar points of wiring systems. The enclosure has</u> ~~with~~ hinged or removable covers for housing and protecting electric wires, cable, and busbars. <u>The enclosure is designed for</u> ~~in which~~ conductors <u>to be</u> ~~are~~ laid <u>or set</u> in place after the <u>enclosures</u> ~~wireway has~~ <u>have</u> been installed as a complete system.

NONMETALLIC AUXILIARY GUTTERS.   A ~~F~~flame retardant, nonmetallic enclosures <u>used to supplement wiring spaces at meter centers, distribution centers, switchboards, and similar points of wiring systems. The enclosure has hinged or</u> ~~with~~ removable covers for housing and protecting electric wires, cable, and busbars. <u>The enclosure is designed for</u> ~~in which~~ conductors <u>to be</u> ~~are~~ laid <u>or set</u> in place after the <u>enclosures</u> ~~wireway has~~ <u>have</u> been installed as a complete system.

Change Significance:

This revision will now resolve the long-standing question of the differences between auxiliary gutters and wireways. The revised text clearly states that an auxiliary gutter is used only to "supplement wiring space." A metal or nonmetallic enclosure with a hinged or removable cover may be an auxiliary gutter or a wireway. The answer is in the application of the enclosure.

If the enclosure is used to supplement wiring space, such as between a ganged meter center and a switchboard, it is an auxiliary gutter. If the enclosure is used solely as a junction box or pull box, it is a wireway. An auxiliary gutter may serve as a pull box.

Auxiliary gutters are not permitted to extend more than 30 feet beyond the equipment supplemented. For example, an enclosure with a hinged or removable cover 25 feet long may be installed as an auxiliary gutter to allow conduits to enter the enclosure and conductors to be routed to a single panelboard.

Auxiliary gutters "supplement wiring space" at meter centers, distribution centers, and similar equipment.

Change Summary:

- Three significant changes have occurred in Article 376 for metal wireways.
  - Clarification of where conductor ampacity requirements exist
  - Clear prohibition on use of power distribution blocks that have exposed live parts
  - New construction requirements

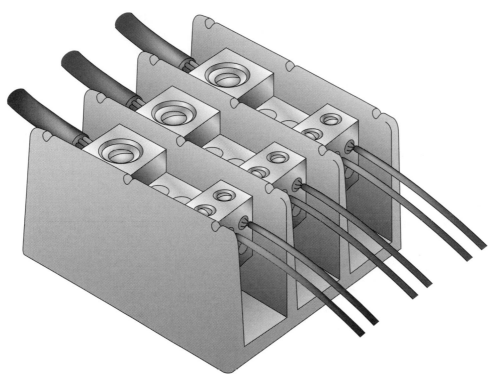

Distribution blocks used in wireways are not permitted to have uninsulated live parts.

Revised 2008 NEC Text:

376.22 NUMBER OF CONDUCTORS <u>AND AMPACITY</u>. <u>The number of conductors and their ampacity shall comply with 376.22(A) and 376.22(B).</u>

(A) <u>CROSS-SECTIONAL AREAS OF WIREWAY.</u> The sum of the cross-sectional areas of all contained conductors at any cross section of a wireway shall not exceed 20 percent of the interior cross-sectional area of the wireway.

(B) ADJUSTMENT FACTORS. The ~~derating~~ <u>adjustment</u> factors in 310.15(B)(2)(a) shall be applied only where the number of current-carrying conductors, including neutral conductors classified as current-carrying under the provisions of 310.15(B)(4), exceeds 30. Conductors for signaling circuits or controller conductors between a motor and its starter and used only for starting duty shall not be considered as current-carrying conductors.

(continued)

Article 376
METAL WIREWAYS

Chapter 3 Wiring Methods and Materials

Article 376 Metal Wireways

Part II Installation

376.22 Number of Conductors and Ampacity

(A) Cross-sectional Areas of Wireways

(B) Adjustment Factors

376.56 Splices, Taps, and Power Distribution Blocks

(B) Power Distribution Blocks

(4) Live Parts

Part III Construction Specifications

376.100 Construction

(A) Electrical and Mechanical Continuity

(B) Substantial Construction

(C) Smooth Rounded Edges

(D) Covers

Change Type: Revision and New

376.56 SPLICES, TAPS AND POWER DISTRIBUTION BLOCKS
(B) POWER DISTRIBUTION BLOCKS
(4) LIVE PARTS.   Power distribution blocks shall not have <u>uninsulated</u> live parts exposed ~~in the~~ <u>within a wireway, whether or not the wireway cover is installed</u> ~~after installation~~.

376.100 CONSTRUCTION
(A) ELECTRICAL AND MECHANICAL CONTINUITY.   <u>Wireways shall be constructed and installed so that adequate electrical and mechanical continuity of the complete system is secured.</u>

(B) SUBSTANTIAL CONSTRUCTION.   <u>Wireways shall be of substantial construction and shall provide a complete enclosure for the contained conductors. All surfaces, both interior and exterior, shall be suitably protected from corrosion. Corner joints shall be made tight, and where the assembly is held together by rivets, bolts, or screws, such fasteners shall be spaced not more than 300 mm (12 in.) apart.</u>

(C) SMOOTH ROUNDED EDGES.   <u>Suitable bushings, shields, or fittings having smooth, rounded edges shall be provided where conductors pass between wireways, through partitions, around bends, between wireways and cabinets or junction boxes, and at other locations where necessary to prevent abrasion of the insulation of the conductors.</u>

(D) COVERS.   <u>Covers shall be securely fastened to the wireway.</u>

Change Significance:

Three significant changes in the 2008 NEC have occurred in Article 376.

- 376.22 has been renamed for clarity and usability. In the 2005 NEC, this section was titled "Number of Conductors." However, this section addressed both the permitted fill of wireways and requirements for calculating conductor ampacity in wireways. This revision now titles this section "Number of Conductors and Ampacity." For clarity and usability, the requirements are subdivided in two first-level subdivisions: (A) "Cross-Sectional Areas of Wireway" and (B) "Adjustment Factors."

- 376.56(B)(4) has been revised to prohibit the use of power distribution blocks in wireways where any uninsulated exposed live parts exist, regardless of whether the covers are on or not.

- New requirements now exist in 376.100 for the construction of metal wireways. These new requirements address (A) Electrical and Mechanical Continuity, (B) Substantial Construction, (C) Smooth Rounded Edges, and (D) Covers for All-metal Wireways.

Notes:

Change Summary:

- Article 382 now recognizes concealed nonmetallic extensions that are permitted to be concealed with paint, texture, joint/concealing compound, plaster, wallpaper, tile, wall paneling, or other similar materials.
- A concealed nonmetallic extension is required to be listed and have five types of protection, as required in 382.6.
- A new Part III has been added for construction specifications of concealable nonmetallic extensions.

Article 382
NONMETALLIC EXTENSIONS

Chapter 3 Wiring Methods and Materials

Article 382 Nonmetallic Extensions

Part I General

382.2 Definition of Concealable Nonmetallic Extension

Part II Installation

382.6 Listing Requirements

382.15 Exposed

Part III Construction Specifications (Concealable Nonmetallic Extensions Only)

Change Type: Revision and New

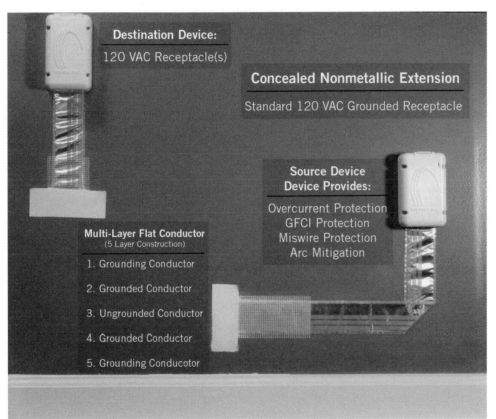

Destination Device:
120 VAC Receptacle(s)

Concealed Nonmetallic Extension
Standard 120 VAC Grounded Receptacle

Source Device
Device Provides:
Overcurrent Protection
GFCI Protection
Miswire Protection
Arc Mitigation

Multi-Layer Flat Conductor
(5 Layer Construction)
1. Grounding Conductor
2. Grounded Conductor
3. Ungrounded Conductor
4. Grounded Conductor
5. Grounding Conducotor

Courtesy of Southwire Company.

Revised 2008 NEC Text:

382.2 DEFINITION

CONCEALABLE NONMETALLIC EXTENSION. A listed assembly of two, three, or four insulated circuit conductors within a nonmetallic jacket, an extruded thermoplastic covering, or a sealed nonmetallic covering. The classification includes surface extensions intended for mounting directly on the surface of walls or ceilings, and concealed with paint, texture, joint compound, plaster, wallpaper, tile, wall paneling, or other similar materials.

(continued)

382.6 LISTING REQUIREMENTS. Concealable nonmetallic extensions and associated fittings and devices shall be listed. The starting/source tap device for the extension shall contain and provide the following protection for all load-side extensions and devices:

(1) Supplementary overcurrent protection

(2) Level of protection equivalent to a Class A GFCI

(3) Level of protection equivalent to a portable GFCI

(4) Line and load-side miswire protection

(5) Provide protection from the effects of arc faults

382.15 EXPOSED

(A) NONMETALLIC EXTENSIONS. One or more extensions shall be permitted to be run in any direction from an existing outlet, but not on the floor or within 50 mm (2 in.) from the floor.

(B) CONCEALABLE NONMETALLIC EXTENSIONS. Where identified for the use, nonmetallic extensions may be concealed with paint, texture, concealing compound, plaster, wallpaper, tile, wall paneling, or other similar materials and installed per 382.15(A).

382.42 DEVICES   (New)

PART III. CONSTRUCTION SPECIFICATIONS (CONCEALABLE NONMETALLIC EXTENSIONS ONLY)   (The construction, insulation and marking of concealable nonmetallic extensions are covered in the new Part III)

Change Significance:

This revision now permits a "concealed nonmetallic extension" to be installed under paint, texture, concealing compound, plaster, wallpaper, tile, wall paneling, or other similar materials on walls or ceilings. A new 382.6 requires that the "concealed nonmetallic extension" and associated fittings be listed and that the starting/source tap device for the extension provide the following protection to all downstream conductors and devices:

(1) Supplementary overcurrent protection

(2) Protection equal to a class A GFCI

(3) Protection equal to a portable type GFCI

(4) Line and load-side miswire protection

(5) Provide protection from the effects of arc faults

A class A GFCI trips when the current to ground is 6 mA or higher and will not trip when the current to ground is less than 4 mA. A portable-type GFCI provides open neutral protection, which means that if the neutral is open or opens the device will open—disallowing energized conductors downstream. AFCI and miswiring protection are also required. These concealed nonmetallic extensions may be installed in accordance with 382.15 behind paint, texture, concealing compound, plaster, wallpaper, tile, wall paneling, or other similar materials.

A new 382.42 provides requirements for all receptacles and housings used as part of the concealed nonmetallic extension. A new Part III provides construction specifications for concealable nonmetallic extensions only.

388.30 and 388.56
SECURING AND SUPPORTING AND SPLICES AND TAPS

Chapter 3 Wiring Methods and Materials

Article 388 Surface Nonmetallic Raceways

Part II Installation

388.30 Securing and Supporting

388.56 Splices and Taps

Change Type: Revision and New

Change Summary:
- New 388.30 requires surface nonmetallic raceways be supported in accordance with manufacturers' instructions.
- Splices and taps are now clearly permitted within a surface nonmetallic raceway, provided there is a cover (capable of being opened) in place after installation.

Revised 2008 NEC Text:
388.30 SECURING AND SUPPORTING. Surface nonmetallic raceways shall be supported at intervals in accordance with the manufacturer's installation instructions.

388.56 SPLICES AND TAPS. Splices and taps shall be permitted in surface nonmetallic raceways having a ~~removable~~ cover capable of being opened in place that is accessible after installation. The conductors, including splices and taps, shall not fill the raceway to more than 75 percent of its area at that point. Splices and taps in surface nonmetallic raceways without ~~removable~~ covers capable of being opened in place shall be made only in boxes. All splices and taps shall be made by approved methods.

Change Significance:
New text in 388.30 requires that surface nonmetallic raceways be supported at intervals in accordance with the manufacturer's installation instructions. This revision establishes a correlation between Article 386 (Surface Metal Raceways) and Article 388 (Surface Nonmetallic Raceways). Listing information for both products requires that they be provided with installation instructions indicating the methods to secure and support.

Listing information for surface nonmetallic raceways requires that the cover for a raceway or fitting shall be constructed and installed so that it is capable of being removed or opened.

For the purposes of the application of 388.56, there is no difference between a raceway that has a cover that can be removed and one that has a cover that can be opened. Splices and taps are now permitted in surface nonmetallic raceways having a cover that is capable of being opened in place after installation. The text prohibiting splices in surface nonmetallic raceways has been removed. The requirement is that the splice or tap be made only where there is a cover that is capable of being opened in place after the installation.

Surface nonmetallic raceways must be secured in accordance with manufacturers' instructions.

392.3(A)
WIRING METHODS

Chapter 3 Wiring Methods and Materials

Article 392 Cable Trays

392.3 Uses Permitted

(A) Wiring Methods

Change Type: Revision

Change Summary:
- Table 392.3(A) is revised to recognize additional wiring methods permitted to be installed in cable trays.

Revised 2008 NEC Text:
392.3 USES PERMITTED

(A) WIRING METHODS.   The wiring methods in Table 392.3(A) shall be permitted to be installed in cable tray systems under the conditions described in their respective articles and sections.

(The following have been added in Table 392.3(A))

Change Significance:
392.3(A) Wiring Methods recognizes only "wiring methods" listed in Table 392.3(A) as being permitted to be installed in cable trays. There were multiple raceways and wiring methods omitted from the table but permitted elsewhere in the NEC to be installed in cable tray. This user-friendly revision will allow the code user to quickly and accurately determine what is permitted in cable trays.

It is important to note that a "cable tray" is not a wiring method. It is a "support system," as detailed in the definition as follows:

392.2 DEFINITION

CABLE TRAY SYSTEM.   A unit or assembly of units or sections and associated fittings forming a structural system used to securely fasten or support cables and raceways.

Wiring Method	Article
CATV Cables	820
CATV raceways	820
Class 2 and Class 3 cables	725
Communications cables	800
Network-powered broadband communications cables	830
Non-power limited fire alarm cable	760
Polyvinyl Chloride PVC conduit	352
Power-limited fire alarm cable	760
RTRC	355
Signaling raceway	725

Notes:

Cable tray cables with conductors is not a wiring method; it is a support system.

Change Summary:
- New text has been added in 392.9(A) to help the code user correlate requirements for cable tray fill with conductor ampacity.

Revised 2008 NEC Text:
392.9 NUMBER OF MULTICONDUCTOR CABLES, RATED 2000 VOLTS OR LESS, IN CABLE TRAYS. The number of multiconductor cables, rated 2000 volts or less, permitted in a single cable tray shall not exceed the requirements of this section. The conductor sizes herein apply to both aluminum and copper conductors.

(A) ANY MIXTURE OF CABLES. Where ladder or ventilated trough cable trays contain multiconductor power or lighting cables, or any mixture of multiconductor power, lighting, control, and signal cables, the maximum number of cables shall conform to the following:

(1) Where all of the cables are 4/0 AWG or larger, the sum of the diameters of all cables shall not exceed the cable tray width, and the cables shall be installed in a single layer. Where cable ampacity is determined according to 392.11(A)(3), the cable tray width shall not be less than the sum of the diameters of the cables and the sum of the required spacing widths between the cables.

Change Significance:
This revision correlates the requirements of 392.9(A) for the maximum cable tray fill with the requirements of 392.11(A)(3) for ampacity. When conductor ampacity is determined by using 392.11(A)(3), cables must maintain a spacing of not less than one cable diameter between cables. This new last sentence informs the code user that the cable tray fill requirements may be modified by the requirements for conductor ampacity.

This new text will inform the code user that cables previously installed in a cable tray with a maintained spacing must continue to have the spacing or they must have their ampacity recalculated.

392.9
NUMBER OF MULTICONDUCTOR CABLES, RATED 2000 VOLTS OR LESS, IN CABLE TRAYS

Chapter 3 Wiring Methods and Materials

Article 392 Cable Trays

392.9 Number of Multiconductor Cables, Rated 2000 Volts or Less, in Cable Trays

(A) Any Mixture of Cables

Change Type: Revision

Article 392 provides limits on the number of multiconductor and single-conductor cables permitted in a cable tray.

392.11(C)
COMBINATIONS OF MULTI-CONDUCTOR AND SINGLE-CONDUCTOR CABLES

Chapter 3 Wiring Methods and Materials

Article 392 Cable Trays

392.11 Ampacity of Cables, Rated 2000 Volts or Less, in Cable Trays

(C) Combinations of Multiconductor and Single-Conductor Cables

Change Type: New

Change Summary:
- New text in 392.11(C) provides the code user with guidance for cable tray fill and conductor ampacity where a combination of multi-conductor and single-conductor cables exists in the same tray.

Revised 2008 NEC Text:
392.11 AMPACITY OF CABLES, RATED 2000 VOLTS OR LESS, IN CABLE TRAYS

(C) COMBINATIONS OF MULTICONDUCTOR AND SINGLE-CONDUCTOR CABLES. Where a cable tray contains a combination of multiconductor and single-conductor cables, the allowable ampacities shall be as given in 392.11(A) for multiconductor cables and 392.11B) for single-conductor cables, provided that:

(1) The sum of the multiconductor cable fill area as a percentage of the allowable fill area for the tray calculated per 392.9, and the single-conductor cable fill area as a percentage of the allowable fill area for the tray calculated per 392.10, totals not more than 100 percent.

(2) Multiconductor cables are installed according to 392.9 and single-conductor cables are installed according to 392.10 and 392.8(D) and (E).

Change Significance:
The 2005 NEC did not provide the code user with guidance for cable tray fill requirements where a combination of multiconductor and single-conductor cables exists in the same tray. This revision now clarifies that multiconductor and single-conductor cable installed in the same tray shall not exceed the fill requirements for each type of cable (as if installed in its own tray) and that all installation requirements and ampacity limits for each cable type apply.

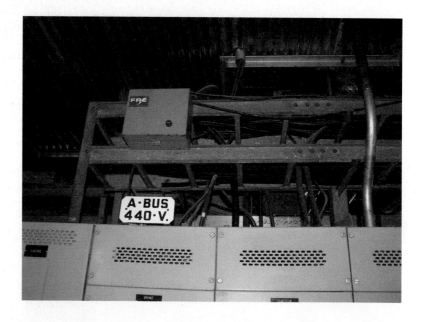

There are many different types of cable tray, including ladder, ventilated trough, solid bottom, and others.

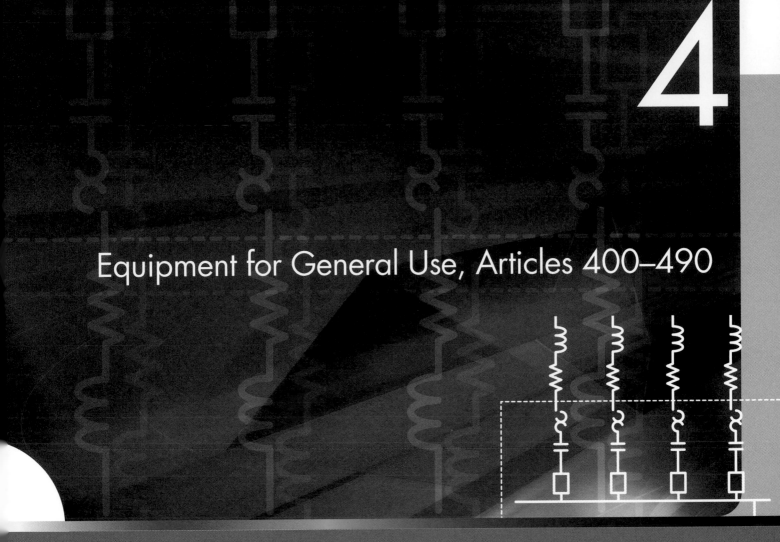

Equipment for General Use, Articles 400–490

Change Summary:
- All surface-mounted switches or circuit breakers in damp or wet locations are now required to be in a weatherproof enclosure.
- All flush-mounted switches or circuit breakers in damp or wet locations are now required to be equipped with a weatherproof cover.

Revised 2008 NEC Text:
404.4 <u>DAMP OR</u> WET LOCATIONS. A <u>surface-mounted</u> switch or circuit breaker in a <u>damp or</u> wet location ~~or outside of a building~~ shall be enclosed in a weatherproof enclosure or cabinet that shall comply with 312.2~~(A)~~. <u>A flush-mounted switch or circuit breaker in a damp or wet location shall be equipped with a weatherproof cover.</u> Switches shall not be installed within wet locations in tub or shower spaces unless installed as part of a listed tub or shower assembly.

Change Significance:
This revision significantly revises the title of 404.4 to include damp locations. The language "or outside of a building" has been deleted. The qualifiers are "damp" or "wet" locations, which is all-inclusive—covering indoor and outdoor installations. Further clarification is provided in this change for surface-mounted and flush-mounted switches and circuit breakers. All surface-mounted switches or circuit breakers in damp or wet locations are now required to be in a weatherproof enclosure. All flush-mounted switches or circuit breakers in damp or wet locations are now required to be equipped with a weatherproof cover.

404.4
DAMP OR WET LOCATIONS

Chapter 4 Equipment for General Use

Article 404 Switches

Part I Installation

404.4 Damp or Wet Locations

Change Type: Revision

Surface- and flush-mounted switches and circuit breaker enclosures/covers installed in damp or wet locations must provide protection that is weatherproof.

404.8(C)
MULTIPOLE SNAP SWITCHES

Chapter 4 Equipment for General Use

Article 404 Switches

Part I Installation

404.8 Accessibility and Grouping

(C) Multipole Snap Switches

Change Type: New

Change Summary:

- A new requirement prohibits a multipole snap switch from switching two separate circuits, unless it is listed and marked for the purpose or unless the voltage rating of the switch is not less than the nominal line-to-line voltage of the system supplying the circuits.

Revised 2008 NEC Text:

404.8 ACCESSIBILITY AND GROUPING
(C) MULTIPOLE SNAP SWITCHES. A multipole, general use snap switch shall not be permitted to be fed from more than a single circuit unless it is listed and marked as a two-circuit or three-circuit switch, or unless its voltage rating is not less than the nominal line-to-line voltage of the system supplying the circuits.

Change Significance:

This revision now requires that where a multipole snap switch is used to switch two separate circuits the switch shall meet one of the following two requirements:

(1) The multipole switch is to be listed and marked as "2-circuit" or "3-circuit."

(2) The multipole switch shall have a voltage rating not less than the nominal line-to-line voltage of the system supplying the circuits.

The UL product category title/code, Snap Switches (WJQR), states that multipole general-use snap switches have not been investigated for more than single-circuit operation unless marked as "2-circuit" or "3-circuit."

It is important to note that as written this new requirement will allow two separate circuits in a 120/240-volt single-phase system (regardless of the voltage between the circuit conductors) to be switched with a multipole snap switch rated at 277 volts. Note that the new text in 404.8(C) does not refer to the voltage between conductors. Only the nominal voltage of the system is referenced. This would prohibit two 277-volt circuits supplied from the same phase to be switched by a multipole snap switch rated at 277 volts.

Types of snap switches include the single-pole single-throw type, the single-pole double-throw type, and the multipole single-throw type.

Change Summary:

- A new 406.4(G) requires identified barriers to be installed between devices where the voltage between the devices is greater than 300 volts.

Revised 2008 NEC Text:

406.4 RECEPTACLE MOUNTING

(G) VOLTAGE BETWEEN ADJACENT DEVICES. A receptacle shall not be grouped or ganged in enclosures with other receptacles, snap switches, or similar devices, unless they are arranged so that the voltage between adjacent devices does not exceed 300 volts, or unless they are installed in enclosures equipped with identified, securely installed barriers between adjacent devices.

Change Significance:

This new first-level subdivision recognizes hazards currently addressed in 404.8(B) for switches. When a switch and any other device are installed next to each other, the NEC has in the past required a barrier between a switch and any other device if the voltage between them was greater than 300 volts.

The same hazard exists if only receptacles are installed. Where receptacles are grouped and the voltage between receptacles is greater than 300 volts, identified barriers must be securely installed for safety. It is logical and user friendly for this requirement to be located in Article 406. Although the provisions of 404.8(B) address the same hazard, they apply only when a switch is involved. In the comments from panel members it was noted that 15-amp 277-volt duplex receptacles are manufactured and sold for lighting and other applications. If two of these devices are installed in the same box and supplied from different phases in the wye-connected system, the voltage between devices is 480 volts.

406.4(G)
VOLTAGE BETWEEN ADJACENT DEVICES

Chapter 4 Equipment for General Use

Article 406 Receptacles, Cord Connectors, and Attachment Plugs (Caps)

406.4 Receptacle Mounting

(G) Voltage Between Adjacent Devices

Change Type: New

406.4(G) requires that voltage between adjacent *devices* (this includes all devices such as switches and receptacles) must not exceed 300 volts. This new text provides correlation.

406.8
RECEPTACLES IN DAMP OR WET LOCATIONS

Chapter 4 Equipment for General Use

Article 406 Receptacles, Cord Connectors, and Attachment Plugs (Caps)

406.8 Receptacles in Damp or Wet Locations

(A) Damp Locations

(B) Wet Locations

(1) 15- and 20-Ampere Receptacles in a Wet Location

(2) Other Receptacles

Change Type: Revision and New

Change Summary:
- All nonlocking 15- and 20-ampere, 125- and 250-volt receptacles installed in damp or wet locations are now required to be listed as "weather-resistant type."

Weather-resistant-type receptacles are required in wet or damp locations.

Revised 2008 NEC Text:
406.8 RECEPTACLES IN DAMP OR WET LOCATIONS

(A) DAMP LOCATIONS. (A new last sentence and FPN are added as follows:)

All nonlocking 15- and 20- ampere, 125- and 250-volt receptacles shall be a listed weather-resistant type.

FPN: The types of receptacles covered by this requirement are identified as 5-15, 5-20, 6-15, and 6-20 in ANSI/NEMA WD 6-2002, National Electrical Manufacturers Association *Standard for Dimensions of Attachment Plugs and Receptacles.*

(B) WET LOCATIONS
(1) 15- AND 20-AMPERE RECEPTACLES IN A WET LOCATION. (A new last sentence, FPN and exception are added as follows:)

All 15- and 20- ampere, 125- and 250-volt nonlocking receptacles shall be listed weather-resistant type.

FPN: The types of receptacles covered by this requirement are identified as 5-15, 5-20, 6-15, and 6-20 in ANSI/NEMA WD 6-2002, National Electrical

(continued)

Manufacturers Association *Standard for Dimensions of Attachment Plugs and Receptacles.*

Exception: 15- and 20-ampere, 125- through 250-volt receptacles installed in a wet location and subject to routine high-pressure spray washing shall be permitted to have an enclosure that is weatherproof when the attachment plug is removed.

(2) OTHER RECEPTACLES.   All other receptacles installed in a wet location shall comply with (B)(2)(a) or (B)(2)(b).

   (a) (No change)

   (b) (No change)

Change Significance:

This revision will now require that all receptacles installed in a damp or wet location be listed as "weather resistant." This requirement is delayed until January 1, 2011 to allow manufacturers to begin making these devices and incorporating them into all equipment and assemblies installed in damp or wet locations.

Receptacles installed in damp or wet locations are exposed to moisture at varying degrees depending on location; cover type; exposure to water from being exposed to rain, sleet, and snow; or presence in a wash-down area. Enclosures for receptacles in damp or wet locations are subject to damage from physical contact or exposure to the elements outdoors. Misapplications occur regularly. An enclosure installed in a wet location for use only while attended may become used for other than the original intent and be subject to moisture while unattended. The addition of a receptacle that is weather resistant will ensure that the device will not be damaged by exposure to moisture.

This requirement applies to all receptacles located in damp locations (as addressed in 406.8(A)) and to all wet locations, as addressed in 406.8(B)(1) and (2). It is interesting to note that the results of a joint National Electrical Manufacturers Association (NEMA) and UL investigation revealed that the greatest number of inoperable ground-fault circuit interrupter (GFCI) receptacles were located outdoors. The failure rate was more than double any other location. Weather-resistant GFCI receptacles will be designed for exposure to moisture in both indoor and outdoor locations.

A new exception is added to 406.8(B)(1) to address enclosures located indoors in food processing plants, where they may be subject to high-pressure washing—which, incidentally, may occur on a daily basis. 406.8(B)(1) requires that all 15- and 20-ampere 125- and 250-volt receptacles installed in a wet locations have an enclosure that is weatherproof whether or not the attachment plug cap is inserted.

By design, this type of enclosure may have an opening on the bottom to allow cords to exit. Where these types of enclosures are located indoors in food processing plants, they may be subject to high-pressure washing—which, incidentally, may occur on a daily basis. Detergents and water may be forced into the enclosure, causing damage to the receptacle. This new exception will now permit an enclosure that is weatherproof when an attachment plug is not inserted for indoor wet locations.

406.11

TAMPER-RESISTANT RECEPTACLES IN DWELLING UNITS

Chapter 4 Equipment for General Use

Article 406 Receptacles, Cord Connectors, and Attachment Plugs (Caps)

406.11 Tamper-Resistant Receptacles in Dwelling Units

Change Type: New

Change Summary:

- All 125-volt 15- and 20-amp receptacles installed in dwelling units must be listed as "tamper resistant." This revision will provide protection, designed to prevent injury to children tampering with the device.

Revised 2008 NEC Text:

406.11 TAMPER-RESISTANT RECEPTACLES IN DWELLING UNITS. In all areas specified in 210.52, all 125-volt, 15- and 20-ampere receptacles shall be listed tamper resistant receptacles.

Change Significance:

Over the 10-year period from 1991 to 2001, more than 24,000 children were injured when they inserted objects into energized receptacles. This is an average of more than 2,400 children per year suffering electrical burn and shock injuries while tampering with receptacles. Children will use anything they can get their hands on to insert into receptacles, including hairpins, keys, needles, screws, nails, paper clips, knifes, and the like.

Most incidents result in shock and burns, but multiple fatalities have been recorded. Preventative measures such as plastic caps inserted into the receptacle are effective only where used and will not prevent a child from unplugging a cord cap to tamper with an unprotected outlet.

The use of tamper-resistant receptacles represents the most cost-effective means of protecting children from electrical accidents. These devices will cost .50 cents more than a standard receptacle. In an average-size home with 75 receptacles, the additional cost is only $37.50. This minor increase in cost represents a tremendous increase in safety. These devices are readily available and have for many years been used to comply with 517.18(C) in pediatric wards.

Although not all homes are purchased by families with children, this requirement is for all dwelling units. Dwelling units are occupied transiently. Although a couple without children may purchase a home, when it is sold a family with children may move in.

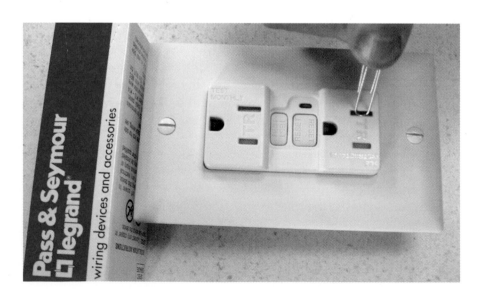

Tamper-resistant receptacles will prevent injury to children.

Change Summary:
- Marking is now required on all switchboards and panelboards that contain a "high leg."

Revised 2008 NEC Text:
408.3 SUPPORT AND ARRANGEMENT OF BUSBARS AND CONDUCTORS (F) HIGH-LEG IDENTIFICATION. A switchboard or panelboard containing a 4-wire, delta-connected system where the midpoint of one phase winding is grounded shall be legibly and permanently field marked as follows: "Caution ___ __ Phase Has ___ __ Volts to Ground"

Change Significance:
This new first level subdivision will require that all switchboards or panelboards containing a four-wire delta-connected system in which the midpoint of one phase winding is grounded be marked to identify the high leg and the voltage to ground of the high leg. This safety-driven change will now provide a clear marking on all switchboards and panelboards to inform persons servicing or upgrading the system that one leg has a higher voltage to ground.

This should help prevent common mistakes in which an installer connects a load from the grounded conductor to the high leg. The purpose of this change is to eliminate some of the hazards of accidentally connecting outlets to the high leg and causing injury to people and the destruction of equipment. It is important to note that 408.3(E) requires that the B phase in a four-wire delta-connected system in which the midpoint of one phase winding is grounded shall be the phase having the higher voltage to ground.

408.3(F)
HIGH-LEG IDENTIFICATION

Chapter 4 Equipment for General Use

Article 408 Switchboards and Panelboards

Part I General

408.3 Support and Arrangement of Busbars and Conductors

(F) High-Leg Identification

Change Type: New

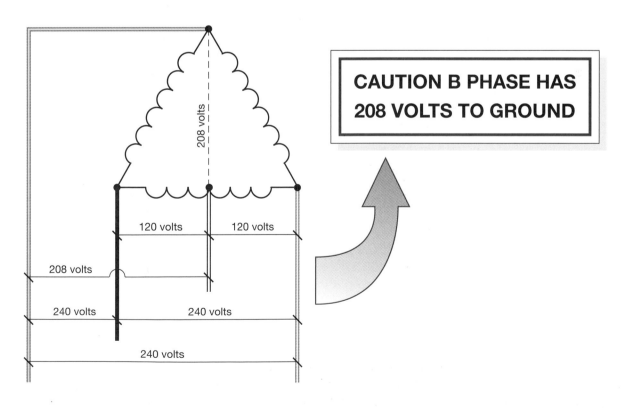

Field marking of high leg voltage ratings in four-wire delta systems is required.

408.4

CIRCUIT DIRECTORY OR CIRCUIT IDENTIFICATION

Chapter 4 Equipment for General Use

Article 408 Switchboards and Panelboards

Part I General

408.4 Circuit Directory or Circuit Identification

Change Type: Revision

Change Summary:

• Spare overcurrent devices in switchboards or panelboards are required to be marked as such.

• Circuits are not permitted to be described in a manner that is transient, such as "Bob's office" or "Bob's bedroom."

PANEL DIRECTORY

200 - AMP   120/208 – VOLT

Circuit		Circuit	
1	AC ROOFTOP	2	OUTDOOR ⏀
3		4	OUTDOOR LIGHTING
5		6	FIRE ALARM SYSTEM
7	⏀ BASEMENT	8	SECURITY SYSTEM
9	⏀ LOADING DOCK	10	PHONE CLOSET ⏀
11	⏀ LOBBY AREA	12	KITCHEN ⏀ COUNTER
13	⏀ RMS 101, 103, 105	14	KITCHEN ⏀
15	⏀ RMS 102, 104, 106	16	REFRIGERATOR ⏀
17	COPIER RM 101	18	MICROWAVE ⏀
19		20	LIGHTING - NORTH 1ST FL
21	WATER HEATER	22	LIGHTING - SOUTH 1ST FL
23		24	LIGHTING RMS 101, 103, 105
25	SPARE	26	LIGHTING RMS 102, 104, 106
27	SPARE	28	LIGHTING LOAD DOCK
29	SPARE	30	LIGHTING MECH AREA
31	SPARE	32	⏀ MECH AREA
33	SPARE	34	OIL HEATER
35	SPARE	36	COMPUTER SERVER ⏀
37	SPARE	38	SPARE
39	SPARE	40	SPARE
41	SPARE	42	SPARE

CES ELECTRIC   5/14/2006

Circuit directories must identify all spare positions.

Revised 2008 NEC Text:

408.4 CIRCUIT DIRECTORY OR CIRCUIT IDENTIFICATION.   Every circuit and circuit modification shall be legibly identified as to its clear, evident, and specific purpose or use. The identification shall include sufficient detail to allow each circuit to be distinguished from all others. Spare positions that contain unused overcurrent devices or switches shall be described accordingly. The identification shall be included in a circuit directory that is located on the face or inside of the panel door in the case of a panelboard, and located at each switch on a switchboard. No circuit shall be described in a manner that depends on transient conditions of occupancy.

(continued)

Change Significance:

This revision requires that any spare overcurrent devices installed in a switchboard or panelboard be identified as such on the circuit directory in a panelboard or at each overcurrent device in a switchboard. This information is necessary for the installer/maintainer, who will service the system after the initial installation.

A new last sentence will now prohibit the use of circuit markings that would be considered transient. For example, circuits in a commercial office space may be identified in a circuit directory as Mr. Kilroy's office. When Mr. Kilroy moves out and Mr. Schaffer moves in, the installer/maintainer cannot identify circuits properly. The room would need to be marked in a manner such as "room 412" or "4th-floor corner office southwest."

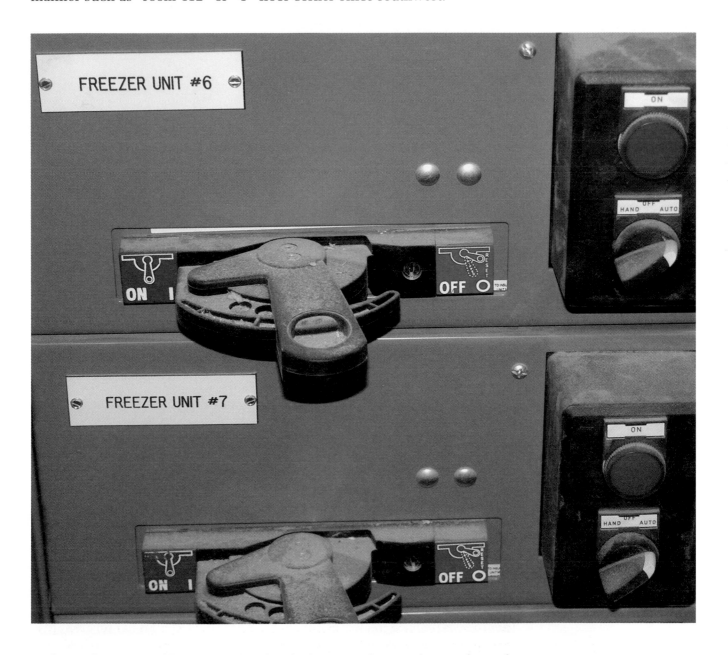

Each switch on a switchboard shall be identified as to its clear, evident, and specific purpose or use.

408.36
OVERCURRENT PROTECTION

Chapter 4 Equipment for General Use

Article 408 Switchboards and Panelboards

Part III Panelboards

408.36 Overcurrent Protection

Exceptions

Change Type: Revision

Change Summary:

- The categories "lighting appliance branch circuit panelboard" and "power panelboard" have been deleted.
- In general, all panelboards are now required to be protected by a single overcurrent protective device located within or upstream of the panelboard.
- Panelboards are no longer limited to 42 overcurrent protective devices.

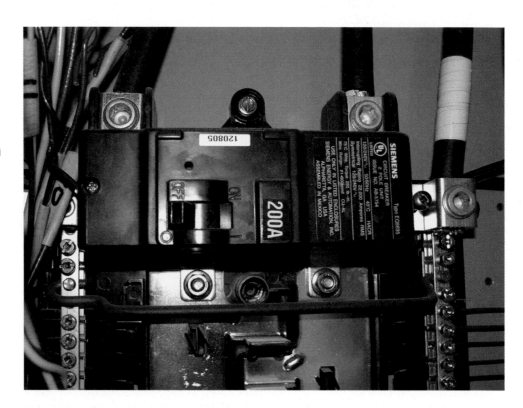

In general, all panelboards must be provided with overcurrent protection rated not greater than the panelboard.

Revised 2008 NEC Text:

~~408.34 CLASSIFICATION OF PANELBOARDS~~
~~(A) LIGHTING AND APPLIANCE BRANCH CIRCUIT PANELBOARD~~
~~(B) POWER PANELBOARD~~
~~408.35 NUMBER OF OVERCURRENT DEVICES ON ONE PANELBOARD~~
408.36 OVERCURRENT PROTECTION.
~~(A) LIGHTING AND APPLIANCE BRANCH CIRCUIT PANELBOARD.~~
~~(B) POWER PANELBOARD PROTECTION.~~
408.36 OVERCURRENT PROTECTION.   In addition to the requirements of 408.30, a panelboard shall be protected by an overcurrent protective device having a rating not greater than that of the panelboard. This overcurrent protective device shall be located within or at any point on the supply side of the panelboard.

(continued)

Exception No. 1: Individual protection shall not be required for a panelboard used as service equipment with multiple disconnecting means in accordance with 230.71. In panelboards protected by three or more main circuit breakers or sets of fuses, the circuit breakers or sets of fuses shall not supply a second bus structure within the same panelboard assembly.

Exception No. 2: Individual protection shall not be required for a panelboard protected on its supply side by two main circuit breakers or two sets of fuses having a combined rating not greater than that of the panelboard. A panelboard constructed or wired under this exception shall not contain more than 42 overcurrent devices. For the purposes of determining the maximum of 42 overcurrent devices, a 2-pole or a 3- pole circuit breaker shall be considered as two or three overcurrent devices, respectively.

Exception No. 3: For existing panelboards, individual protection shall not be required for a panelboard used as service equipment for an individual residential occupancy.

~~(C)~~ **(A)** **Snap Switches Rated at 30 Amperes or Less.** (No change)

~~(D)~~ **(B)** **Supplied Through a Transformer.** (No change)

~~(E)~~ **(C)** **Delta Breakers.** (No change.)

~~(F)~~ **(D)** **Back-Fed Devices.** (No change)

Change Significance:

Note that although the revised 2008 NEC text cited previously is underlined as new this revision incorporates editorial and substantive changes to existing 2005 NEC text. This revision provides clarity and usability by simplifying the rules for overcurrent protection of panelboards. The panelboard categories "lighting and appliance branch circuit panelboard" and "power panelboard" have been deleted. The parent text of 408.36 now requires that all panelboards be protected by an overcurrent protective device having a rating not greater than that of the panelboard. This general rule further incorporates a previous exception by allowing the overcurrent protective device to be located within or at any point on the supply side of the panelboard.

Three exceptions are added to modify the general rule. Exception No. 1 is based on the 2005 exception in 408.36(B) and addresses a panelboard used as service equipment. This exception mirrors the rule for the maximum number of disconnects for a service in 230.71, and permits up to six disconnects. Additional text in this exception prohibits any of the devices permitted by 230.71 to supply a split buss in the same panelboard.

Exception No. 2 is based on the 2005 text of 408.36(A). This addresses split-bus panelboards and continues to limit these panelboards to 42 overcurrent protective devices.

Exception No. 3 is based on the 2005 Exception No. 2 to 408.36(A). This exception has been editorially revised to continue to allow existing panelboards without protection where they are used as service equipment in individual residential occupancies.

Notes:

409.2
DEFINITION OF *INDUSTRIAL CONTROL PANEL*

Chapter 4 Equipment for General Use

Article 409 Industrial Control Panels

Part I General

409.2 Definitions, Industrial Control Panel

Change Type: Revision

Change Summary:
- This definition is revised to clarify that an industrial control panel may consist of power circuit components only, or control circuit components only, or a combination of power and control circuit components.

Revised 2008 NEC Text:
409.2 DEFINITIONS

INDUSTRIAL CONTROL PANEL. An assembly of ~~a systematic and standard arrangement of~~ two or more components consisting of one of the following:

Power circuit components only, such as motor controllers, overload relays, fused disconnect switches, and circuit breakers

Control circuit components only, ~~and related control devices~~ such as pushbuttons ~~stations~~, pilot lights, selector switches, timers, switches, control relays

(3) A combination of power and control circuit components These components, ~~the like~~ with associated wiring~~,~~ and terminals, are mounted on or contained within an enclosure or mounted on a sub-panel. ~~blocks, pilot lights, and similar components.~~ The industrial control panel does not include the controlled equipment.

Change Significance:
This revision clarifies that an industrial control panel may serve to supply power, to supply control, or to supply both. This is a significant revision that will impact the construction and protection of an industrial control panel depending on how it is applied. A new exception in 409.110 recognizes that an industrial control panel used only for control circuit components is not required to be marked with a short-circuit current rating. The approved method for determining the short-circuit current rating of an industrial control panel is based only on power circuit components.

An industrial control panel may contain power components, control components, or both.

Change Summary:
- The marking requirements for an industrial control panel have been revised to clearly identify information on each incoming supply circuit, as well as the identification/location of wiring diagrams.
- A new exception has been added that clarifies that short-circuit current rating markings are not required for industrial control panels containing only control circuit components.

409.110
MARKING

Chapter 4 Equipment for General Use

Article 409 Industrial Control Panels

Part III Construction Specifications

409.110 Marking

Change Type: Revision

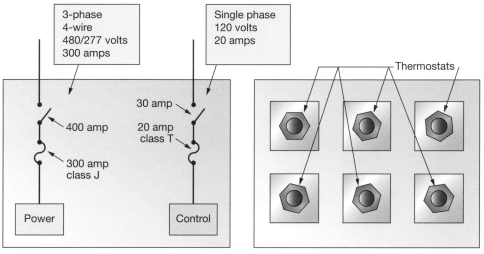

Industrial control panel

Marking must include both the
3-phase, 4-wire, 480/277-volt supply
and the single phase, 120-volt supply

Industrial control panel

Short-circuit current rating not required
because load current does not flow through
components in this industrial control panel

Each incoming supply circuit in an industrial control panel shall be marked and plainly visible.

Revised 2008 NEC Text:
409.110 MARKING. An industrial control panel shall be marked with the following information that is plainly visible after installation:
(1) Manufacturer's name, trademark, or other descriptive marking by which the organization responsible for the product can be identified.
(2) Supply voltage, <u>number of</u> phase<u>s</u>, frequency, and full-load current <u>for each incoming supply circuit</u>.
(3) Short-circuit current rating of the industrial control panel based on one of the following:
   a. Short-circuit current rating of a listed and labeled assembly
   b. Short-circuit current rating established utilizing an approved method
   FPN: UL 508A ~~2001~~, Supplement SB, is an example of an approved method.
   <u>*Exception to (3): Short-circuit current rating markings are not required for industrial control panels containing only control circuit components.*</u>
(4) If the industrial control panel is intended as service equipment, it shall be marked to identify it as being suitable for use as service equipment.

(continued)

(5) Electrical wiring diagram or the <u>identification</u> number of ~~the index to the~~ ~~electrical drawings showing the~~ <u>a separate</u> electrical wiring diagram <u>or a</u> <u>designation referenced in a separate wiring diagram.</u>

(6) An enclosure type number shall be marked on the industrial control panel enclosure.

Change Significance:

This revision clarifies the required markings for an industrial control panel. 409.110(2) is revised to mark information for each incoming supply circuit and to identify the number of phases. 409.110(3) is revised to include a new exception that will not require a short-circuit current rating for industrial control panels that contain only control circuit components. 409.110(5) is revised to require an identification number of an applicable wiring diagram for the individual industrial control panel.

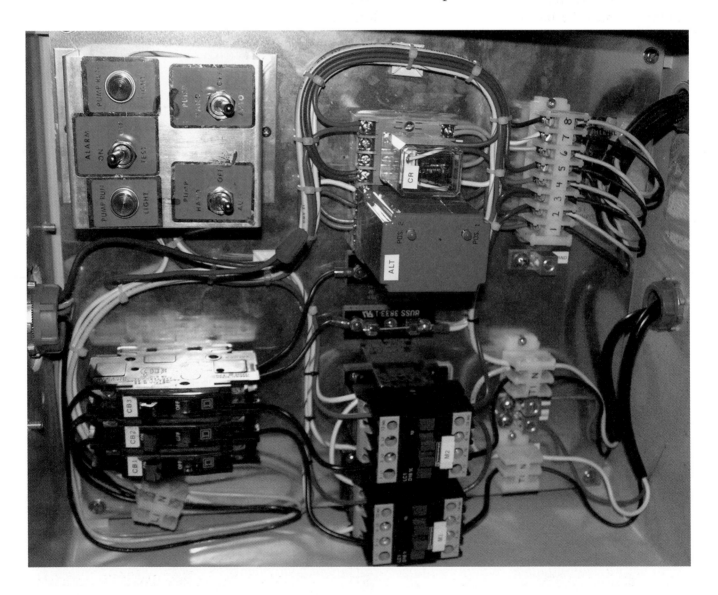

Industrial control panels typically include line voltage supply and control systems for pumps, motors, and other equipment.

Change Summary:

- A new section is added in Part I of Article 410 to require that all luminaries and lampholders be listed.

Revised 2008 NEC Text:

410.6 LISTING REQUIRED.   All luminaries and lampholders shall be listed.

Change Significance:

This revision now requires that all luminaries and lampholders be listed. The technical committee states in their substantiation that requirements for interior wiring, construction, clearances, suitability for specific locations and conditions, and other safety measures are more properly contained within the product standards.

It is important to note that as required in 110.3(B) all listed and labeled equipment must be installed in accordance with the listing and labeling. Where listing requirements existed throughout Article 410 in the 2005 NEC, they are deleted due to this new requirement located in "Part I: General" (which applies to all luminaires).

410.6
LISTING REQUIRED

Chapter 4 Equipment for General Use

Article 410 Luminaires, Lampholders, and Lamps

Part I General

410.6 Listing Required

Change Type: New

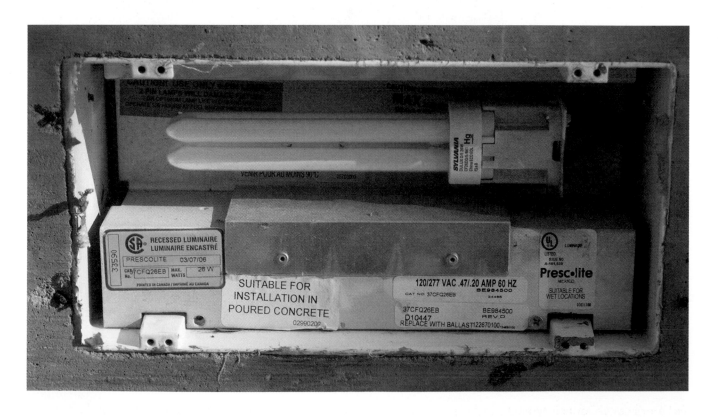

A fixture installed in concrete must be listed for the purpose.

410.16
LUMINAIRES IN CLOTHES CLOSETS

Chapter 4 Equipment for General Use

Article 410 Luminaires, Lampholders, and Lamps

Part II Luminaire Locations

410.16 Luminaires in Clothes Closets

(A) Luminaire Types Permitted

(C) Location

Change Type: Revision

Change Summary:
- Surface-mounted LED luminaries identified for the use are now permitted to be installed in a clothes closet.
- 410.8 is editorially moved to 410.16. The remainder of this section has been editorially revised for clarity.

LED-type fixtures where identified for the use may be installed in a clothes closet.

(continued)

Revised 2008 NEC Text:

410.~~1~~68 LUMINAIRES ~~(FIXTURES)~~ IN CLOTHES CLOSETS

(A)~~(B)~~ LUMINAIRE ~~(FIXTURE)~~ TYPES PERMITTED. Listed luminaires ~~(fixtures)~~ of the following types shall be permitted to be installed in a closet:

(1) A surface-mounted or recessed incandescent luminaire ~~(fixture)~~ with a completely enclosed lamp

(2) A surface-mounted or recessed fluorescent luminaire ~~(fixture)~~

(3) Surface mounted fluorescent or LED luminaires identified as suitable for installation within the storage area.

(C)~~(D)~~ LOCATION. The minimum clearance between luminaires installed in clothes closets and the nearest point of a storage space shall be ~~permitted to be installed~~ as follows:

(1) 300 mm (12 in.) for surface-mounted incandescent or LED luminaires with a completely enclosed light source installed on the wall above the door or on the ceiling

(2) 150 mm (6 in.) for surface-mounted fluorescent luminaires installed on the wall above the door or on the ceiling

(3) 150 mm (6 in.) for recessed incandescent or LED luminaires with a completely enclosed light source installed in the wall or the ceiling

(4) 150 mm (6 in.) for recessed fluorescent luminaires installed in the wall or the ceiling.

(5) Surface-mounted fluorescent or LED luminaires shall be permitted to be installed within the storage space where identified for this use.

Change Significance:

This revision will now permit a surface-mounted LED luminaire identified for the use to be installed in a clothes closet. This revision will now allow any LED fixture (such as a surface-mounted or wall-mounted clothes rod luminaire), provided it is "identified" and installed in accordance with the manufacturer's installation instructions. One method of being recognized as "identified" is for a product to be listed for the purpose.

IDENTIFIED (AS APPLIED TO EQUIPMENT). Recognizable as suitable for the specific purpose, function, use, environment, application, and so forth where described in a particular code requirement.

FPN: Some examples of ways to determine suitability of equipment for a specific purpose, environment, or application include investigations by a qualified testing laboratory (listing and labeling), an inspection agency, or other organizations concerned with product evaluation.

The technical committee noted the following in their panel statement to include LEDs: "LEDs are a relatively new and developing light source technology. Individual LEDs typically operate at low temperatures but some LEDs can operate at elevated temperatures. As LED technology advances even the operating temperatures of individual LEDs are not certain. Revising the Code as proposed recognizes the advent of LEDs as a general light source while maintaining the risk of fire at or below existing levels."

410.130(G)
DISCONNECTING MEANS

Chapter 4 Equipment for General Use

Article 410 Luminaires, Lampholders, and Lamps

Part XIII Special Provisions for Electric-Discharge Lighting Systems of 1000 Volts or Less

410.130 General

(G) Disconnecting Means

(1) General

(2) Multiwire Branch Circuits

(3) Location

Change Type: Revision

Change Summary:
- Requirements for disconnection means in ballasted fluorescent luminaries are revised to clarify only luminaries with double-ended lamps are included.
- The grounded conductor is required to be disconnected only where the luminaire is supplied from a multiwire branch circuit.
- The location of the disconnect when it is external of the luminaire is required to be within sight of the luminaire.

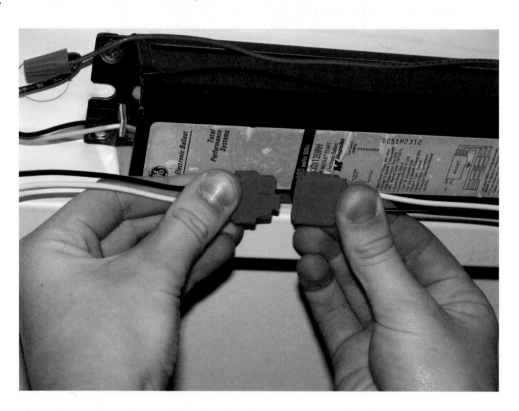

The required disconnecting means must be accessible to qualified persons before servicing the ballast.

Revised 2008 NEC Text:
410.130~~73~~ GENERAL
(G) DISCONNECTING MEANS
(1) GENERAL.   In indoor locations, other than dwellings and associated accessory structures, fluorescent luminaries ~~(fixtures)~~ that utilize double-ended lamps and contain ballast(s) that can be serviced in place ~~or ballasted luminaires that are supplied from multiwire branch circuits and contain ballast(s) that can be serviced in place~~ shall have a disconnecting means either internal or external to each luminaire. The line side terminals of the disconnecting means shall be guarded.

   Exception No. 1 through 4: (No change)

   Exception No. 5: Where more than one luminaire is installed and supplied by other than a multiwire branch circuit, a disconnecting means shall

(continued)

not be required for every luminaire when the design of the installation in-cludes ~~locally accessible disconnects~~ disconnecting means, *such that the il-luminated space cannot be left in total darkness.*

(2) MULTIWIRE BRANCH CIRCUITS. When connected to multiwire branch circuits, the disconnect shall simultaneously break all the ~~from the source of~~ supply ~~all~~ conductors ~~of~~ to the ballast, including the grounded conductor ~~if any~~.

(3) LOCATION. The disconnecting means shall be located so as to be accessible to qualified persons before servicing or maintaining the ballast. Where the disconnecting means is external to the luminaire, it shall be a single device, and shall be attached to the luminaire or the luminaire shall be located within sight of the disconnecting means. ~~This requirement shall become effective January 1, 2008.~~

Change Significance:

This requirement is editorially relocated in Article 410 and logically separated into three second-level subdivisions for clarity and usability. This revision clarifies in (G)(1) that the requirement for a disconnecting means is limited to double-ended fluorescent luminaires. The previous text included the language "or ballasted luminaires that are supplied from multiwire branch circuits and contain ballast(s) that can be serviced in place," which included any type of luminaire utilizing a ballast connected to a multiwire branch circuit.

It is now clarified that only fluorescent luminaires that utilize double-ended lamps and that contain ballast(s) are included in this rule. This will include all ballasted fixtures using straight or U-shape fluorescent tubes. High-hat–type fluorescent fixtures with a single-ended lamp are not affected by this rule.

The requirement for disconnecting the grounded conductor in addition to the ungrounded conductor is now limited to luminaries supplied from a multiwire branch circuit. Where a luminaire is supplied from a two-wire branch circuit, a dangerous "open neutral" situation will not occur.

The location of the disconnect, when it is not internal to the luminaire, is clarified to be a single device—which must be located in sight of the luminaire. This provides clarity and usability of this requirement, aiding proper installation and enforcement. The text that delayed this requirement to allow manufacturers to include these disconnects in luminaries is removed.

Notes:

_____

_____

_____

_____

410.141(B)
WITHIN SIGHT OR LOCKED TYPE

Chapter 4 Equipment for General Use

Article 410 Luminaires, Lampholders, and Lamps

Part XIV Special Provisions for Electric-Discharge Lighting Systems of More Than 1000 Volts

410.141 Control

(B) Within Sight or Locked Type

Change Type: Revision

Change Summary:
- The disconnecting means of electric-discharge lighting systems of more than 1000 volts are now required to remain in place with or without a lock installed. Portable means of adding the lock are not permitted.

Revised 2008 NEC Text:
430.~~141~~181 CONTROL

(B) WITHIN SIGHT OR LOCKED TYPE. The switch or circuit breaker shall be located within sight from the luminaires (~~fixtures~~) or lamps, or it shall be permitted elsewhere if it is provided with a means for locking in the open position. <u>The provisions for locking or adding a lock to the disconnecting means must remain in place at the switch or circuit breaker whether the lock is installed or not. Portable means for adding a lock to the switch or circuit breaker shall not be permitted.</u>

Change Significance:
This safety-driven revision will now strengthen and clarify the intent of this rule. Where a disconnecting means is not within sight of electric-discharge lighting systems of more than 1000 volts, disconnects must be capable of being locked in the open position. This revision requires that the provisions for locking or adding the lock remain in place with or without the lock installed.

When these lighting systems are maintained or repaired, the installer/maintainer must deenergize the luminaire. The text added in this revision will allow persons to "lock out" the disconnect with the use of only a lock. No special device will be needed. Where the disconnect is a circuit breaker, an identified accessory device readily available from manufacturers must remain in place for maintenance purposes.

Part XIV of Article 410 provides the code user with requirements for electric-discharge lighting systems of more than 1000 volts.

Change Summary:

- The definition of *Lighting Systems Operating at 30 Volts or Less* is revised to clarify that the luminaries are included and the output of the power supply is "rated" for not more than 25 amps.

Revised 2008 NEC Text:

411.2 DEFINITION

LIGHTING SYSTEMS OPERATING AT 30 VOLTS OR LESS.   A lighting system consisting of an isolating power supply, the low voltage luminaires, and associated equipment that are all identified for the use. The output circuits of the power supply are rated for not more than 25 amperes and operate at 30 volts (42.4 volts peak) or less under all load conditions. (Remainder of deleted text is revised and relocated in the definition.)

Change Significance:

This revised definition clarifies that "Lighting Systems Operating at 30 Volts or Less" are made up of (1) an isolating power supply, (2) associated equipment, and (3) the luminaires. As previously written in the 2005 NEC, the definition did not include the luminaires. It is important to note that as stated in the definition all of these parts must be identified for the use. This revision also revises the requirement for the system to be "rated" instead of "limited to" not more than 25 amps. The use of the term *rated* will ensure safe operation because these systems are required to be identified for the use. Product standards, including UL 1838 and UL 2108, require the power supply to limit the output current.

411.2
DEFINITION OF *LIGHTING SYSTEMS OPERATING AT 30 VOLTS OR LESS*

Chapter 4 Equipment for General Use

Article 411 Lighting Systems Operating at 30 Volts or Less

411.2 Definition, Lighting Systems Operating at 30 Volts or Less

Change Type: Revision

Low-voltage lighting systems are available in a wide range of styles that may look like line voltage lighting.

411.3
LISTING REQUIRED

Chapter 4 Equipment for General Use

Article 411 Lighting Systems Operating at 30 Volts or Less

411.3 Listing Required

(A) Listed System

(B) Assembly of Listed Parts

Change Type: Revision

Listed parts of a low-voltage lighting assembly will be identified by the listing agency.

Change Summary:
* Listing requirements for lighting systems operating at 30 volts or less have been clarified to permit installation of a single listed system, or a system assembled from listed parts.

Revised 2008 NEC Text:
411.3 LISTING REQUIRED.   Lighting systems operating at 30 volts or less shall ~~be listed~~ comply with 411.3(A) or 411.3(B).

(A) LISTED SYSTEM.   Lighting systems operating at 30 volts or less shall be listed as a complete system. The luminaires, power supply and luminaire fittings (including the exposed bare conductors) of an exposed bare conductor lighting system shall be listed for the use as part of the same identified lighting system.

(B) ASSEMBLY OF LISTED PARTS.   A lighting system assembled from the following listed parts shall be permitted.
(1) Low voltage luminaires
(2) Low voltage luminaire power supply
(3) Class 2 power supply
(4) Low voltage luminaire fittings
(5) Cord (secondary circuit) that the luminaires and power supply are listed for use with
(6) Cable, conductors in conduit, or other fixed wiring method for the secondary circuit. The luminaires, power supply, and luminaire fittings (including the exposed bare conductors) of an exposed bare conductor lighting system shall be listed for use as part of the same identified lighting system

Change Significance:
The previous text of 411.3 inferred that all parts of a lighting system operating at 30 volts or less must be part of a single listed entire lighting system. Lighting systems operating at 30 volts or less have long been field assembled from individually listed low-voltage luminaires, listed luminaire power units, listed cord, and any involved listed luminaire fittings.

This revision clarifies that a lighting system operating at 30 volts or less may be a complete listed lighting system (as outlined in 410.3(A)) or be assembled from listed parts as outlined in 410.3(B). The revised text provides clarity and usability for the code user, clarifying that systems may be in a listed complete set or be assembled with listed parts. It is interesting to note that the technical committee requests that the product standards provide instructions with the power supply specifying the type, length, and size of conductors permitted for use with the system.

A new last sentence requires that systems incorporating exposed bare conductors shall be listed for use as part of the same identified lighting system. This includes the luminaires, power supply, and luminaire fittings of a system incorporating bare conductors.

Change Summary:

- 411.4 has been renamed to clarify that this requirement addresses "specific location requirements."
- The installation of conductors is clarified where they are concealed or extended through walls, floors, or ceilings.
- Installations near pools, spas, fountains, and similar locations are clarified to be installed a minimum of 10 feet horizontally from the nearest edge of water unless permitted in Article 680.

Revised 2008 NEC Text:

411.4 <u>SPECIFIC LOCATION REQUIREMENTS</u> ~~LOCATIONS NOT PERMITTED~~. ~~Lighting systems operating at 30 volts or less shall not be installed in the locations described in 411.4(A) and 411.4(B).~~

(A) <u>WALLS, FLOORS, AND CEILINGS.</u> ~~Where~~ <u>Conductors</u> concealed or extended through a ~~building~~ wall<u>, floor, or ceiling shall be in accordance with</u> ~~unless permitted in~~ (1) or (2):

(1) Installed using any of the wiring methods specified in Chapter 3

(2) Installed using wiring supplied by a listed Class 2 power source and installed in accordance with 725.130

(B) <u>POOLS, SPAS, FOUNTAINS, AND SIMILAR LOCATIONS.</u> ~~Where installed within~~ <u>Lighting systems shall not be installed less than</u> 3.0 m (10 ft) <u>horizontally from the nearest edge of the water</u> ~~of pools, spas, fountains, or similar locations~~, unless permitted by Article 680.

Change Significance:

The previous title and text of 411.4 was not user friendly, creating confusion for the code user. The previous title of this section was "Locations Not Permitted," but the text was permissive in nature. This section has been renamed to clarify that this requirement addresses "specific location requirements."

First-level subdivision (A) is titled "Walls, Floors, and Ceilings" for clarity. The requirement is further clarified to address "conductors" that pass through walls, floors, or ceilings.

First-level subdivision (B) is titled "Pools, Spas, Fountains, and Similar Locations" for clarity. The requirement is further clarified to prohibit these systems from being installed closer than 10 feet horizontally from the nearest edge of water unless permitted in Article 680.

411.4
SPECIFIC LOCATION REQUIREMENTS

Chapter 4 Equipment for General Use

Article 411 Lighting Systems Operating at 30 Volts or Less

411.4 Specific Location Requirements

(A) Walls, Floors, and Ceilings

(B) Pools, Spas, Fountains, and Similar Locations

Change Type: Revision

New text in 411.4(B) prohibits lighting systems operating at 30-volts or less within 10-feet of the nearest edge of a swimming pool.

422.51
CORD-AND-PLUG-CONNECTED VENDING MACHINES

Chapter 4 Equipment for General Use

Article 422 Appliances

Part IV Construction

422.51 Cord-and-Plug-Connected Vending Machines

Change Type: Revision

A vending machine is now qualified as a self-service device that dispenses products or merchandise.

Change Summary:
- The requirements for GFCI protection of vending machines, regardless of voltage or ampere rating, have been revised for clarity and usability.
- A description of a "vending machine" is included to aid the code user in proper application of this requirement.

Revised 2008 NEC Text:
422.51 CORD-AND-PLUG-CONNECTED VENDING MACHINES. Cord-and-plug-connected vending machines manufactured or re-manufactured on or after January 1, 2005, shall include a ground-fault circuit-interrupter as an integral part of the attachment plug or <u>be</u> located ~~in the power supply cord~~ within 300 mm (12 in.) of the attachment plug. ~~Cord-and-plug-connected~~ <u>Older</u> vending machines <u>manufactured or remanufactured prior to January 1, 2005,</u> ~~not incorporating integral GFCI protection~~ shall be connected to a GFCI protected outlet. <u>For the purpose of this section, the term "vending machine" means any self-service device that dispenses products or merchandise without the necessity of replenishing the device between each vending operation and designed to require insertion of a coin, paper currency, token, card, key, or receipt of payment by other means.</u>

FPN: For further information, see ANSI/UL 541-2005, *Standard for Refrigerated Vending Machines*, or ANSI/UL 751-2005, *Standard for Vending Machines.*

Change Significance:
422.51 requires that all vending machines manufactured or remanufactured on or after January 1, 2005 include a ground-fault circuit interrupter as an integral part of the attachment plug, or that they be within 300 mm (12 inches) of the attachment plug. This requirement has been editorially revised for clarity.

A description of "vending machine" is now included to clarify the application of this requirement. This description includes "any self-service device that dispenses products or merchandise without the necessity of replenishing the device between each vending operation and designed to require insertion of a coin, paper currency, token, card, key, or receipt of payment by other means." Only machines that "dispense products or merchandise" are considered vending machines within 422.51. This would exclude washers and dryers in a laundromat and arcade games in an arcade. A picture-taking booth, however, would be considered a "vending machine" because it dispenses your picture.

Notes:
_____

_____

_____

_____

Change Summary:
*   A new Section 422.52 now requires that electric drinking fountains be "protected" by a ground-fault circuit interrupter.

Revised 2008 NEC Text:
422.52 ELECTRIC DRINKING FOUNTAINS.   Electric drinking fountains shall be protected with ground-fault circuit-interrupter protection.

Change Significance:
This new section will now require that the installation of all electric drinking fountains ensure that GFCI protection is provided. It is interesting that this requirement is located in "Part IV, Construction" of Article 422. This would lead one to believe that all electric drinking fountains will be supplied with a GFCI device in the cord or elsewhere to achieve the level of protection required. However, the new text does not mirror language in 422.41 for immersion detection circuit interrupter (IDCI) protection that states "shall be constructed to provide protection."

This new requirement is safety driven and is similar to recent revisions mandating GFCI protection of vending machines. Where GFCI receptacles are used to provide this protection, the installer may need to locate the device upstream of the fountain to permit testing and resetting the device without removing the fountain enclosure.

422.52
ELECTRIC DRINKING FOUNTAINS

Chapter 4 Equipment for General Use

Article 422 Appliances

Part IV Construction

422.52 Electric Drinking Fountains

Change Type: New

An electric drinking fountain must have GFCI protection.

424.19
DISCONNECTING MEANS

Chapter 4 Equipment for General Use

Article 424 Fixed Electric Space-Heating Equipment

Part III Control and Protection of Fixed Electric Space-Heating Equipment

424.19 Disconnecting Means

Change Type: Revision

Change Summary:

- All disconnects for electric space-heating equipment are now required to be sized at not less than 125% of the total load of the motors and the heaters.
- Provisions for locking or adding a lock to the disconnecting means for electric space-heating equipment must remain in place with or without the lock installed.

Revised 2008 NEC Text:

424.19 DISCONNECTING MEANS.   Means shall be provided to <u>simultaneously</u> disconnect the heater, motor controller(s), and supplementary overcurrent protective device(s) of all fixed electric space-heating equipment from all ungrounded conductors. Where heating equipment is supplied by more than one source, the disconnecting means shall be grouped and marked. <u>The disconnecting means specified in 424.19(A) and 424.19(B) shall have an ampere rating not less than 125 percent of the total load of the motors and the heaters. The provision for locking or adding a lock to the disconnecting means shall be installed on or at the switch or circuit breaker used as the disconnecting means and shall remain in place with or without the lock installed.</u>

Change Significance:

The disconnecting means for all electric space-heating equipment shall have an ampere rating not less than 125% of the total load of the motors and the heaters. This revision correlates the requirements for sizing branch circuits (in 424.3(B)) and requirements for sizing overcurrent protection (in 210.20(A)) at 125%.

A new last sentence requires that disconnects for electric space-heating equipment be installed with provisions for locking or adding a lock to the disconnecting means. These provisions must remain in place with or without the lock installed.

Provisions for locking this disconnecting means remain in place with or without the lock installed.

Change Summary:
- Electrically heated pipelines, vessels, or both are now required to be marked with caution signs or markings at intervals not exceeding 20 feet.
- In addition to marking along the pipeline or vessel, markings must also now be placed on or adjacent to equipment in the piping system that requires periodic servicing.

Revised 2008 NEC Text:
427.13 IDENTIFICATION.   The presence of electrically heated pipelines, vessels, or both, shall be evident by the posting of appropriate caution signs or markings at ~~frequent~~ intervals <u>not exceeding 6 m (20 ft)</u> along the pipeline or vessel <u>and on or adjacent to equipment in the piping system that requires periodic servicing</u>.

Change Significance:
The previous requirement of 427.13 for identification required the markings to be at "frequent intervals." The use of the term *frequent* is subjective as well as unenforceable. The NEC style manual requires that technical committees refrain from the use of vague and unenforceable terms. The implementation of markings for electrically heated pipelines and vessels is now clear and enforceable at intervals not to exceed 20 feet.

This revision also requires that these markings be applied on or adjacent to equipment in the piping system that requires periodic servicing. This safety-driven revision will warn those who service or maintain such equipment that under the insulation on the equipment exist electric heat elements that may be damaged and that if energized pose a serious shock hazard.

427.13
IDENTIFICATION

Chapter 4 Equipment for General Use

Article 427 Fixed Electric Heating Equipment for Pipelines and Vessels

Part II Installation

427.13 Identification

Change Type: Revision

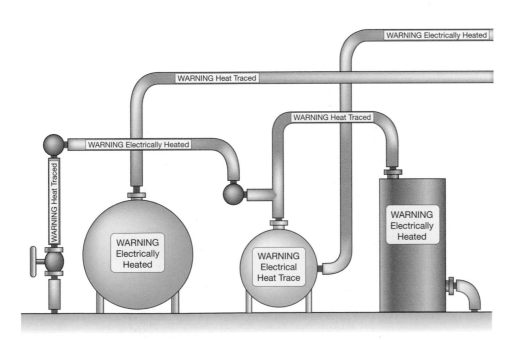

Electrically heated pipelines and vessels must be marked as such at least every 20 feet.

430.32(C)
SELECTION OF OVERLOAD DEVICE

Chapter 4 Equipment for General Use

Article 430 Motors, Motor Circuits, and Controllers

Part III Motor and Branch-circuit Overload Protection

430.32 Continuous-duty Motors

(C) Selection of Overload Device

Change Type: Revision

Change Summary:

- 430.32(C) has been revised to more accurately describe acceptable types of overload protection as a device. Although a relay can be used, a properly sized overcurrent device is also permitted

Revised 2008 NEC Text:

430.32 CONTINUOUS-DUTY MOTORS

(C) SELECTION OF OVERLOAD DEVICE ~~RELAY~~. Where the sensing element or setting <u>or sizing</u> of the overload <u>device</u> ~~relay~~ selected in accordance with 430.32(A)(1) and 430.32(B)(1) is not sufficient to start the motor or to carry the load, higher size sensing elements or incremental settings <u>or sizing</u> shall be permitted to be used, provided the trip current of the overload <u>device</u> ~~relay~~ does not exceed the following percentage of motor nameplate full-load current rating:

Motors with marked service factor 1.15 or greater 140%
Motors with a marked temperature rise 40°C or less 140%
All other motors 130%

If not shunted during the starting period of the motor as provided in 430.35, the overload device shall have sufficient time delay to permit the motor to start and accelerate its load.

FPN: A Class 20 or Class 30 overload relay will provide a longer motor acceleration time than a Class 10 or Class 20, respectively. Use of a higher class overload relay may preclude the need for selection of a higher trip current.

Change Significance:

This revision clarifies that overload protection is not required to be in the form of a relay. Although an overload relay may be used, a properly sized overcurrent protective device is also permitted to provide overload protection.

In many cases, general motor applications are provided with overcurrent protection in the form of two devices. The first is sized in accordance with 430.52 (using table current values) to provide branch-circuit, short-circuit, and ground-fault protection. The second is sized with nameplate current values to provide overload protection. It is possible (and in many cases very practical) to provide both levels of protection with a single properly sized overcurrent protective device.

A properly sized overcurrent protective device can serve as both the overload and the branch-circuit, short-circuit, ground-fault device.

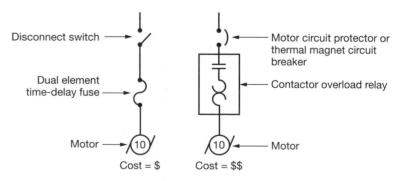

Complete motor and motor circuit protection can be provided by a properly rated disconnecting means and properly sized dual-element time-delay fuses

Change Summary:
- 430.73 is editorially renamed "Protection of Conductor from Physical Damage" and separated to create a new section, "430.74, Electrical Arrangement of Control Circuits."

Revised 2008 NEC Text:
430.73 ~~MECHANICAL~~ PROTECTION OF CONDUCTOR FROM PHYSICAL DAMAGE. Where damage to a motor control circuit would constitute a hazard, all conductors of such a remote motor control circuit that are outside the control device itself shall be installed in a raceway or be otherwise suitably protected from physical damage.

430.74 ELECTRICAL ARRANGEMENT OF CONTROL CIRCUITS. Where one side of the motor control circuit is grounded, the motor control circuit shall be arranged so that an accidental ground in the control circuit remote from the motor controller will (1) not start the motor and (2) not bypass manually operated shutdown devices or automatic safety shutdown devices.

Change Significance:
This revision provides clarity and usability for the code user by separating the existing requirements of 430.73 in two logical sections. "430.73, Protection of Conductor from Physical Damage" provides requirements for the physical protection of control circuits. "430.74, Electrical Arrangement of Control Circuits" provides requirements for the arrangement of motor control circuits. This requirement has not been substantively revised. However, significant clarity and usability are provided by placing this requirement in an individual properly titled section. The user of the NEC using the codeology method will be able to quickly and accurately locate this information.

430.73
PROTECTION OF CONDUCTOR FROM PHYSICAL DAMAGE

430.74
ELECTRICAL ARRANGEMENT OF CONTROL CIRCUITS

Chapter 4 Equipment for General Use

Article 430 Motors, Motor Circuits, and Controllers

Part VI Motor Control Circuits

430.73 Protection of Conductor From Physical Damage

430.74 Electrical Arrangement of Control Circuits

Change Type: Revision

Motor control circuits must be protected from physical damage.

430.102(A)
CONTROLLER

Chapter 4 Equipment for General Use

Article 430 Motors, Motor Circuits, and Controllers

Part IX Disconnecting Means

430.102 Location

(A) Controller

Exception No. 3

Change Type: New

Change Summary:

- A new Exception No. 3 is added to 430.102(A) to address disconnect requirements of "valve-actuated motor" controllers where the installation of a disconnect means could introduce additional or increased hazards to persons or property.

Revised 2008 NEC Text:

430.102 LOCATION

(A) CONTROLLER.   An individual disconnecting means shall be provided for each controller and shall disconnect the controller. The disconnecting means shall be located in sight from the controller location.

Exception No. 1 and No. 2: (No change)

Exception No. 3: The disconnecting means shall not be required to be in sight from valve actuator motor (VAM) assemblies containing the controller where such a location introduces additional or increased hazards to persons or property and conditions (a) and (b) are met.

(a) *The valve actuator motor assembly is marked with a warning label giving the location of the disconnecting means.*

(b) *The provision for locking or adding a lock to the disconnecting means shall be installed on or at the switch or circuit breaker used as the disconnecting means and shall remain in place with or without the lock installed.*

Change Significance:

This new exception permits the required disconnecting means for the controller of a valve actuator motor to be located other than "in sight from" the controller location. Where the installation of a disconnecting means for the valve actuator motor controller could introduce additional or increased hazards to persons or property, Exception No. 3 provides an option.

This exception allows the disconnect to be installed other than "in sight from" provided the following two conditions are met: (1) the valve actuator motor assembly must be marked with a warning label giving the location of the disconnecting means; and (2) provisions for locking or adding a lock to the disconnecting means must be installed on or at the switch or circuit breaker used as the disconnecting means and must remain in place with or without the lock installed. A new definition of "valve actuator motor" has been added to Article 430 as follows:

430.2 VALVE ACTUATOR MOTOR (VAM) ASSEMBLIES.   A manufactured assembly used to operate a valve consisting of an actuator motor, and other components such as controllers, torque switches, limit switches, and overload protection.

FPN: VAMs typically have short-time duty and high torque characteristics.

Notes:

_____

_____

A valve actuated motor is utilized to operate valves in automatic or remote applications.

Change Summary:
- 430.102(B) has been editorially revised to provide clarity in the application of requirements for motor disconnecting means.

Revised 2008 NEC Text:
430.102 LOCATION

(B) MOTOR. <u>A disconnecting means shall be provided for a motor in accordance with (B)(1) or (B)(2).</u>

(1) SEPARATE MOTOR DISCONNECT. A disconnecting means <u>for the motor</u> shall be located in sight from the motor location and the driven machinery location.

(2) CONTROLLER DISCONNECT. The <u>controller</u> disconnecting means required in accordance with 430.102(A) shall be permitted to serve as the disconnecting means for the motor if it is located in sight from the motor location and the driven machinery location.

Exception <u>to (1) and (2)</u>: The disconnecting means <u>for the motor</u> shall not be required to be in sight from the motor and the driven machinery location under either condition (a) or (b), provided the <u>controller</u> disconnecting means required in accordance with 430.102(A) is individually capable of being locked in the open position. The provision for locking or adding a lock to the <u>controller</u> disconnecting means shall be installed on or at the switch or circuit breaker used as the disconnecting means and shall remain in place with or without the lock installed.

(a) Where such a location of the disconnecting means <u>for the motor</u> is impracticable or introduces additional or increased hazards to persons or property

(b) In industrial installations, with written safety procedures, where conditions of maintenance and supervision ensure that only qualified persons service the equipment

FPN No. 1: (No change)
FPN No. 2: (No change)

Change Significance:
This revision provides the code user with clear and easy-to-read text for the requirements of motor disconnecting means. The parent text of 430.102(B) now clearly permits compliance with 430.102(B)(1) or (B)(2).

This requirement is editorially separated in two second-level subdivisions:
- 430.102(B)(1) addresses a "Separate Motor Disconnect" to meet the requirement
- 430.102(B)(2) addresses the use of a "Controller Disconnect" to meet this requirement.

The exception applies to both 430.102(B)(1) and (B)(2). The exception is editorially revised to describe how the "controller" disconnecting means is permitted to serve as the "motor disconnecting means" under the provisions of the exception.

430.102(B)
MOTOR

Chapter 4 Equipment for General Use

Article 430 Motors, Motor Circuits, and Controllers

Part IX Disconnecting Means

430.102 Location

(B) Motor

(1) Separate Motor Disconnect

(2) Controller Disconnect

Exception

Change Type: Revision

A disconnect within sight of motors is required for safety reasons.

430.103
OPERATION

Chapter 4 Equipment for General Use

Article 430 Motors, Motor Circuits, and Controllers

Part IX Disconnecting Means

430.103 Operation

Change Type: Revision

Change Summary:
- 430.103 is revised to require that the disconnecting means for motors and controllers shall be designed so that it cannot be closed automatically.

Revised 2008 NEC Text:
430.103 OPERATION. The disconnecting means shall open all ungrounded supply conductors and shall be designed so that no pole can be operated independently. The disconnecting means shall be permitted in the same enclosure with the controller. <u>The disconnecting means shall be designed so that it cannot be closed automatically.</u>

FPN: See 430.113 for equipment receiving energy from more than one source.

Change Significance:
This revision prohibits the use of any device that may be closed automatically from serving as a disconnecting means for a motor or controller. A time clock, for example, could not be used as a disconnecting means for a controller or a motor. 430.103 requires that the disconnecting means open all ungrounded supply conductors and be designed so that no pole can be operated independently.

The purpose of the disconnecting means for a controller or motor is to isolate the motor and/or the controller for the safety of persons performing maintenance and/or repair. Any disconnecting means that can be closed automatically is not permitted to be installed to comply with 430.102.

A time clock is not permitted to serve as a motor disconnecting means.

Change Summary:

- Exceptions that permit a disconnecting means for a motor or compressor to be rated in horsepower and have a value less than 115% of the specified current values have been revised. The exceptions now specifically address the use of an "unfused" motor-circuit switch.

430.110(A)
GENERAL

440.12
RATING AND INTERRUPTING CAPACITY

Chapter 4 Equipment for General Use

Article 430 Motors, Motor Circuits, and Controllers

Part IX Disconnecting Means

430.110 Ampere Rating and Interrupting Capacity

(A) General

Article 440 Air Conditioning and Refrigeration Equipment

Part II Disconnecting Means

440.12 Rating and Interrupting Capacity

(A) Hermetic Refrigerant Motor-Compressor

(1) Ampere Rating

(B) Combination Loads

(2) Full-Load Current Equivalent

Change Type: Revision

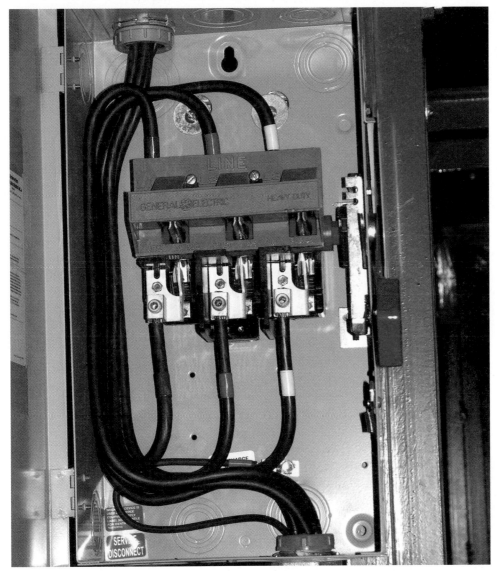

The ampere rating and interrupting capacity of disconnects for motors are addressed in 430.110.

Revised 2008 NEC Text:
430.110 AMPERE RATING AND INTERRUPTING CAPACITY
(A) GENERAL. The disconnecting means for motor circuits rated 600 volts, nominal, or less shall have an ampere rating not less than 115 percent of the full-load current rating of the motor.

(continued)

Exception: A listed ~~nonfused~~ <u>unfused</u> motor-circuit switch having a horsepower rating not less than the motor horsepower shall be permitted to have an ampere rating less than 115 percent of the full-load current rating of the motor.

440.12 RATING AND INTERRUPTING CAPACITY
(A) HERMETIC REFRIGERANT MOTOR-COMPRESSOR
(1) AMPERE RATING.   The ampere rating shall be at least 115 percent of the nameplate rated-load current or branch-circuit selection current, whichever is greater.

Exception: A listed ~~nonfused~~ <u>unfused</u> motor circuit switch, <u>without fuseholders,</u> having a horsepower rating not less than the equivalent horsepower determined in accordance with 440.12(A)(2) shall be permitted to have an ampere rating less than 115 percent of the specified current.

(B) COMBINATION LOADS
(2) FULL-LOAD CURRENT EQUIVALENT.   The ampere rating of the disconnecting means shall be at least 115 percent of the sum of all currents at the rated-load condition determined in accordance with 440.12(B)(1).

Exception: A listed ~~nonfused~~ <u>unfused</u> motor circuit switch, <u>without fuseholders,</u> having a horsepower rating not less than the equivalent horsepower determined in accordance with 440.12(B)(1) shall be permitted to have an ampere rating less than 115 percent of the sum of all currents.

Change Significance:
430.6 and 430.110(A) require the disconnecting means for motors be rated at 115% of the table current value of the motor. 440.12(A)(1) and (B)(2) require the disconnecting means for compressors and combination loads to be rated at 115% of the compressor nameplate/branch circuit, short circuit (BCSC) values, or for (B)(2) (equivalent horsepower) to be rated as the sum of all currents.

Exceptions to these rules permit a switch properly rated in horsepower, with a current value less than 115% to be used. These exceptions now clearly require an unfused switch. Using a conductive element of any type to jump out a fused switch is prohibited.

Notes:

_____

_____

_____

_____

_____

Change Summary:
- The requirements for motor overtemperature protection of motors in an adjustable-speed drive system are clarified to be application dependent and are not required in all cases.

430.126
MOTOR OVERTEMPERATURE PROTECTION

Chapter 4 Equipment for General Use

Article 430 Motors, Motor Circuits, and Controllers

Part X Adjustable-speed Drive Systems

430.126 Motor Overtemperature Protection

(A) General

Change Type: Revision

The overload requirements of 430.126 are in addition to those in 430.32.

Revised 2008 NEC Text:
430.126 MOTOR OVERTEMPERATURE PROTECTION
(A) GENERAL. Adjustable speed drive systems shall protect against motor overtemperature conditions~ <u>where the motor is not rated to operate at</u>

(continued)

the nameplate rated current over the speed range required by the application. This ~~overtemperature~~ protection shall be provided ~~is~~ in addition to the conductor protection required in 430.32. Protection shall be provided by one of the following means.

(1) Motor thermal protector in accordance with 430.32

(2) Adjustable speed drive system ~~controller~~ with load and speed-sensitive overload protection and thermal memory retention upon shutdown or power loss

Exception to 2: Thermal memory retention upon shutdown or power loss is not required for continuous duty loads.

(3) Overtemperature protection relay utilizing thermal sensors embedded in the motor and meeting the requirements of 430.32(A)(2) or (B)(2)

(4) Thermal sensor embedded in the motor whose communications are ~~that is~~ received and acted upon by an adjustable speed drive system

FPN: The relationship between motor current and motor temperature changes when the motor is operated by an adjustable speed drive. In certain applications, overheating of motors can occur ~~W~~when operated at reduced speed, ~~overheating of motors may occur~~ even at current levels less than ~~or equal to~~ a motor's rated full load current. ~~This is~~ The overheating can be the result of reduced motor cooling when its shaft-mounted fan is operating less than rated nameplate RPM. As part of the analysis to determine whether overheating will occur, it is necessary to consider the continuous torque capability curves for the motor given the application requirements. This will assist in determining whether the motor overload protection will be able, on its own, to provide protection against overheating. These overheating protection requirements are only intended to apply to applications where an adjustable speed drive, as defined in 430.2, is used.

For motors that utilize external forced air or liquid cooling systems, overtemperature can occur if the cooling system is not operating. Although this issue is not unique to adjustable speed applications, externally cooled motors are most often encountered with such applications. In these instances, overtemperature protection using direct temperature sensing is recommended (i.e. 430.126(A)(1), (A)(3) or (A)(4)) or additional means should be provided to assure that the cooling system is operating (flow or pressure sensing, interlocking of adjustable speed drive system and cooling system, etc.).

~~(B) MOTORS WITH COOLING SYSTEMS~~   (Delete 430.126 and FPN)

Change Significance:

The motor overtemperature requirements in 430.126 are in addition to the overload requirements in 430.32. This revision clarifies that the additional protection required in 430.126 applies only when the motor is not rated to operate at the nameplate-rated current over the speed range required by the application. An exception has been added to 430.126(2) exempting continuous-duty loads from incorporating thermal memory retention. The existing FPN in 430.126(C) is revised and relocated in 430.126(A). This FPN provides the code user with valuable information regarding the application of these requirements. Because 430.126(A) covers all types of motors, including those with an external source of cooling, this subdivision is deleted.

Change Summary:

- The disconnecting means for controllers of motors operating at greater than 600 volts are now required to have provisions for "lockout," which must remain in place with or without the lock installed.

Revised 2008 NEC Text:

430.227 DISCONNECTING MEANS. The controller disconnecting means shall be capable of being locked in the open position. <u>The provision for locking or adding a lock to the disconnecting means shall be installed on or at the switch or circuit breaker used as the disconnecting means and shall remain in place with or without the lock installed.</u>

Change Significance:

This safety-driven revision modifies the disconnect requirement for controllers of motors operating at greater than 600 volts nominal. In addition to being capable of being locked in the open position, the means to apply a lock must remain in place—with or without the lock installed. Where motors at greater than 600 volts are installed, we know that maintenance will take place. This change ensures that only a lock is needed by an installer/maintainer to work safely.

Without provisions to apply a lock at the controller, personnel safety is dependent on whether or not an installer has the proper portable locking device. This revision will allow the installer/maintainer to "lock out" the controller with only a lock. The ability to lock out will not be dependent on a locking-type device that just happens to fit the circuit breaker or switch on a given controller.

430.227
DISCONNECTING MEANS

Chapter 4 Equipment for General Use

Article 430 Motors, Motor Circuits, and Controllers

Part XI Over 600 Volts, Nominal

430.227 Disconnecting Means

Change Type: Revision

Disconnects for high-voltage motors must provide a means of locking the switch in the open position.

445.18
DISCONNECTING MEANS REQUIRED FOR GENERATORS

Chapter 4 Equipment for General Use

Article 445 Generators

445.18 Disconnecting Means Required for Generators

Change Type: Revision

Change Summary:

- Disconnecting means required for generators are now required to be "lockable in the open position."

Revised 2008 NEC Text:

445.18 DISCONNECTING MEANS REQUIRED FOR GENERATORS. Generators shall be equipped with disconnect(s), lockable in the open position, by means of which the generator and all protective devices and control apparatus are able to be disconnected entirely from the circuits supplied by the generator except where both of the following conditions apply:

(1) The driving means for the generator can be readily shut down.

(2) The generator is not arranged to operate in parallel with another generator or other source of voltage.

Change Significance:

This revision will now clearly require that the disconnecting means for a generator be capable of being locked in the open position. This change illustrates the global effort within the NEC to provide installers and maintainers of electrical distribution systems with adequate means of isolating themselves from an electrical source when working on electrical equipment.

The electrical industry is presently being revamped with respect to electrical safety. Owners, customers, and insurance companies are demanding safe electrical work practices. The ability to properly apply a lock and tag is dependent on the type of disconnecting means and whether or not it is capable of being locked in the open position.

The required disconnect for a generator must be capable of being locked in the open position.

Change Summary:

- 445.19 is added to clarify that where a single generator supplies more than one load or where multiple generators are in parallel a switchboard or single feeder may be used to supply multiple overcurrent protective devices.

Revised 2008 NEC Text:

445.19 GENERATORS SUPPLYING MULTIPLE LOADS. A single generator supplying more than one load, or multiple generators operating in parallel, shall be permitted to supply:

(1) A vertical switchboard with separate sections, or

(2) Individual enclosures with overcurrent protection tapped from a single feeder for load separation and distribution.

Change Significance:

This revision recognizes that standby generators are not supplied with or capable of containing multiple disconnects and overcurrent protection to supply separate loads such as emergency loads, fire pumps, legally required standby loads, and optional standby loads. This revision clarifies that conductor separation of normal and emergency loads is not required until after termination at the switchboard or individual feeder taps.

Where more than one generator is used in parallel, they are permitted to serve a single switchboard. The separation of normal and emergency loads is accomplished in different vertical sections of the switchboard.

445.19
GENERATORS SUPPLYING MULTIPLE LOADS

Chapter 4 Equipment for General Use

Article 445 Generators

445.19 Generators Supplying Multiple Loads

Change Type: New

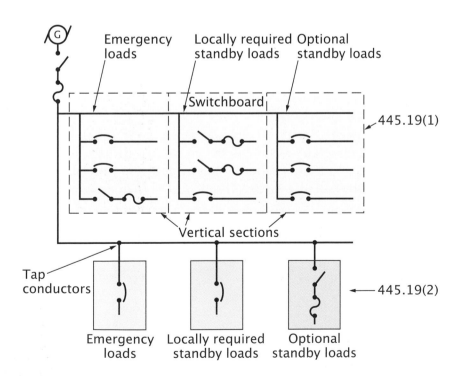

Generators supplying multiple loads may employ one of the two methods shown in the diagram.

480.5
DISCONNECTING MEANS

Chapter 4 Equipment for General Use

Article 480 Storage Batteries

480.5 Disconnecting Means

Change Type: New

Change Summary:

- A new Section 480.5 has been added to require a disconnecting means for all ungrounded conductors (derived from a stationary battery system) that operate at greater than 30 volts. The disconnecting means is required to be readily accessible and located within sight of the battery system.

Revised 2008 NEC Text:

480.5 DISCONNECTING MEANS. A disconnecting means shall be provided for all ungrounded conductors derived from a stationary battery system over 30 volts. A disconnecting means shall be readily accessible and located within sight of the battery system.

Change Significance:

All battery systems require maintenance if they are to remain functional. In some cases, such as in 700.3(C) and (D), the NEC requires battery system maintenance and documentation of the maintenance. To safely perform maintenance on a stationary battery system, a disconnect means should be provided within sight of the battery system. This new section will now provide installers and the enforcement community with clear text requiring a disconnecting means for all ungrounded conductors in a system that operates at greater than 30 volts. This requirement also mandates that the disconnecting means be "readily accessible" and be "within sight" of the battery system.

Although this requirement will apply only to battery systems operating at greater than 30 volts, new requirements for overcurrent protection will now mandate overcurrent protection for all battery systems. The general provisions of 240.21 require an overcurrent protective device at the point of supply for conductors. This is impractical in battery systems. The installation of battery systems and conductors will be impacted by a new first-level subdivision in 240.21 as follows:

240.21 LOCATION IN CIRCUIT
(H) BATTERY CONDUCTORS. Overcurrent protection shall be permitted to be installed as close as practicable to the storage battery terminals in a non-hazardous location. Installation of the overcurrent protection within a hazardous location shall also be permitted.

The 2008 NEC contains new requirements for battery installations including, the location of overcurrent protection in 240.21(H) and a disconnecting means in 480.5.

Change Summary:
- A new last sentence in 490.44(C) requires that the provisions for locking the switching mechanism of a fused interrupter switch remain in place with or without the lock installed.
- A new Section 490.46 provides requirements for locking circuit breakers greater than 600 volts in the open position.

Revised 2008 NEC Text:
490.44 FUSED INTERRUPTER SWITCHES
(C) SWITCHING MECHANISM. The switching mechanism shall be arranged to be operated from a location outside the enclosure where the operator is not exposed to energized parts and shall be arranged to open all ungrounded conductors of the circuit simultaneously with one operation. Switches shall be capable of being locked in the open position. The provisions for locking shall remain in place with or without the lock installed.

490.46 CIRCUIT BREAKER LOCKING. Circuit breakers shall be capable of being locked in the open position or, if they are installed in a drawout mechanism, that mechanism shall be capable of being locked in such a position that the mechanism cannot be moved into the connected position. In either case, the provision for locking shall remain in place with or without the lock.

Change Significance:
These safety-driven revisions are designed to provide the installer/maintainer with the ability to lock out energy sources for systems greater than 600 volts nominal with the use of only a lock. A new last sentence in 490.44(C) requires that the provisions for locking the switching mechanism of a fused interrupter switch remain in place with or without the lock installed.

A new section, "490.46, Circuit Breaker Locking," requires that circuit breakers greater than 600 volts nominal be capable of being locked in the open position—or if they are installed in a drawout mechanism that mechanism shall be capable of being locked in such a position that the mechanism cannot be moved into the connected position. Furthermore, the provision for locking shall remain in place with or without the lock.

490.44
FUSED INTERRUPTER SWITCHES

490.46
CIRCUIT BREAKER LOCKING

Chapter 4 Equipment for General Use

Article 490 Equipment, Over 600 Volts, Nominal

Part III Equipment—Metal-Enclosed Power Switchgear and Industrial Control Assemblies

490.44 Fused Interrupter Switches

(C) Switching Mechanism

490.46 Circuit Breaker Locking

Change Type: Revision and New

A circuit breaker as well as a racked-out device must be capable of being locked in the open position with or without the lock installed.

Special Occupancies, Articles 500 – 590

O U T L I N E

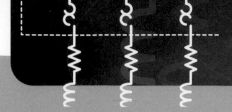

Change Summary:
- The term *flammable liquids* is deleted and the new terms *flammable liquid-produced vapors* and *combustible liquid-produced vapors* are added for clarity. The language "fibers or flyings" is replaced editorially with "fibers/flyings."

Revised 2008 NEC Text:
500.1 SCOPE—ARTICLES 500 THROUGH 504. Articles 500 through 504 cover the requirements for electrical and electronic equipment and wiring for all voltages in Class I, Divisions 1 and 2; Class II, Divisions 1 and 2; and Class III, Divisions 1 and 2 locations where fire or explosion hazards may exist due to flammable gases, ~~or~~ flammable liquid-produced vapors, ~~flammable liquids,~~ combustible liquid-produced vapors, combustible dust, or ignitable fibers/flyings ~~fibers or flyings~~.

Change Significance:
This revision correctly focuses on "flammable liquid-produced vapors" and "combustible liquid-produced vapors" instead of "flammable liquids." Correlation with other NFPA standards is achieved by adding the qualifier "liquid-produced." The language "fibers or flyings" is replaced editorially with "fibers/flyings." These revisions are repeated throughout Articles 500, 503, 504, and 506.

It is important to note that the use of the terms *flammable* and *combustible* within these articles is dependent on the flash point of the vapors, gases, or dusts involved. Where the flash point of a liquid-produced vapor (such as gasoline) is below 100°F, it is considered flammable. Where the flash point of a liquid-produced vapor (such as No. 1-D diesel fuel) is above 100°F, it is considered combustible.

500.1
SCOPE—ARTICLES 500 THROUGH 504

Chapter 5 Special Occupancies

Article 500 Hazardous (Classified) Locations, Classes I, II, and III; Divisions 1 and 2

500.1 Scope—Articles 500 Through 504

Change Type: Revision

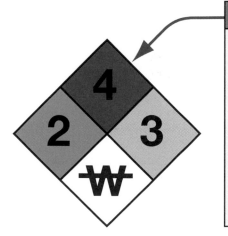

NFPA Hazard Identification System
0 Material will not burn.
1 Material must be pre-heated before ignition can occur.
2 Material must be moderately heated or exposed to relatively high ambient temperature conditions.
3 Liquids and solids that can be ignited under almost all ambient temperature conditions.
4 Materials that will rapidly or completely vaporize at atmospheric pressure and normal ambient temperature, or that are readily dispersed in air and that will burn readily.

The NFPA Hazard Identification System is one method in which materials such as liquids that produce combustible vapors are identified.

500.7(K)
COMBUSTIBLE GAS DETECTION SYSTEM

Chapter 5 Special Occupancies

Article 500 Hazardous (Classified) Locations, Classes I, II, and III; Divisions 1 and 2

500.7 Protection Techniques

(K) Combustible Gas Detection System

(1) Inadequate Ventilation

(2) Interior of a Building

(3) Interior of a Control Panel

Change Type: Revision

Change Summary:
- Gas detection equipment is now required to be listed for the location in which it is installed, for the appropriate material group, and for the detection of the specific gas or vapor to be encountered.

Revised 2008 NEC Text:
(K) COMBUSTIBLE GAS DETECTION SYSTEM.   (Only the deleted text is shown.) ~~Gas detection equipment shall be listed for detection of the specific gas or vapor to be encountered.~~

(1) INADEQUATE VENTILATION.   In a Class I, Division 1 location that is so classified due to inadequate ventilation, electrical equipment suitable for Class I, Division 2 locations shall be permitted. <u>Combustible gas detection equipment shall be listed for Class I, Division 1, the appropriate material group, and for the detection of the specific gas or vapor to be encountered.</u>

(2) INTERIOR OF A BUILDING.   In a building located in, or with an opening into, a Class I, Division 2 location where the interior does not contain a source of flammable gas or vapor, electrical equipment for unclassified locations shall be permitted. <u>Combustible gas detection equipment shall be listed for Class I, Division 1 or Class I, Division 2, for the appropriate material group, and for the detection of the specific gas or vapor to be encountered.</u>

(3) INTERIOR OF A CONTROL PANEL.   In the interior of a control panel containing instrumentation utilizing or measuring flammable liquids, gases, or vapors, electrical equipment suitable for Class I, Division 2 locations shall be permitted. <u>Combustible gas detection equipment shall be listed for Class I, Division 1, for the appropriate material group, and for the detection of the specific gas or vapor to be encountered.</u>

Change Significance:
The second sentence of 500.7(K) is deleted. This sentence previously required that the gas detection equipment had to be listed only for detection of the specific gas or vapor to be encountered. A new last sentence is added to each of the three second-level subdivisions to specifically require that the gas detection equipment be listed for (1) the class and division in which it is located, (2) the appropriate material group, and (3) the specific gas or vapor it is designed to detect. A similar revision has occurred in 505.8(K)(1), (2), and (3) as a result of action on proposal 14-120.

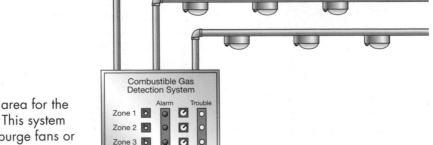

A gas detection system monitors a given area for the accumulation of combustible gas/vapor. This system will activate an alarm and may activate purge fans or interrupt a process.

Change Summary:

- Suitability requirements for hazardous locations are relocated as a new first-level subdivision for clarity, for enforcement, and for correlation with Articles 505 and 506.
- A new informational fine print note (FPN) is included describing the practice of issuing certificates.

Revised 2008 NEC Text:

500.8 EQUIPMENT

(A) SUITABILITY. Suitability of identified equipment shall be determined by one of the following:

(1) Equipment listing or labeling

(2) Evidence of equipment evaluation from a qualified testing laboratory or inspection agency concerned with product evaluation

(3) Evidence acceptable to the authority having jurisdiction such as a manufacturer's self-evaluation or an owner's engineering judgment.

FPN: Additional documentation for equipment may include certificates demonstrating compliance with applicable equipment standards, indicating special conditions of use, and other pertinent information.

Change Significance:

Previous suitability requirements existed in 500.8(A), Approval for Class and Properties. Relocation of this requirement into a new first-level subdivision titled "500.8(A), Suitability" is user friendly and correlates with existing requirements in 505.9(A) and 506.9(A). A new FPN is added to all three locations: 500.8(A), 505.9(A), and 506.9(A). This new informational FPN informs the code user of the common practice of having the testing laboratory, inspection agency, manufacturer, or engineer issue a certificate to document equipment suitability.

Suitability of equipment in hazardous locations is critical. Without equipment that is listed and labeled for the purpose, the code user must document suitability through a third party.

500.8(A)
SUITABILITY

Chapter 5 Special Occupancies

Article 500 Hazardous (Classified) Locations, Classes I, II, and III; Divisions 1 and 2

500.8 Equipment

(A) Suitability

Change Type: New

Hazardous locations come in all shapes and sizes with different hazards. Electrical equipment installed in these areas must be suitable for the purpose.

501.10(B)(1)
GENERAL

Chapter 5 Special Occupancies

Article 501 Class I Locations

Part II Wiring

501.10 Wiring Methods

(B) Class I, Division 2

(1) General

Change Type: New

Change Summary:
- A new list item (7) is added to 501.10(B)(1), expanding the wiring methods permitted in Class I, Division 2, locations to include RTRC and schedule 80 PVC conduits under specified conditions.

Revised 2008 NEC Text:
501.10 WIRING METHODS
(B) CLASS I, DIVISION 2
(1) GENERAL.   In Class I, Division 2 locations, the following wiring methods shall be permitted:

(7) In industrial establishments with restricted public access where the conditions of maintenance and supervision ensure that only qualified persons service the installation and where metallic conduit does not provide sufficient corrosion resistance, reinforced rhermosetting resin conduit (RTRC), factory elbows, and associated fittings, all marked with the suffix-XW, and Schedule 80 PVC conduit, factory elbows, and associated fittings, in accordance with 352.6, shall be permitted.

Where seals are required for boundary conditions as defined in 501.15 (A) (4), the Division 1 wiring method shall extend into the Division 2 area to the seal, which shall be located on the Division 2 side of the Division 1 - Division 2 boundary.

Change Significance:
This revision now permits reinforced thermosetting resin conduit (RTRC) and schedule 80 PVC conduits in Class I, Division 2, locations in industrial establishments under specified conditions. These conditions include (1) restricted public access, (2) conditions of maintenance and supervision that ensure that only qualified persons service the installation, and (3) where metallic conduit does not provide sufficient corrosion resistance.

Throughout the NEC, exceptions are granted to "industrial installations where conditions of maintenance and supervision ensure that only qualified persons service the installation." Although this new list item is written in positive text, it is an exception for an industrial installation. Without a clear definition of *industrial installation,* the implementation of this permissive text is subjective. The enforcement community is also challenged with this type of permissive text. How does one determine if conditions of maintenance and supervision ensure that only qualified persons service the installation?

Industrial establishments with restricted public access and qualified maintenance personnel are permitted to relax many stringent NEC requirements.

Change Summary:
- Flexible metal conduit and liquidtight flexible metal conduit are now clearly prohibited from serving as a sole ground-fault path in 501.30(B).

Revised 2008 NEC Text:
501.30 GROUNDING AND BONDING, CLASS I, DIVISIONS 1 AND 2
(B) TYPES OF EQUIPMENT GROUNDING CONDUCTORS. ~~Where flexible metal conduit or liquidtight flexible metal conduit is used as permitted in 501.10(B) and is to be relied on to complete a sole equipment grounding path, it shall be installed with internal or external bonding jumpers in parallel with each conduit and complying with 250.102.~~

<u>Flexible metal conduit and liquidtight flexible metal conduit shall not be used as the sole ground-fault current path. Where equipment bonding jumpers are installed, they shall comply with 250.102.</u>

Exception: (No change)

502.30(B) TYPES OF EQUIPMENT GROUNDING CONDUCTORS
(Similar revision)

505.25(B) TYPES OF EQUIPMENT GROUNDING CONDUCTORS
(Similar revision)

506.25(B) TYPES OF EQUIPMENT GROUNDING CONDUCTORS
(Similar revision)

Change Significance:
This revision clarifies that flexible metal conduit and liquid-tight flexible metal conduit are not permitted to serve as the sole ground-fault path. Where equipment bonding jumpers are installed, they are required to meet the provisions of 250.102 for sizing and installation. Flexible metal conduit and liquid-tight flexible metal conduit where used as an equipment grounding conductor are limited by 250.118(5) and (6). These flexible raceways are never permitted to serve as the sole ground-fault return path in a hazardous location. These changes have occurred in the following four separate articles:
- **Article 501: Class I Locations**
- **Article 502: Class II Locations**
- **Article 505: Class I, Zones 0, 1, and 2 Locations**
- **Article 506: Zones 20, 21, and 22 Locations for Combustible Dusts or Ignitable Fibers/Flyings**

501.30(B)
TYPES OF EQUIPMENT GROUNDING CONDUCTORS

Chapter 5 Special Occupancies

Article 501 Class I Locations

Part II Wiring

501.30 Grounding and Bonding, Class I, Divisions 1 and 2

(B) Types of Equipment Grounding Conductors

Change Type: Revision

The use of flexible metal conduit in an Article 501 installation requires an equipment grounding conductor in all situations.

502.115(A)
CLASS II, DIVISION 1

Chapter 5 Special Occupancies

Article 502 Class II Locations

Part III Equipment

502.115 Switches, Circuit Breakers, Motor Controllers, and Fuses

(A) Class II, Division 1

(1) Type Required

(2) Metal Dusts

Change Type: Revision

Article 502 provides special modifications and supplemental requirements for installations in which fire or explosion hazards exist due to combustible dust.

Change Summary:
- 502.115 is revised to require that all switches, circuit breakers, motor controllers, fuses, pushbuttons, relays, and similar devices installed in a Class II, Division 1, location be provided with identified dust–ignition-proof enclosures.

Revised 2008 NEC Text:
502.115 SWITCHES, CIRCUIT BREAKERS, MOTOR CONTROLLERS, AND FUSES

(A) CLASS II, DIVISION 1.   In Class II, Division 1 locations, switches, circuit breakers, motor controllers, and fuses shall comply with 502.115(A)(1) ~~through~~ and (A)(2)~~(3)~~.

(1) TYPE REQUIRED.   Switches, circuit breakers, motor controllers, and fuses, including pushbuttons, relays, and similar devices ~~that are intended to interrupt current during normal operation or that are installed where combustible dusts of an electrically conductive nature may be present,~~ shall be provided with identified dust-ignitionproof enclosures.

~~(2) ISOLATING SWITCHES.~~   ~~Disconnecting and isolating switches containing no fuses and not intended to interrupt current and not installed where dusts may be of an electrically conductive nature shall be provided with tight metal enclosures that shall be designed to minimize the entrance of dust and that shall (1) be equipped with telescoping or close-fitting covers or with other effective means to prevent the escape of sparks or burning material and (2) have no openings (such as holes for attachment screws) through which, after installation, sparks or burning material might escape or through which exterior accumulations of dust or adjacent combustible material might be ignited.~~

~~(3)~~ (2) METAL DUSTS.   (No change)

Change Significance:
All switches, circuit breakers, motor controllers, fuses, pushbuttons, relays, and similar devices installed in a Class II, Division 1, location are now required to be provided with identified dust–ignition-proof enclosures. The previous text in 502.115(A)(1) required that only those items that were "intended to interrupt current during normal operation or that are installed where combustible dusts of an electrically conductive nature may be present" were required to be installed in an enclosure identified as a dust–ignition-proof type.

Second-level subdivision "(2) Isolating Switches" is deleted. This subdivision previously permitted "switches containing no fuses and not intended to interrupt current and not installed where dusts may be of an electrically conductive nature" to be installed in enclosures not identified as a dust–ignition-proof type.

Change Summary:

- In Class II, Division 2, locations, coils and windings are required to utilize enclosures that are "dusttight" beginning January 1, 2011.

Revised 2008 NEC Text:

502.120 CONTROL TRANSFORMERS AND RESISTORS
(B) CLASS II, DIVISION 2
(2) COILS AND WINDINGS.   Where not located in the same enclosure with switching mechanisms, control transformers, solenoids, and impedance coils shall be provided with tight metal housings without ventilating openings <u>or shall be installed in dusttight enclosures. Effective January 1, 2011, only dusttight enclosures shall be permitted</u>.

Change Significance:

This revision provides consistency with the requirements of 502.10(B)(4), which requires all boxes and fittings in a Class II, Division 2, location to be dusttight. Recognizing the impact on the industry, the technical committee retained the original text with an option of a "dusttight enclosure" and requires that as of January 1, 2011 all such enclosures shall be "dusttight." Several revisions have occurred throughout Article 502 to consistently require dusttight enclosures for all electrical distribution in a Class II, Division 2, location.

502.120
CONTROL TRANSFORMERS AND RESISTORS

Chapter 5 Special Occupancies

Article 502 Class II Locations

Part III Equipment

502.120 Control Transformers and Resistors

(B) Class II, Division 2

(2) Coils and Windings

Change Type: Revision

The storage of grain and other materials that create dust may be hazardous locations subject to the rules in Article 502.

502.130 (B)(2)
FIXED LIGHTING

Chapter 5 Special Occupancies

Article 502 Class II Locations

Part III Equipment

502.130 Luminaires

(B) Class II, Division 2

(2) Fixed Lighting

Change Type: Revision

Change Summary:
- All luminaires installed in Class II, Division 2, locations not identified for Class II locations are now required to be provided with "dusttight" enclosures.

Revised 2008 NEC Text:
502.130 LUMINAIRES
(B) CLASS II, DIVISION 2
(2) FIXED LIGHTING.   Luminaires for fixed lighting, where not of a type identified for Class II locations, shall <u>be</u> provide<u>d with dusttight</u> enclosures. ~~for lamps and lampholders that shall be designed to minimize the deposit of dust on lamps and to prevent the escape of sparks, burning material, or hot metal.~~ Each fixture shall be clearly marked to indicate the maximum wattage of the lamp that shall be permitted without exceeding an exposed surface temperature in accordance with 500.8~~(C)~~<u>(D)</u>(2) under normal conditions of use.

Change Significance:
Luminaires in a Class II, Division 2 location are not required to be identified for Class II locations—as is required in a Class II, Division 1 location in 502.115(A)(1). Previous code text required that these luminaires in a Class II, Division 2 location not identified as Class II be designed to minimize the deposit of dust on lamps and to prevent the escape of sparks, burning material, or hot metal. All luminaires in a Class II location must now be identified as suitable for a Class II location or identified as dusttight. This revision also achieves consistency with 502.10(B)(4), which requires that all boxes and fittings used in a Class II, Division 2 location be dusttight.

Equipment for use in Class II installations must be marked as *dusttight*. Labels are provided by the manufacturer with information necessary for installation and enforcement.

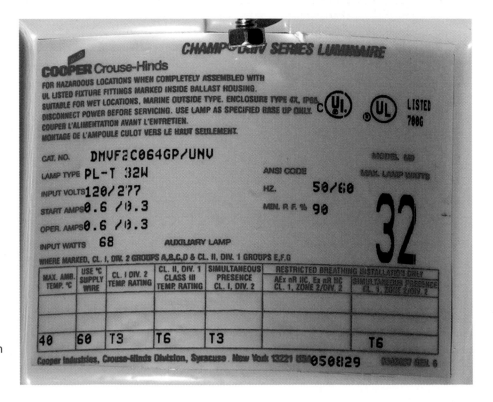

Change Summary:

- In Class II, Division 2, locations, signaling, alarm, remote-control, and communications systems (and meters, instruments, and relays) are required to utilize enclosures that are "dusttight" beginning January 1, 2011.

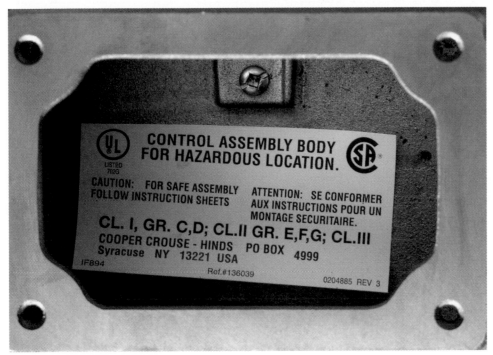

Conduit bodies and enclosures of all types that are identified for use in hazardous locations must be marked with the appropriate information.

502.150(B)
CLASS II, DIVISION 2

Chapter 5 Special Occupancies

Article 502 Class II Locations

Part III Equipment

502.150 Signaling, Alarm, Remote-Control, and Communications Systems; and Meters, Instruments, and Relays

(B) Class II, Division 2

(1) Contacts

(2) Transformers and Similar Equipment

(3) Resistors and Similar Equipment

(4) Rotating Machinery

Change Type: Revision

Revised 2008 NEC Text:

502.150 SIGNALING, ALARM, REMOTE-CONTROL, AND COMMUNICATIONS SYSTEMS; AND METERS, INSTRUMENTS, AND RELAYS

(B) CLASS II, DIVISION 2. In Class II, Division 2 locations, signaling, alarm, remote-control, and communications systems; and meters, instruments, and relays shall comply with 502.150(B)(1) through (B)(4)(5).

(1) CONTACTS. ~~Enclosures~~ Contacts shall comply with 502.150(A)(2), or contacts shall have tight metal enclosures designed to minimize the entrance of dust and shall have telescoping or tight-fitting covers and no openings through which, after installation, sparks or burning material might escape <u>or shall be installed in dusttight enclosures, effective January 1, 2011, only dustight enclosures shall be permitted.</u>

Exception: In nonincendive circuits, enclosures shall be permitted to be of the general-purpose type.

(2) TRANSFORMERS AND SIMILAR EQUIPMENT. The windings and terminal connections of transformers, choke coils, and similar equipment shall

(continued)

comply with 502.120(B)(2). ~~be provided with tight metal enclosures without ventilating openings.~~

(3) RESISTORS AND SIMILAR EQUIPMENT. Resistors, resistance devices, thermionic tubes, rectifiers, and similar equipment shall comply with 502.120(B)(3) ~~130(A)(3)~~.

~~Exception: Enclosures for thermionic tubes, nonadjustable resistors, or rectifiers for which maximum operating temperature will not exceed 120°C (248°F) shall be permitted to be of the general-purpose type.~~

(4) ROTATING MACHINERY. Motors, generators, and other rotating electric machinery shall comply with 502.125(B).

~~(5) WIRING METHODS.~~ ~~The wiring method shall comply with 502.10(B).~~

Change Significance:

This revision provides consistency with the requirements of 502.10(B)(4), which requires all boxes and fittings in a Class II, Division 2, location to be dusttight. Recognizing the impact on the industry, the technical committee retained the original text with an option of a "dusttight enclosure" and requires that as of January 1, 2011 all such enclosures shall be "dusttight." Several revisions have occurred throughout Article 502 to consistently require dusttight enclosures for all electrical distribution in a Class II, Division 2, location.

502.150(B)(1) and (2) are revised to clearly require that all enclosures be "dusttight" by January 1, 2011. 502.150(B)(3) is revised to reference 502.120(B)(3), which requires that all resistors and resistance devices be installed in dust–ignition-proof enclosures identified for Class II locations. The exception for thermionic tubes, nonadjustable resistors, or rectifiers for which maximum operating temperature will not exceed 120°C (248°F) is deleted. 502.150(B)(5) is deleted because the reference to 502.10(B) is redundant.

Notes:

_____

_____

_____

_____

_____

_____

_____

Change Summary:
- Classification of areas and selection of equipment and wiring methods in Article 505 installations is no longer required to be under the supervision of a qualified registered professional engineer.

Revised 2008 NEC Text:
505.7 SPECIAL PRECAUTION
(A) IMPLEMENTATION OF ZONE CLASSIFICATION SYSTEM. Classification of areas, engineering and design, selection of equipment and wiring methods, installation, and inspection shall be performed by qualified persons.

~~**(A) SUPERVISION OF WORK.** Classification of areas and selection of equipment and wiring methods shall be under the supervision of a qualified Registered Professional Engineer.~~

Change Significance:
This revision removes the requirement that a registered professional engineer supervise the classification of areas and selection of equipment and wiring methods in a Class I Zone 0, 1, or 2 location. The revised text is modeled after 506.6(A) for Zones 20, 21, and 22 locations. Article 506 addresses the zone concept for combustible dusts fibers and flyings. Article 506 was new in the 2005 NEC, and 506.6(A) did not require a registered professional engineer. 506.6(A), however, mandates that classification of areas, engineering, design, equipment selection, wiring methods, and inspections is required by "qualified persons."

The revised text in 505.7(A) mirrors the intent found in Article 506. Enforcement of 506.6(A) and 505.7(A) will be difficult for the authority having jurisdiction. There are no qualifiers to determine who is and who is not qualified to classify areas, engineer, design, select equipment, and select wiring methods in these hazardous locations.

505.7(A)
IMPLEMENTATION OF ZONE CLASSIFICATION SYSTEM

Chapter 5 Special Occupancies

Article 505 Class I, Zones 0, 1, and 2 Locations

505.7 Special Precaution

(A) Implementation of Zone Classification System

Change Type: Revision

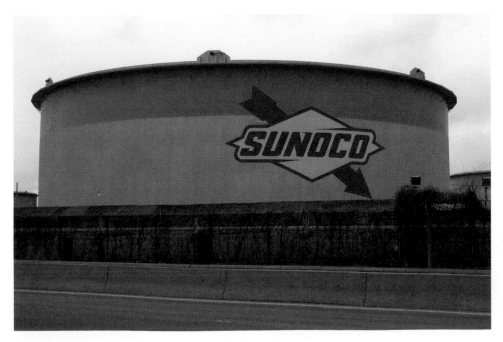

The zone classification system and methods of installation may be used in a hazardous location created by the storage of flammable liquids.

506.2

DEFINITIONS

Chapter 5: Special Occupancies

Article 506 Zones 20, 21, and 22 Locations for Combustible Dusts or Ignitable Fibers/ Flyings

506.2 Definitions

Change Type: New

Change Summary:

- Four new definitions are added to 506.2 to describe protection systems unique to Zones 20, 21, and 22 classified areas.

Revised 2008 NEC Text:

506.2 DEFINITIONS

PROTECTION BY ENCAPSULATION "mD". Type of protection where electrical parts that could cause ignition of a mixture of combustible dust, fibers/flyings in air are protected by enclosing them in a compound in such a way the explosive atmosphere cannot be ignited.

FPN No. 1: For additional information see ISA-61241-18 (12.10.07)-2006 *Electrical Apparatus for Use in Zone 20, Zone 21 and Zone 22 Hazardous (Classified) Locations-Protection by Encapsulation "mD".*

FPN No. 2: Encapsulation is designated level of protection "maD" for use in Zone 20 locations. Encapsulation is designated level of protection "mbD" for use in Zone 21 locations.

Table 506.9(C)(2)(2) Types of Protection Designation

Designation	Technique	Zone
iaD	Protection by intrinsic safety	20
ibD	Protection by intrinsic safety	21
iaD	Associated apparatus	Unclassified**
ibD	Associated apparatus	Unclassified**
maD	Protection by encapsulation	20
mbD	Protection by encapsulation	21
pD	Protection by pressurization	21
tD	Protection by enclosures	21

* Does not address use where a combination of techniques is used.

**Associated apparatus is permitted to be installed in a hazardous (classified) location if suitably protected using another type of protection.

Table 506.9(C)(2)(2): Types of Protection Designation.

(continued)

PROTECTION BY ENCLOSURE "tD". Type of protection for explosive dust atmospheres where electrical apparatus is provided with an enclosure providing dust ingress protection and a means to limit surface temperatures.

FPN: For additional information see ISA 61241-0 (12.10.02) *Electrical Apparatus for Use in Zone 20, Zone 21 and Zone 22 Hazardous (Classified) Locations—General Requirements* (IEC 61241-0 Mod), and ISA 61241-1 (12.10.03) *Electrical Apparatus for Use in Zone 21 and Zone 22 Hazardous (Classified) Locations—Protection by Enclosure "tD"* (IEC 61241-1 Mod).

PROTECTION BY INTRINSIC SAFETY "iD". Type of protection where any spark or thermal effect is incapable of causing ingition of a mixture of combustible dust or fibers/flyings in air under prescribed test conditions.

FPN: For additional information see ISA 61241-11 (12.10.06), *Electrical Apparatus for use in Zone 20, Zone 21 and Zone 22 Hazardous (Classified) Locations- Protection by Intrinsic Safety "iD".*

PROTECTION BY PRESSURIZATION "pD". Type of protection that guards against the ingress of a mixture of combustible dust or fibers/flyings in air into an enclosure containing electrical equipment by providing and maintaining a protective gas atmosphere inside the enclosure at a pressure above that of the external atmosphere.

FPN: For additional information see , ISA 61241-2 (12.10.04), *Electrical Apparatus for use in Zone 21 and Zone 22 Hazardous (Classified) Locations- Protection by Pressurization "pD".*

Change Significance:

Throughout Article 506, revisions have occurred to clarify permitted protection techniques. These new definitions provide the code user with the necessary information to identify the acronyms used for these protection techniques and the differences between methods of protection. The four basic methods of protection include encapsulation, types of enclosures, intrinsically safe conductors, and pressurization.

Notes:

_____

_____

_____

_____

_____

_____

511.2
DEFINITIONS

511.3
AREA CLASSIFICATION, GENERAL

Change Summary:
- Article 511 now defines "minor" and "major" repair garages to provide clarity. 511.3 has been completely reorganized, providing a clear and comprehensive outline to classify Article 511 locations.

Where service and repair is performed on self-propelled vehicles in which volatile flammable liquids or gas are used as fuels, the occupancy is considered a commercial garage.

Revised 2008 NEC Text:
511.2 DEFINITIONS
MAJOR REPAIR GARAGE.   A building or portions of a building where major repairs, such as engine overhauls, painting, body and fender work, and repairs that require draining of the motor vehicle fuel tank are performed on motor vehicles, including associated floor space used for offices, parking, or showrooms. [NFPA 30A-2003, 3.3.12.1]

MINOR REPAIR GARAGE.   A building or portions of a building used for lubrication, inspection, and minor automotive maintenance work, such as engine tune-ups, replacement of parts, fluid changes (e.g., oil, antifreeze, transmission fluid, brake fluid, air conditioning refrigerants, etc.), brake system repairs, tire rotation, and similar routine maintenance work, including associated floor space used for offices, parking, or showrooms. [NFPA 30A-2003, 3.3.12.2]

511.3 AREA CLASSIFICATION, GENERAL
(A) PARKING GARAGES
(B) REPAIR GARAGES, WITH DISPENSING

(continued)

(C) MAJOR REPAIR GARAGES
(1) FLOOR AREAS.   (a) *Ventilation Provided,* (b) *Ventilation Not Provided*
(2) CEILING AREAS.   (a) *Ventilation Provided,* (b) *Ventilation Not Provided*
(3) PIT AREAS IN LUBRICATION OR SERVICE ROOM.   (a) *Ventilation Provided,* (b) *Ventilation Not Provided.*
(D) MINOR REPAIR GARAGES
(1) FLOOR AREAS.   (a) *Ventilation Provided,* (b) *Ventilation Not Provided*
(2) CEILING AREAS
(3) PIT AREAS IN LUBRICATION OR SERVICE ROOM.   (a) *Ventilation Provided,* (b) *Ventilation Not Provided*
(E) MODIFICATIONS TO CLASSIFICATION
(1) SPECIFIC AREAS ADJACENT TO CLASSIFIED LOCATIONS
(2) ALCOHOL-BASED WINDSHIELD WASHER FLUID

Change Significance:

This revision includes two new definitions that provide the code user with clear guidelines to determine when an occupancy is a "minor" or "major" repair garage. This revision includes a complete reorganization of 511.3. This section is now written in a user-friendly format that provides increased clarity of the requirements. The provisions of NFPA 30A Code for Motor Fuel Dispensing Facilities and Repair Garages have been incorporated through reference as extract material to provide the code user with a comprehensive approach to classifying Article 511 locations.

(D) Minor Repair Garages

(1) Floor Areas

(a) Ventilation Provided

(b) Ventilation Not Provided

(2) Ceiling Areas

(3) Pit Areas in Lubrication or Service Room

(a) Ventilation Provided

(b) Ventilation Not Provided

(E) Modifications to Classification

(1) Specific Areas Adjacent to Classified Locations

(2) Alcohol-Based Windshield Washer Fluid

Change Type: Revision and New

511.2 now contains definitions to aid the code user to qualify an area as a *major repair garage* or a *minor repair garage.*

513.2
DEFINITION OF *AIRCRAFT PAINTING HANGAR*

513.3(C)
VICINITY OF AIRCRAFT

Change Type: Revision and New

Change Summary:
* A new definition of *aircraft painting hangar* is included in 513.2, along with prescriptive requirements to outline classified areas around the airplane to be painted.

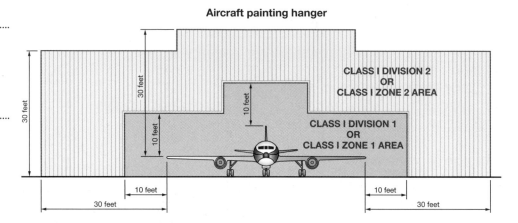

Aircraft painting hanger

Class and zone designations are now clearly defined for hangars in which aircraft are painted.

Revised 2008 NEC Text:
513.2 DEFINITIONS
AIRCRAFT PAINTING HANGAR.   An aircraft hangar constructed for the express purpose of spray/coating/dipping applications and provided with dedicated ventilation supply and exhaust.

513.3 CLASSIFICATION OF LOCATIONS
(C) VICINITY OF AIRCRAFT
(1) AIRCRAFT MAINTENANCE AND STORAGE HANGARS.   The area within 1.5 m (5 ft) horizontally from aircraft power plants or aircraft fuel tanks shall be classified as a Class I, Division 2 or Zone 2 location that shall extend upward from the floor to a level 1.5 m (5 ft) above the upper surface of wings and of engine enclosures.

(2) AIRCRAFT PAINTING HANGARS.   The area within 3 m (10 ft) horizontally from aircraft surfaces from the floor to 3 m (10 ft) above the aircraft shall be classified as Class I, Division 1 or Class I, Zone 1. The area horizontally from aircraft surfaces between 3.0 m (10 ft) and 9.0 m (30 ft) from the floor to 9.0 m (30 ft) above the aircraft surface shall be classified as Class I, Division 2 or Class I, Zone 2.
    FPN: See NFPA 33-2003, *Standard for Spray Application Using Flammable or Combustible Materials for information on ventilation and grounding for static protection in spray painting areas.*

Change Significance:
This revision provides a clear definition of an "aircraft painting hangar," which must have been built for the express purpose of painting aircraft and

(continued)

must be provided with dedicated ventilation supply and exhaust. A typical aircraft hangar will not be considered an "aircraft painting hangar."

513.3(C) is separated in two second-level subdivisions to classify hazardous locations in aircraft hangars. 513.3(C)(1) addresses a typical maintenance and storage hangar with the previous code text of 513.3(C). New text is provided in 513.3(C)(2) for classifying locations in an "aircraft painting hangar."

A typical aircraft hangar (unless built for the express purpose of painting) will not be considered as an aircraft painting hangar.

517.2
DEFINITIONS

Chapter 5 Special Occupancies

Article 517 Health Care Facilities

Part I General

517.2 Definitions

Change Type: Revision

Change Summary:
- Twelve definitions in 517.2 have been revised to correlate with preferred definitions in other NFPA standards. A reference in brackets ([]) is located at the end of each definition to inform the code user of the document from which they are extracted in accordance with 90.5(C).

Revised 2008 NEC Text:

AMBULATORY HEALTH CARE ~~FACILITY~~ OCCUPANCY. A building or <u>portion</u> ~~part~~ thereof used to provide services or treatment <u>simultaneously</u> to four or more patients ~~at the same time and meeting either (1) or (2)~~ <u>that provides, on an outpatient basis, one or more of the following:</u> (Delete existing list items (1) and (2))

<u>(1) Treatment for patients that renders the patients incapable of taking action for self-preservation under emergency conditions without the assistance of others;</u>

<u>(2) Anesthesia that renders the patients incapable of taking action for self-preservation under emergency conditions without the assistance of others;</u>

<u>(3) Emergency or urgent care for patients who, due to the nature of their injury or illness, are incapable of taking action for self-preservation under emergency conditions without the assistance of others. [NFPA: 101:3.3.168.1]</u>

CRITICAL BRANCH. [NFPA 99:3.3.26] (No change to definition)

ELECTRICAL LIFE SUPPORT EQUIPMENT. <u>[NFPA 99: 3.3.37]</u> (No change to definition)

EMERGENCY SYSTEM. <u>[NFPA 99: 3.3.41]</u> (No change to definition)

ESSENTIAL ELECTRICAL SYSTEM. <u>[NFPA 99: 3.3.44]</u> (No change to definition)

HOSPITAL. A building or ~~part~~ <u>portion</u> thereof used <u>on a 24-hour basis</u> for the medical, psychiatric, obstetrical, or surgical care, ~~on a 24-hour basis~~, of four or more inpatients. <u>[NFPA 101: 3.3.124]</u>

~~Hospital, wherever used in this Code, shall include general hospitals, mental hospitals, tuberculosis hospitals, children's hospitals, and any such facilities providing inpatient care.~~

LIFE SAFETY BRANCH. <u>[NFPA 99: 3.3.96]</u> (No change to definition)

LIMITED CARE FACILITY. A building or ~~part~~ <u>portion</u> thereof used on a 24-hour basis for the housing of four or more persons who are incapable of self-preservation because of age; physical limitation due to accident or illness; or ~~mental~~ limitations such as mental retardation/developmental disability, mental illness, or chemical dependency. <u>[NFPA 99:3.3.97]</u>"

NURSING HOME. A building or ~~part thereof~~ <u>portion of a building</u> used <u>on a 24-hour basis</u> for the ~~lodging, boarding,~~ <u>housing</u> and nursing care, ~~on a 24-~~

Article 517 contains 39 definitions in 517.2 that are critical to proper application of these installation requirements.

(continued)

~~hour basis~~, of four or more persons who, because of mental or physical incapacity, ~~may~~ might be unable to provide for their own needs and safety without the assistance of another person. [NFPA 99: 3.3.129] ~~Nursing home, wherever used in this Code, shall include nursing and convalescent homes, skilled nursing facilities, intermediate care facilities, and infirmaries of homes for the aged.~~

PATIENT BED LOCATION.   The location of an inpatient sleeping bed~~:~~, or the bed or procedure table ~~used in~~ of a critical ~~patient~~ care area. [NFPA 99:3.3.137]

PATIENT EQUIPMENT GROUNDING POINT. [NFPA 99:3.3.141] (No change to definition)

PATIENT CARE VICINITY.   [NFPA 99: 3.3.140] (No change to definition)

Change Significance:

These revisions are primarily editorial in nature. These changes recognize the preferred NFPA definitions from other standards. Consistent definitions throughout NFPA codes and standards provide the code user with improved clarity, usability, ease of implementation, and uniform enforcement. The definition *of ambulatory health care occcupancy* is revised to correlate with NFPA 101 and provide additional clarity and prescriptive elements for proper classification. The revisions correlate with the preferred definitions in:

*   **NFPA 99: Standard for Health Care Facilities**
*   **NFPA 101: Life Safety Code**

Requirements for health care facilities in the NEC are heavily influenced by NFPA 99 the Standard for Health Care Facilities and NFPA 101 the Life Safety Code.

517.2

DEFINITION OF PATIENT CARE AREA

Chapter 5 Special Occupancies

Article 517 Health Care Facilities

Part I General

517.2 Definitions

Change Type: Revision

The electrical installation in a patient care area is modified significantly in Article 517 from the general requirements in Chapters 1 through 4.

Change Summary:

- The definition of *patient care area* is revised to clarify general care areas and wet procedure locations.

Revised 2008 NEC Text:

517.2 DEFINITIONS

PATIENT CARE AREA.   Any portion of a health care facility wherein patients are intended to be examined or treated. Areas of a health care facility in which patient care is administered are classified as general care areas or critical care areas~~, either of which may be classified as a wet location.~~ The governing body of the facility designates these areas in accordance with the type of patient care anticipated and with the following definitions of the area classification.

FPN: Business offices, corridors, lounges, day rooms, dining rooms, or similar areas typically are not classified as patient care areas.

General Care Areas. Patient bedrooms, examining rooms, treatment rooms, clinics, and similar areas in which it is intended that the patient will come in contact with ordinary appliances such as a nurse-call system, electric beds, examining lamps, telephones, and entertainment devices. ~~In such areas, it may also be intended that patients be connected to electromedical devices (such as heating pads, electrocardiographs, drainage pumps, monitors, otoscopes, opthalmoscopes, intraveneous lines etc.)~~ [99:3.3.138.2]

Critical Care Areas. (No change)

Wet <u>Procedure</u> Locations. Those <u>spaces within</u> patient care areas <u>where a procedure is performed and</u> that are normally subject to wet conditions while patients are present. These include standing fluids on the floor or drenching of the work area, either of which condition is intimate to the patient or staff. Routine housekeeping procedures and incidental spillage of liquids do not define a wet location.

Change Significance:

Article 517 contains 39 definitions in 517.2 that are critical to proper application of the installation requirements in this article. The definition of *patient care area* has a serious impact on wiring methods and electrical systems that serve these areas. This definition is further separated into two variations of a "patient care area." These include "general care" and "critical care." The definition of *general care area* is revised by removing the laundry list of examples to classify a "general care" area. No change has occurred in "critical care areas."

The definition of *wet location* has always applied to both "general care" and "critical care" areas. This definition has been revised as a "wet *procedure* location." This provides necessary clarity for the code user.

Notes:

_____

_____

Change Summary:

- Wiring methods for luminaires in patient care areas must qualify in 250.118 as equipment grounding conductors (EGCs). Clarification is provided to require that the provisions of 517.13(A) be met.

Revised 2008 NEC Text:

517.13 GROUNDING OF RECEPTACLES AND FIXED ELECTRICAL EQUIPMENT IN PATIENT CARE AREAS

(B) INSULATED EQUIPMENT GROUNDING CONDUCTOR. (Editorial revision only)

Exception No. 2: Luminaires more than 2.3 m (7½ ft.) above the floor and switches located outside of the patient vicinity shall ~~not be required to be grounded by an insulated equipment grounding conductor.~~ *be permitted to be connected to an equipment grounding return path complying with 517.13(A).*

Change Significance:

The previous permissive text of Exception No. 2 to 517.13 relieved the requirement for an "insulated equipment grounding conductor." This exception does not provide any relief from the provisions of 517.13(A). The wiring method used must qualify as an equipment grounding conductor based on the metal raceway system or a cable with a metallic armor meeting the requirements of 250.118. The revision of Exception No. 2 to 517.13(B) now clearly prohibits the use of type NM cable or the standard type MC cable with an interlocking metal jacket.

517.13(B)
INSULATED EQUIPMENT GROUNDING CONDUCTOR

Chapter 5 Special Occupancies

Article 517 Health Care Facilities

Part II Wiring and Protection

517.13(B) Insulated Equipment Grounding Conductor

Exception No. 2

Change Type: Revision

250.118 Types of Equipment Grounding Conductors
(2) Rigid metal conduit
(3) Intermediate metal conduit
(4) Electrical metallic tubing
(5) Listed flexible metal conduit meeting all the following conditions: *(four conditions must be met)*
(6) Listed liquidtight flexible metal conduit meeting all the following conditions: *(four conditions must be met)*
(7) Flexible metallic tubing where the tubing is terminated in fittings listed for grounding and meeting the following conditions: *(two conditions must be met)*
(8) Armor of type AC cable as provided in 320.108
(9) The copper sheath of mineral-insulated, metal-sheathed cable
(10) Type MC cable where listed and identified for grounding in accordance with the following: (A) The combined metallic sheath and grounding conductor of interlocked metal tape-type MC cable (B) The metallic sheath or the combined metallic sheath and grounding conductors of the smooth or corrugated tube-type MC cable
(13) Other listed electrically continuous metal raceways and listed auxiliary gutters
(14) Surface metal raceways listed for grounding

Table 250.118 provides the code user with a prescriptive list of recognized *equipment grounding conductors.*

517.32
LIFE SAFETY BRANCH

Chapter 5 Special Occupancies

Article 517 Health Care Facilities

Part III Essential Electrical System

517.32 Life Safety Branch

(C) Alarm and Alerting Systems

(E) Generator Set and Transfer Switch Locations

(F) Generator Set Accessories

Change Type: Revision and New

517.30 of the NEC outlines the essential electrical system. The life safety and critical branches of the essential electrical system are part of the emergency system.

Change Summary:

- Functions required to be supplied by the "life safety branch" are expanded to include HVAC controls, dampers, task illumination at transfer switch locations, and required generator accessories.

Revised 2008 NEC Text:

517.32 LIFE SAFETY BRANCH

(A) & (B)   (No change)

(C) ALARM AND ALERTING SYSTEMS.   Alarm and alerting systems including the following:
(1) & (2) (No change)
<u>(3) Mechanical, control, and other accessories required for effective life safety systems operation shall be permitted to be connected to the life safety branch.</u>

(D)   (No change)

(E) GENERATOR SET <u>AND TRANSFER SWITCH</u> LOCATIONS.   Task Illumination battery charger for battery-powered lighting units(s) and selected receptacles at the generator set <u>and essential transfer switch locations.</u> <u>[NFPA 99:4.4.2.2.2.2(5)]</u>

(F) GENERATOR SET ACCESSORIES.   <u>Generator set accessories as required for generator performance.</u>
(Rename existing first level subdivisions (F) and (G), as (G) and (H))

Change Significance:

517.32 lists the equipment, lighting, and receptacles required to be supplied by the "life safety branch" of the emergency portion of the essential electrical system addressed in Part III of Article 517. This revision now recognizes the need for supplying necessary loads from the "life safety branch" and the list of required loads is expanded.

517.32(C) is expanded to include mechanical and control functions that would include HVAC controls and required damper activation. 517.32(E) is expanded to include lighting at essential transfer switch locations to facilitate troubleshooting and repair of automatic transfer switches during a power outage. 517.32(F) is new and now requires that generator accessories (such as a day tank) necessary to generator performance be connected to the life safety branch.

Notes:

_____

_____

Change Summary:
* Requirements for delayed automatic and/or manual connection to an alternate source to the equipment and critical branch are clarified.

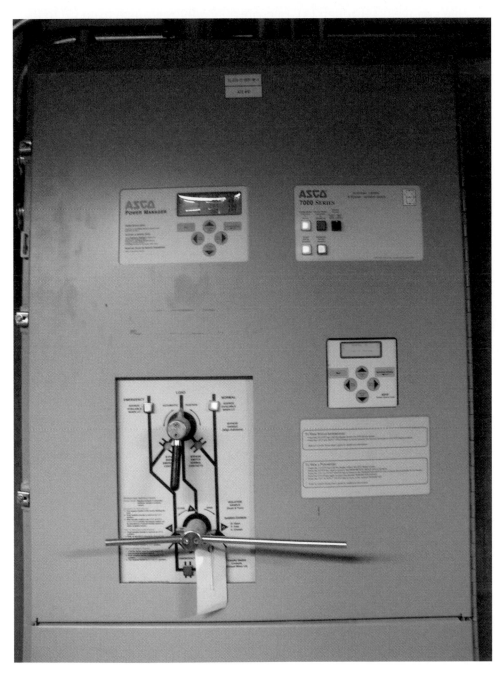

517.34
EQUIPMENT SYSTEM CONNECTION TO ALTERNATE POWER SOURCE

517.43
CONNECTION TO CRITICAL BRANCH

Chapter 5 Special Occupancies

Article 517 Health Care Facilities

Part III Essential Electrical System

517.34 Equipment System Connection to Alternate Power Source

(A) Equipment for Delayed Automatic Connection

(B) Equipment for Delayed Automatic or Manual Connection

517.43 Connection to Critical Branch

(A) Delayed Automatic Connection

(B) Delayed Automatic or Manual Connection

Change Type: Revision

An automatic transfer switch can be designed and installed to provide an automatic transfer of power, or a delayed transfer may be designed to allow the alternate source to come online.

(continued)

Revised 2008 NEC Text:

517.34 EQUIPMENT SYSTEM CONNECTION TO ALTERNATE POWER SOURCE

(A) EQUIPMENT FOR DELAYED AUTOMATIC CONNECTION.   The following equipment shall be <u>permitted to be</u> arranged for delayed automatic connection to the alternate power source:

(No change (1) through (4), relocate NFPA 99 reference from (5) to (6))

<u>(7) Supply, return, and exhaust ventilating systems for operating and delivery rooms.</u>

Exception: (No change)

(B) EQUIPMENT FOR DELAYED AUTOMATIC OR MANUAL CONNECTION.   The following equipment shall be <u>permitted to be</u> arranged for either delayed automatic or manual connection to the alternate power source:

(No change to exception or list items, add NFPA 99 reference to list item #8)

517.43 CONNECTION TO CRITICAL BRANCH

(A) DELAYED AUTOMATIC CONNECTION.   The following equipment shall be <u>permitted to be</u> connected to the critical branch and shall be arranged for delayed automatic connection to the alternate power source:

(No change (1) through (4), add NFPA 99 reference to (5))

(B) DELAYED AUTOMATIC OR MANUAL CONNECTION.   The following equipment shall be <u>permitted to be</u> connected to the critical branch and shall be arranged for either delayed automatic or manual connection to the alternate power source. <u>[NFPA 99: 4.5.2.2.34(A and B)]</u>

(No change (1) & (2), expand NFPA 99 reference to (3))

Change Significance:

The parent text of 517.34 requires that equipment described in 517.34(A) and (B) be automatically restored to operation by the alternate source. This revision clearly permits "delayed" or "automatic" connection to alternate sources for the equipment identified in 517.34(A) and (B).

The parent text of 517.43 provides requirements for connections to the critical branch of the essential electrical system. This text requires that equipment described in 517.43(A) and (B) be automatically restored to operation by the alternate source. This revision permits the listed equipment in 517.43(A) to be connected to the critical branch.

The parent text in 517.43 also requires that equipment listed in 517.43(B) be connected to the alternate source by delayed automatic or manual operation. This revision permits the equipment in 517.43(B) to be connected to the critical branch.

Modern standby generator systems can handle automatic or manual transfer of the full load when designed for the purpose. Requiring delayed connection is no longer necessary and may necessitate the need for multiple transfer switches when a single device is adequate. NFPA 99 references were added, expanded, and relocated throughout these sections. 517.34(A) now includes supply, return, and exhaust ventilating systems for operating and delivery rooms to ensure automatic transfer for these loads.

Change Summary:

- The term *inpatient hospital care* is deleted in 517.40(B). Clarification is provided by the qualifier of "patients who need to be sustained by electrical life support equipment."

Revised 2008 NEC Text:

517.40 ESSENTIAL ELECTRICAL SYSTEMS FOR NURSING HOMES AND LIMITED CARE FACILITIES

(B) INPATIENT HOSPITAL CARE FACILITIES. For those ~~N~~nursing homes and limited care facilities that ~~provide inpatient hospital care~~ <u>admit patients who need to be sustained by electrical life support equipment, the essential electrical system from the source to the portion of the facility where such patients are treated</u> shall comply with requirements of Part III, 517.30 through 517.35.

Change Significance:

This revision provides clarity for the classification of nursing homes and limited care facilities. 517.40(B) provides the threshold for when a nursing home or limited care facility needs to comply with the following sections.

517.30 ESSENTIAL ELECTRICAL SYSTEMS FOR HOSPITALS
517.31 EMERGENCY SYSTEM
517.32 LIFE SAFETY BRANCH
517.33 CRITICAL BRANCH
517.34 EQUIPMENT SYSTEM CONNECTION TO ALTERNATE POWER SOURCE
517.35 SOURCES OF POWER

517.40(B) now clearly requires that nursing homes and limited care facilities comply with 517.30 through 517.35 when those venues *admit patients who need to be sustained by electrical life support equipment.*

517.40(B)
INPATIENT HOSPITAL CARE FACILITIES

Chapter 5: Special Occupancies

Article 517 Health Care Facilities

Part III Essential Electrical System

517.40 Essential Electrical Systems for Nursing Homes and Limited Care Facilities

(B) Inpatient Hospital Care Facilities

Change Type: Revision

Nursing homes, limited care facilities, and hospices that do not provide patients with electrical life support equipment are not an *inpatient hospital care facility.*

517.80
PATIENT CARE AREAS

Chapter 5 Special Occupancies

Article 517 Health Care Facilities

Part VI Communications, Signaling Systems, Data Systems, Fire Alarm Systems, and Systems Less Than 120 Volts, Nominal

517.80 Patient Care Areas

Change Type: Revision

Change Summary:
- The installation of communications and signaling systems supplied by a transformer secondary below 120 volts is clarified. A raceway is not required unless specified in Chapter 7 or 8.

Revised 2008 NEC Text:
517.80 PATIENT CARE AREAS.   Equivalent insulation and isolation to that required for the electrical distribution systems in patient care areas shall be provided for communications, signaling systems, data system circuits, fire alarm systems, and systems less than 120 volts, nominal.

FPN: An acceptable alternate means of providing isolation for patient/nurse call systems is by the use of nonelectrified signaling, communications, or control devices held by the patient or within reach of the patient.

Secondary circuits of transformer-powered communications or signaling systems shall not be required to be enclosed in raceways unless otherwise specified by Chapters 7 or 8. [NFPA 99-2005, 4.4.2.2.4.6.]

Change Significance:
This revision provides clear direction to the code user with reference to the installation of low-voltage communications, fire alarm, or other signaling system conductors. The existing text seems to require a raceway for these Chapters 7 and 8 conductor installations. The new second paragraph clearly permits communications and signaling system conductors supplied by a transformer secondary below 120 volts to be installed without a raceway. Only where specified in Chapter 7 or 8 is a raceway required.

In a similar revision, 517.30(C)(3)(5) now permits "Secondary circuits of Class 2 or Class 3 communications or signaling systems with or without raceways."

Raceways are not required in patient care areas for communications or signaling circuits meeting the requirements of 517.80.

Change Summary:

*   Identification requirements for conductors in an isolated power system are revised to require distinctive colored stripes on the required brown, orange, and yellow conductors.

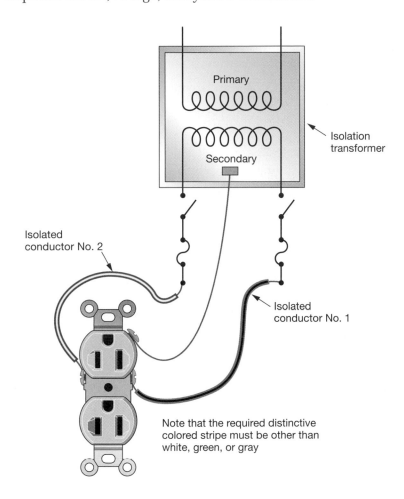

Primary

Secondary

Isolation transformer

Isolated conductor No. 2

Isolated conductor No. 1

Note that the required distinctive colored stripe must be other than white, green, or gray

517.160 (A)(5)
CONDUCTOR IDENTIFICATION

Chapter 5 Special Occupancies

Article 517 Health Care Facilities

Part VII Isolated Power Systems

517.160 Isolated Power Systems

(A) Installations

(5) Conductor Identification

Change Type: Revision

The installation of conductors for isolated power systems in a health care facility requires that a color code system in accordance with 517.160(A)(5) be implemented.

Revised 2008 NEC Text:
517.160 ISOLATED POWER SYSTEMS
(A) INSTALLATIONS
(5) CONDUCTOR IDENTIFICATION. The isolated circuit conductors shall be identified as follows:

(1) Isolated Conductor No. 1—Orange <u>with a distinctive colored stripe other than white, green or gray</u>

(2) Isolated Conductor No. 2—Brown <u>with a distinctive colored stripe other than white, green or gray</u>

For 3-phase systems, the third conductor shall be identified as yellow <u>with a distinctive colored stripe other than white, green or gray</u>. Where isolated circuit conductors supply 125-volt, single-phase, 15- and 20-ampere receptacles, the <u>striped</u> orange conductor(s) shall be connected to the

(continued)

terminal(s) on the receptacles that are identified in accordance with 200.10(B) for connection to the grounded circuit conductor.

Change Significance:

In the panel statement to accept this proposal in principle, the technical committee indicates concern over the clear identification of conductors in isolated power systems as follows: "The critical nature of these circuits demands that their conductors be identified in a manner that leaves no question as to the nature of their service."

The color orange is required in 110.15 to identify the "high leg" in a three-phase four-wire delta-connected system. The colors brown and orange are typically used for "A-phase, brown and B-phase, orange" in a three-phase four-wire wye-connected 277-/480-volt system. Due to the common use of the colors brown, orange, and yellow, this revision is designed to eliminate confusion between isolated power system conductors and those of another system. Colored stripes (other than white, green, or gray) on brown, orange, and (in three-phase applications) yellow conductors are now required for isolated power system conductors.

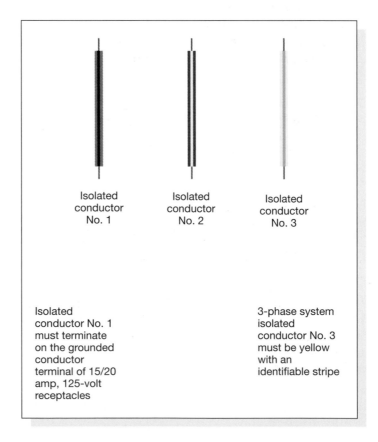

Isolated
conductor
No. 1

Isolated
conductor
No. 2

Isolated
conductor
No. 3

Isolated
conductor No. 1
must terminate
on the grounded
conductor
terminal of 15/20
amp, 125-volt
receptacles

3-phase system
isolated
conductor No. 3
must be yellow
with an
identifiable stripe

Isolated Conductor No. 1 is orange with a distinctive stripe other than white, green, or gray.
Isolated Conductor No. 2 is brown with a distinctive stripe other than white, green, or gray.
Isolated Conductor No. 3 (for three phase systems) is yellow with a distinctive stripe other than white, green or gray.

Change Summary:

- The general wiring methods required in an assembly occupancy are clarified to require that the wiring method itself qualify as an EGC in 250.118, or the wiring method must contain an insulated EGC.

Revised 2008 NEC Text:

518.4 WIRING METHODS

(A) GENERAL.   The fixed wiring methods shall be metal raceways, flexible metal raceways, nonmetallic raceways encased in not less than 50 mm (2 in.) of concrete, Type MI, MC, or AC cable. ~~containing~~ <u>The wiring method shall itself qualify as an equipment grounding conductor according to 250.118, or shall contain</u> an insulated equipment grounding conductor sized in accordance with Table 250.122.

Change Significance:

This revision now requires that wiring methods in an assembly occupancy qualify as an equipment grounding conductor in accordance with 250.118, or the wiring method is required to contain an insulated equipment grounding conductor. Type MI cable, for example, will qualify as an EGC if the outer sheath is copper. Stainless-steel-jacketed MI cable would be required to contain an insulated EGC. EMT and the jacket of AC cable are recognized as EGCs in 250.118 and an insulated EGC would not be required.

518.4(A)
GENERAL

Chapter 5 Special Occupancies

Article 518 Assembly Occupancies

518.4 Wiring Methods

(A) General

Change Type: Revision

All buildings or portions of buildings or structures designed or intended for the gathering together of 100 or more persons are considered by the NEC as an *assembly occupancy.*

518.5
SUPPLY

Chapter 5 Special Occupancies

Article 518 Assembly Occupancies

518.5 Supply

Change Type: Revision

Types of dimming systems include (1) solid-state sine wave three-phase four-wire and (2) solid-state phase control three-phase four-wire. Courtesy of Electronic Theatre Controls Inc.

Change Summary:

- This revision recognizes that there are two types of dimming systems available for use in assembly occupancies and addresses when the neutral is considered a current-carrying conductor.

Revised 2008 NEC Text:

518.5 SUPPLY.   Portable switchboards and portable power distribution equipment shall be supplied only from listed power outlets of sufficient voltage and ampere rating. Such power outlets shall be protected by overcurrent devices. Such overcurrent devices and power outlets shall not be accessible to the general public. Provisions for connection of an equipment grounding conductor shall be provided. The neutral <u>conductor</u> of feeders supplying solid-state <u>phase control</u>, 3-phase, 4-wire dimmer systems shall be considered a current-carrying conductor <u>for purposes of derating</u>. <u>The neutral conductors of feeders supplying solid-state sine wave, 3-phase, 4-wire dimming systems shall not be considered a current-carrying conductor for purposes of derating.</u>

    <u>*Exception: The neutral conductor of feeders supplying systems that use or may use both phase-control and sine-wave dimmers shall be considered as current-carrying for purposes of derating.*</u>

    <u>FPN: For definitions of solid state dimmer types, see 520.2.</u>

Change Significance:

This revision recognizes that there are two types of dimming systems available for use in assembly occupancies. Traditional "phase control" are nonlinear loads and the neutral conductor must be considered a current-carrying conductor. A new class of listed solid-state dimmer has been introduced to the professional performance lighting market: the solid-state sine wave dimmer.

    This type of dimmer varies the amplitude of the applied voltage wave form, without any of the nonlinear switching found in traditional phase-control solid-state dimmers. Because solid-state sine wave dimmers are linear loads, they do not require the neutral of the feeder to the dimmers to be considered a current-carrying conductor. An informational FPN is included to refer the code user to new definitions of solid-state dimmers in 520.2.

Notes:

_____

_____

_____

_____

_____

Change Summary:

- Two new definitions for dimmers are added to 520.2 for implementation of neutral conductor sizing.
- Neutral conductor sizing requirements are based on the type of solid-state dimmer employed.

Revised 2008 NEC Text:

520.2 DEFINITIONS

SOLID-STATE PHASE-CONTROL DIMMER.   A solid-state dimmer where the wave shape of the steady-state current does not follow the wave shape of the applied voltage, such that the wave shape is nonlinear.

SOLID-STATE SINE-WAVE DIMMER.   A solid-state dimmer where the wave shape of the steady-state current follows the wave shape of the applied voltage, such that the wave shape is linear.

520.27 STAGE SWITCHBOARD FEEDERS

(B) NEUTRAL CONDUCTOR.   For the purpose of derating, the following shall apply:

(1) The neutral conductor of feeders supplying solid-state, phase control 3-phase, 4-wire dimming systems shall be considered a current-carrying conductor.

(2) The neutral conductor of feeders supplying solid-state, sine wave 3-phase, 4-wire dimming systems shall not be considered a current-carrying conductor.

(3) The neutral conductor of feeders supplying systems that use or may use both phase-control and sine-wave dimmers shall be considered as current-carrying.

Change Significance:

These revisions include two new definitions for dimmers. A new type of listed solid-state dimmer has been introduced to the market: the solid-state sine wave dimmer. These devices are a linear load, which varies the amplitude of the applied voltage wave form without any of the nonlinear switching found in traditional phase control solid-state dimmers. Both types of dimmers are defined for clarity.

520.27(B) is revised to address the neutral conductor size where solid-state dimming systems are employed. The neutral conductor of a feeder that supplies any phase control dimmers will be considered current carrying. The neutral conductor of a feeder supplying dimmers of only the sine wave type is not considered current carrying. Similar revisions occurred in 520.51 and 520.53.

Notes:

_____

_____

520.2
DEFINITIONS

520.27(B)
NEUTRAL CONDUCTOR

Chapter 5 Special Occupancies

Article 520 Theaters, Audience Areas of Motion Picture and Television Studios, Performance Areas, and Similar Locations

Part I General

520.2 Definitions

Part II Fixed Stage Switchboards

520.27 Stage Switchboard Feeders

(B) Neutral Conductor

Change Type: Revision

Article 520 occupancies include theaters and the *audience* areas of motion picture and television studios.

Article 522
CONTROL SYSTEMS FOR PERMANENT AMUSEMENT ATTRACTIONS

Chapter 5 Special Occupancies

Article 522 Control Systems for Permanent Amusement Attractions

Part I General

Part II Control Circuits

Part III Control Circuit Wiring Methods

Change Type: New

Change Summary:
- A new Article 522 is added to address control systems for permanent amusement attractions.

Permanent amusement attractions may implement complex control systems now recognized in Article 522.

Revised 2008 NEC Text:
ARTICLE 522 CONTROL SYSTEMS FOR PERMANENT AMUSEMENT ATTRACTIONS
I. GENERAL
522.1 SCOPE.   This article covers the installation of control circuit power sources and control circuit conductors for electrical equipment, including associated control wiring in or on all structures, that are an integral part of a permanent amusement attraction.

522.2 DEFINITIONS
CONTROL CIRCUIT.   For the purposes of this article the circuit of a control system that carries the electric signals directing the performance of the controller but does not carry the main power current.

ENTERTAINMENT DEVICE.   A mechanical or electromechanical device that provides an entertainment experience.

FPN: These devices may include animated props, show action equipment, animated figures, and special effects, coordinated with audio and lighting to provide an entertainment experience.

(continued)

PERMANENT AMUSEMENT ATTRACTION. Ride devices, entertainment devices, or combination thereof, that are installed so that portability or relocation is impracticable.

RIDE DEVICE. A device or combination of devices that carry, convey, or direct a person(s) over or through a fixed or restricted course within a defined area for the primary purpose of amusement or entertainment.

522.3 OTHER ARTICLES
522.5 VOLTAGE LIMITATIONS. *Maximum 150-volts AC to ground or 300-volts DC*

522.7 MAINTENANCE. *Only "qualified persons" may service attractions*

II. CONTROL CIRCUITS
522.10 POWER SOURCES FOR CONTROL CIRCUITS. *(A) Power-Limited Control Circuits & (B) Non-Power-Limited Control Circuits*

III. CONTROL CIRCUIT WIRING METHODS
522.20 CONDUCTORS, BUSBARS, AND SLIP RINGS
522 21 CONDUCTOR SIZING. *30 AWG or larger as listed component or assembly*

522.22 CONDUCTOR AMPACITY. *See Table 522.22*

522 23 OVERCURRENT PROTECTION FOR CONDUCTORS
522.24 CONDUCTORS OF DIFFERENT CIRCUITS IN THE SAME CABLE, CABLE TRAY, ENCLOSURE, OR RACEWAY. *(A) Two or More Control Circuits, (B) Control Circuits with Power Circuits*

522.25 UNGROUNDED CONTROL CIRCUITS
522.28 CONTROL CIRCUITS IN WET LOCATIONS

Change Significance:

As outlined in the scope of Article 522, Control Systems for Permanent Amusement Attractions, this article addresses only (1) control circuit power sources and (2) control circuit conductors for electrical equipment, including associated control wiring in or on all structures that are an integral part of a permanent amusement attraction. The permanent amusement attraction industry has seen a tremendous amount of growth over the last two decades and this new article is designed to increase the safety of the attractions and to provide consistent installation methods and enforcement.

This revision will permit complex control systems for permanent amusement attractions where control reliability principles are employed. The current wiring methods of the NEC limit or prevent the use of newer microdevices, connectors, and computer-based technologies used for the increased monitoring, verification, redundancy, and diagnostics of the apparatus or amusement attraction under control.

525.2
DEFINITIONS

Chapter 5 Special Occupancies

Article 525 Carnivals, Circuses, Fairs, and Similar Events

Part I General Requirements

520.2 Definitions

Change Type: New

Change Summary:
- The term *operator* is now defined to clarify that this person is responsible for control of the ride or concession.
- The term *portable structure* is now defined to clarify equipment within the scope of Article 525.

Revised 2008 NEC Text:
525.2 DEFINITIONS

OPERATOR.   The individual responsible for starting, stopping, and controlling an amusement ride or supervising a concession.

PORTABLE STRUCTURES.   Units designed to be moved including, but not limited to, amusement rides, attractions, concessions, tents, trailers, trucks, and similar units.

Change Significance:
The term *operator* is sometimes used for the owner of amusement rides or concessions. This new definition in 520.2 provides clear guidance to the code user where the term *operator* is used in Article 525. This is the person responsible for starting, stopping, and controlling an amusement ride or supervising a concession.

525.3(B) clearly explains to the code user that "permanent structures" are addressed in Articles 518 and 520. Defining the term *portable structure* provides clarity by covering all "units designed to be moved." Everything "moved" into the field, parking lot, farm, schoolyard, or other area is included. Additional text within the definition specifically recognizes but is not limited to amusement rides, attractions, concessions, tents, trailers, trucks, and similar units. The term *portable structure* has been added to multiple sections throughout Article 525 for clarity.

Carnivals, circuses, fairs, and similar events are portable in nature, which makes them special occupancies addressed in Article 525.

Change Summary:

* The term *portable structure* is added throughout Article 525 for clarity.

Revised 2008 NEC Text:

525.5 OVERHEAD CONDUCTOR CLEARANCES
(B) CLEARANCE TO ~~RIDES AND ATTRACTIONS~~ PORTABLE STRUCTURES
(1) UNDER 600 VOLTS. ~~Amusement rides and amusement attractions~~ Portable structures shall be maintained not less than 4.5 m (15 ft) in any direction from overhead conductors operating at 600 volts or less, except for the conductors supplying the <u>portable structure</u> ~~amusement ride or attraction.~~ <u>Portable structures included in 525.3(D) shall comply with Article 680, Table 680.8.</u>

(2) OVER 600 VOLTS. ~~Amusement rides or attractions~~ <u>Portable structures</u> shall not be located under or within 4.5 m (15 ft) horizontally of conductors operating in excess of 600 volts.

525.6 PROTECTION OF ELECTRICAL EQUIPMENT.   Electrical equipment and wiring methods in or on ~~rides, concessions, or other units~~ <u>portable structures</u> shall be provided with mechanical protection where such equipment or wiring methods are subject to physical damage.

525.11 MULTIPLE SOURCES OF SUPPLY.   Where multiple services or separately derived systems, or both supply ~~rides, attractions, and~~ portable ~~other~~ structures, <u>the equipment grounding conductors of</u> all <u>the</u> sources of supply that serve ~~rides, attractions, or other~~ <u>such</u> structures separated by less than 3.7 m (12 ft) shall be bonded ~~to the same grounding electrode system~~ <u>together at the portable structures. The bonding conductor shall be copper and sized in accordance with Table 250.122 based on the largest overcurrent device supplying the portable structures, but not smaller than No. 6 AWG.</u>

525.21 RIDES, TENTS, AND CONCESSIONS
(A) DISCONNECTING MEANS.   Each ~~ride and concession~~ <u>portable structure</u> shall be provided with a ~~fused~~ disconnect switch ~~or circuit breaker~~ located within sight <u>of</u> and within 1.8 m (6 ft) of the operator's station. The disconnecting means shall be readily accessible to the operator, including when the ride is in operation. Where accessible to unqualified persons, the enclosure for the switch or circuit breaker shall be of the lockable type. A shunt trip device that opens the fused disconnect or circuit breaker when a switch located in the ride operator's console is closed shall be a permissible method of opening the circuit.

525.30 EQUIPMENT BONDING.   The following equipment connected to the same source shall be bonded:
(1) & (2) (No change)
(3) Metal frames and metal parts of ~~rides, concessions, tents~~ <u>portable structures,</u> trailers, trucks, or other equipment that contain or support electrical equipment

(continued)

525.5, 525.6, 525.11, 525.21, and 525.30

Change Type: Revision

The equipment grounding conductor of the circuit supplying the equipment in items (1), (2) or (3) that is likely to energize the metal frame or part shall be permitted to serve as the bonding means.

Change Significance:

These revisions are implemented along with a new definition in 525.2 of *portable structure.* The terms *ride, attraction, concession,* and *tent* are deleted and replaced with *portable structure.* This definition is all-encompassing, covering every "unit designed to be moved."

525.21(A) is revised to require an operator disconnect only. 525.5(B) is separated for clarity and usability in two second-level subdivisions: (1) Under 600 Volts and (2) Over 600 Volts.

Portable structures (such as rides, tents, and concessions) require special consideration of clearances, grounding, and bonding.

Change Summary:

- Revisions have been made in Article 547 to clarify that (1) a distribution point is permitted to supply any building or structure and (2) electrical installations in buildings or structures not addressed in 547.1 must conform to applicable articles in the NEC.

Revised 2008 NEC Text:

547.2 DEFINITIONS

DISTRIBUTION POINT.   An electrical supply point from which service drops, service <u>conductors</u> ~~laterals~~, feeders, or branch circuits to ~~agricultural buildings, associated farm dwelling(s), and associated~~ buildings <u>or structures utilized</u> under single management are supplied.

547.3 OTHER ARTICLES.

For ~~agricultural~~ buildings <u>and structures</u> not having conditions as specified in 547.1, the electrical installations shall be made in accordance with the applicable articles in this Code.

547.9 ELECTRICAL SUPPLY TO BUILDING(S) OR STRUCTURE(S) FROM A DISTRIBUTION POINT.

<u>A distribution point shall be permitted to supply any building or structure located on the same premises. The</u> ~~O~~<u>o</u>verhead electrical supply shall comply with 547.9(A) and 547.9(B), or with 547.9(C). <u>The</u> ~~U~~<u>u</u>nderground electrical supply shall comply with 547.9(C) and 547.9(D).

(A) SITE-ISOLATING DEVICE

(1) WHERE REQUIRED.   A site-isolating device shall be installed at the distribution point where two or more agricultural buildings, <u>or structures</u>~~,~~ ~~associated farm dwelling(s), or other buildings~~ are supplied from the distribution point.

Change Significance:

The definition of *distribution point* is clarified by removing specific types of buildings or structures and qualifying that this is a point from which "buildings and structures" under single management are supplied. The parent text of 547.9 is revised with a new first sentence. This new text clearly permits a distribution point to serve any building or structure located on the same premises.

547.3 is revised to clarify that electrical installations for buildings and structures that do not meet the requirements of 547.1 must comply with other applicable articles in the NEC. Together, these revisions clarify that (1) a "distribution point" may supply more than just agricultural buildings/structures (such as a dwelling unit), as outlined in 547.1; and (2) buildings/structures not meeting the requirements of 547.1 must comply with other applicable articles of the NEC for their electrical installations.

Notes:

_____

_____

547.2, 547.3, and 547.9

Chapter 5 Special Occupancies

Article 547 Agricultural Buildings

547.2 Definitions

547.3 Other Articles

547.9 Electrical Supply to Building(s) or Structure(s) from a Distribution Point

(A) Site-Isolating Device

(1) Where Required

Change Type: Revision

Agricultural buildings may be effected by excessive dust, water, and corrosive atmospheres created by feed, litter, and other materials where poultry, livestock, or fish are involved.

547.9
ELECTRICAL SUPPLY TO BUILDING(S) OR STRUCTURE(S) FROM A DISTRIBUTION POINT

Chapter 5 Special Occupancies

Article 547 Agricultural Buildings

547.9 Electrical Supply to Building(s) or Structure(s) from a Distribution Point

(A) Site-Isolating Device

(10) Marking

(E) Identification

Change Type: New

Change Summary:
- All site-isolating devices must be permanently marked as such and marking must be located on the operating handle or immediately adjacent.
- Where a site is supplied by more than one service with any two services located a distance of 150 m (500 feet) or less apart as measured in a straight line, a permanent plaque or directory is required.

Revised 2008 NEC Text:
547.9 ELECTRICAL SUPPLY TO BUILDING(S) OR STRUCTURE(S) FROM A DISTRIBUTION POINT
(A) SITE-ISOLATING DEVICE
(10) MARKING.   A site-isolating device shall be permanently marked to identify it as a site-isolating device. This marking shall be located on the operating handle or immediately adjacent thereto.

(E) IDENTIFICATION.   Where a site is supplied by more than one service with any two services located a distance of 150 m (500 ft) or less apart, as measured in a straight line, a permanent plaque or directory shall be installed at each of these distribution points denoting the location of each of the other distribution points and the buildings or structures served by each.

Change Significance:
Two new marking requirements have been added to 547.9 where an electrical supply to a building or structure is fed from a distribution point. A site-isolating device is defined in 547.2 as disconnecting means installed at the distribution point for the purpose of isolation, system maintenance, emergency disconnection, or connection of optional standby systems. 547.9(A)(10) now requires all site-isolating devices to be *permanently marked* to identify them as site-isolating devices. Marking is required on the operating handle or immediately adjacent to the site-isolating device.

A new first-level subdivision 547.9(E) requires that where two services supply a site and are located 500 feet or less apart they must be identified in a manner similar to the marking requirements of 230.2(E). A permanent plaque or directory is now required at each distribution point denoting the location of all the other distribution points and the buildings or structures served by each.

The electrical distribution on a farm or other area with agricultural buildings is installed in accordance with the provisions of Article 547.

Change Summary:

- The use of a grounded circuit conductor as a fault return path when supplying a building or structure is now prohibited in other than existing buildings or structures.

The wiring of multiple buildings or structures in an agricultural occupancy must be in accordance with 250.32.

547.9(B)(3)
GROUNDING AND BONDING

Chapter 5 Special Occupancies

Article 547 Agricultural Buildings

547.9 Electrical Supply to Building(s) or Structure(s) From a Distribution Point

(B) Service Disconnecting Means and Overcurrent Protection at the Building(s) or Structure(s)

(3) Grounding and Bonding

Change Type: Revision

Revised 2008 NEC Text:

547.9 ELECTRICAL SUPPLY TO BUILDING(S) OR STRUCTURE(S) FROM A DISTRIBUTION POINT
(B) SERVICE DISCONNECTING MEANS AND OVERCURRENT PROTECTION AT THE BUILDING(S) OR STRUCTURE(S). ~~(a) System with grounded neutral conductor.~~ (Deleted)

~~FPN:~~ (Deleted)

~~(b) System with separate equipment grounding conductor.~~ (Deleted)
~~FPN:~~ (Deleted)

(3) GROUNDING AND BONDING. For each building or structure, <u>grounding and bonding of the supply conductors shall be in accordance with the requirements of 250.32</u>, and the following conditions shall be met:

(1) The equipment grounding conductor shall be the same size as the largest supply conductor if of the same material, or adjusted in size in accordance with the equivalent size columns of Table 250.122 if of different materials.

(2) The equipment grounding conductor is bonded to the grounded circuit conductor and the site-isolating device at the distribution point.
(List items (3) and (4) are deleted)

(continued)

Change Significance:

This revision will now require that all supply conductors to buildings or structures include an EGC. The previous text of 547.9(B)(3) permitted the grounded conductor of the supply to be connected to a building/structure disconnecting means and to the grounding electrode system in accordance with 250.32(B)(2). This means of providing a ground-fault return path is eliminated and an EGC must be installed.

It is important to note that the reference to 250.32 will allow use of the exception for existing premises wiring systems. The intent of this change is to eliminate situations in which the grounded conductor becomes bonded at multiple locations, allowing normal and fault current to flow on buildings, structures, equipment or through the earth—creating serious safety concerns.

547.9(B)(3) list items (3) and (4) are eliminated because these requirements are covered in 250.32. The increased requirement for the equipment grounding conductor remains in 547.9(B)(3)(1), and requirements for bonding to the site-isolating device remain in 547.9(B)(3)(2).

All structures in which poultry, livestock, and fish confinement systems are contained will accumulate litter dust or feed dust including mineral feed particles and are considered as agricultural buildings.

Change Summary:
* The use of a grounded circuit conductor as a fault return path in a feeder supplying mobile or manufactured homes is now prohibited.

Revised 2008 NEC Text:
550.33 FEEDER

(A) FEEDER CONDUCTORS. <u>Feeder conductors shall comply with the following:</u>

<u>(1)</u> Feeder conductors shall consist of either a listed cord, factory installed in accordance with 550.10(B), or a permanently installed feeder consisting of four insulated, color-coded conductors that shall be identified by the factory or field marking of the conductors in compliance with 310.12. Equipment grounding conductors shall not be identified by stripping the insulation.

<u>(2) Feeder conductors shall be installed in compliance with 250.32(B).</u>

~~Exception: Where a feeder is installed between service equipment and a disconnecting means as covered in 550.32(A), it shall be permitted to omit the equipment grounding conductor where the grounded circuit conductor is grounded at the disconnecting means as required in 250.32(B).~~

550.33(A)
FEEDER CONDUCTORS

Chapter 5 Special Occupancies

Article 550 Mobile Homes, Manufactured Homes, and Mobile Home Parks

Part III Services and Feeders

550.33 Feeder

(A) Feeder Conductors

Change Type: Revision

Change Significance:
This revision will now require that all feeder conductors supplying mobile or manufactured homes include an EGC. The exception now deleted previously permitted the grounded conductor of the feeder to be connected to the disconnecting means in a mobile or manufactured home.

The intent of this change is to eliminate situations in which the grounded conductor becomes bonded at multiple locations, allowing normal and fault current to flow on mobile homes, manufactured homes, buildings, structures, and equipment or through the earth—creating serious safety concerns. The separation of requirements in two list items is user friendly. List item (2) now clearly requires installation in accordance with 250.32 and includes an exception for existing buildings/structures.

A mobile home is a *special occupancy* that must meet the requirements of Article 550.

551.4
GENERAL REQUIREMENTS

Chapter 5 Special Occupancies

Article 551 Recreational Vehicles and Recreational Vehicle Parks

Part I General

551.4 General Requirements

(A) Not Covered

(B) Systems

Change Type: Revision

Change Summary:
- Recreational vehicles and recreational vehicle parks are now permitted to be supplied by and to utilize power at 208Y/120 volts in addition to 120/240 volts.

Revised 2008 NEC Text:
551.4 GENERAL REQUIREMENTS

(A) NOT COVERED.   A recreational vehicle not used for the purposes as defined in 551.2 shall not be required to meet the provisions of Part I pertaining to the number or capacity of circuits required. It shall, however, meet all other applicable requirements of this article if the recreational vehicle is provided with an electrical installation intended to be energized from a 120-<u>volt, 208Y/120-volt</u> or 120/240-volt, nominal, ac power-supply system.

(B) SYSTEMS.   This article covers combination electrical systems, generator installations, and 120-<u>volt, 208Y/120</u> or 120/240-volt, nominal, systems.

Change Significance:
Due to capacity requirements, large recreational vehicle parks are typically served from a three-phase four-wire wye-connected system at 208/120 volts. This revision will now allow electrical supply and utilizations at single-phase 120/240 volts or three-phase 208Y/120 volts. The present text of the NEC requires that recreational vehicle parks install transformers to supply power at 120/240 volts.

Many recreational vehicles are equipped with larger loads, such as air conditioners, capable of being connected at 208 or 240 volts. Multiple revisions have occurred in Article 551 to expand permitted power supply and utilization to include 208Y/120 volts in Sections 551.20, 551.31, 551.40, 551.42, 551.44, and 551.46.

A recreational vehicle is a vehicular-type unit primarily designed as temporary living quarters for recreational, camping, or travel use, which either has its own motive power or is mounted on or drawn by another vehicle.

Change Summary:

- This revision permits conductor splices above the deck of a floating pier, above the waterline but below the electrical datum field where sealed wire connector systems and approved junction boxes are employed.

Revised 2008 NEC Text:

555.9 ELECTRICAL CONNECTIONS.   ~~All e~~Electrical connections shall be located at least 305 mm (12 in.) above the deck of a floating pier. <u>Conductor splices, within approved junction boxes, utilizing sealed wire connector systems listed and identified for submersion shall be permitted where located above the waterline, but below the electrical datum field for floating piers.</u>

All electrical connections shall be located at least 305 mm (12 in.) above the deck of a fixed pier but not below the electrical datum plane.

Change Significance:

Electrical connections on a floating pier are generally required to be located 12 inches above the deck of the floating pier. This revision permits conductor splices above the waterline but below the electrical datum field where sealed wire connector systems and approved junction boxes are employed. This will allow splices to be made just above the water line. Note that the sealed wire connector systems permitted by this revision must be listed and identified for "submersion."

Sealed wire connector systems listed for the application, such as salt water exposure, are available in NEMA Type 6P enclosures described as follows:

NEMA Enclosure Type Designation "6P." Indoor or outdoor use primarily to provide a degree of protection against hose-directed water, the entry of water during prolonged submersion at a limited depth and damage from external ice formation.

555.9
ELECTRICAL
CONNECTIONS

Chapter 5 Special Occupancies

Article 555 Marinas and Boatyards

555.9 Electrical Connections

Change Type: Revision

The electrical datum plane is defined in 555.2 and is different for land areas subject to tidal fluctuation, those not subject to fluctuation, and floating piers.

555.21

MOTOR FUEL DISPENSING STATIONS— HAZARDOUS (CLASSIFIED) LOCATIONS

Change Type: Revision and New

Change Summary:
- Requirements for electrical wiring and equipment supplying motor-fuel-dispensing stations on or near floating/fixed docks and piers are revised to provide clear direction for the code user to determine area classifications.

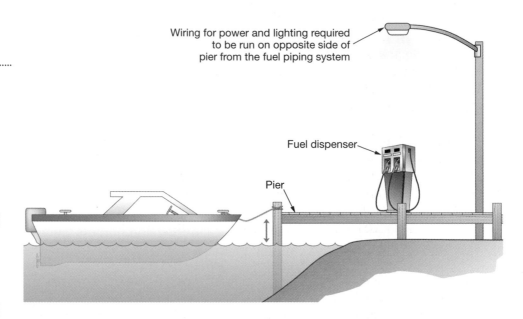

Wiring for power and lighting required to be run on opposite side of pier from the fuel piping system

Fuel dispenser

Pier

Introducing motor fuel to a marina installation requires that multiple special requirements be combined in Article 555.

Revised 2008 NEC Text:
555.21 MOTOR FUEL DISPENSING STATIONS—HAZARDOUS (CLASSIFIED) LOCATIONS
(A) GENERAL. Electrical wiring and equipment located at or serving motor fuel dispensing <u>locations</u> ~~stations~~ shall comply with Article 514 in addition to the requirements of this article. All electrical wiring for power and lighting shall be installed on the side of the wharf, pier, or dock opposite from the liquid piping system.

FPN: For additional information, see NFPA 303-2000, *Fire Protection Standard for Marinas and Boatyards,* and NFPA 30A-2003, *Motor Fuel Dispensing Facilities and Repair Garages.*

(B) CLASSIFICATION OF CLASS I, DIVISION 1 AND 2 AREAS. <u>The following criteria shall be used for the purposes of applying Table 514.3(B)(1) and 514.3(B)(2) to motor fuel dispensing equipment on floating or fixed piers, wharfs or docks.</u>

(1) CLOSED CONSTRUCTION. <u>Where the construction of floating docks, piers or wharfs is closed so that there is no space between the bottom of the dock, pier or wharf and the water, such as concrete enclosed expanded foam or similar construction, and having integral service boxes with supply chases:</u>

(continued)

(a) The space above the surface of the floating dock, pier, or wharf shall be a Class I, Division 2 location with distances as identified in Table 514.3(B)(1) Dispenser and Outdoor.

(b) The space below the surface of the floating dock, pier or wharf having areas or enclosures such as tubs, voids, pits, vaults, boxes, depressions, fuel piping chases, or similar spaces where flammable liquid or vapor can accumulate shall be a Class I, Division 1 location.

Exception No. 1: Dock, pier, or wharf sections that do not support fuel dispensers and abut but are 6.0 m (20 ft) or more from dock sections that support fuel dispenser(s) shall be permitted to be Class I, Division 2 where documented air space is provided between dock sections to permit flammable liquids or vapors to dissipate and not travel to these dock sections. Such documentation shall comply with 500.4(A)

Exception No. 2: Dock, pier, or wharf sections that do not support fuel dispensers and do not directly abut sections that support fuel dispensers shall be permitted to be unclassified where documented air space is provided and where flammable liquids or vapors can not travel to these dock sections. Such documentation shall comply with 500.4(A).

FPN: See 500.4(A) for documentation requirements.

(2) OPEN CONSTRUCTION.   Where the construction of floating piers, wharfs or docks is open, such as decks built on stringers supported by pilings, floats, pontoons or similar construction:

(a) The area 450 mm (18 in) above the surface of the dock, pier or wharf and extending 6.0 m (20 ft) horizontally in all directions from the outside edge of the dispenser and down to the water level shall be Class 1 Division 2.

(b) Enclosures such as tubs, voids, pits, vaults, boxes, depressions, piping chases, or similar spaces where flammable liquids or vapors can accumulate within 6.0 m (20 ft) of the dispenser shall be a Class I, Division 1 location.

Change Significance:

The previous requirement of this section was simply to apply the rules in Article 514 along with Article 555. Motor-fuel-dispensing stations on floating/fixed docks and piers are unique installations, and the construction of these locations and installation methods must be considered when classifying areas.

This revision separates rules into two first-level subdivisions: (A) General and (B) Classification of Class I, Divisions 1 and 2. This revision provides specific prescriptive requirements in Article 555 that correlate with the requirements of Article 514.

Notes:

_____

_____

590.4(D)
RECEPTACLES

Chapter 5 Special Occupancies

Article 590 Temporary Installations

590.4 General

(D) Receptacles

Change Type: Revision

Temporary installations for construction purposes are permitted to remain in place "during the period of construction."

Change Summary:

- Where metal raceways and metal-covered cable assemblies are used as an EGC, in a temporary installation to supply receptacles the metal raceway or metal sheath of the cable must be listed in 250.118 as an EGC. The metal raceway or metal sheath listed in 250.118 must be continuous from the source to the receptacle or a separate EGC shall be installed.

Revised 2008 NEC Text:

590.4 GENERAL

(D) RECEPTACLES.   All receptacles shall be of the grounding type. Unless installed in a continuous ~~grounded~~ metal raceway <u>that qualifies as an equipment grounding conductor in accordance with 250.118</u> or <u>a continuous</u> metal-covered cable <u>that qualifies as an equipment grounding conductor in accordance with 250.118</u>, all branch circuits shall ~~contain~~ <u>include</u> a separate equipment grounding conductor, and all receptacles shall be electrically connected to the equipment grounding conductors. Receptacles on construction sites shall not be installed on branch circuits that supply temporary lighting. Receptacles shall not be connected to the same ungrounded conductor of multiwire circuits that supply temporary lighting.

Change Significance:

The qualifying term *grounded* used to describe the metal raceway or metal-covered cable has been deleted. The requirement has been clarified by mandating that where the metal raceway or metal-covered cable is used as an EGC the raceway or cable shall be listed as such in 250.118. The metal raceway or metal sheath listed in 250.118 must be continuous from the source to the receptacle or a separate EGC shall be installed.

The most popular wiring method for branch circuits in temporary installations is type NM cable. There are many reasons NM is used over other wiring methods. Type NM is less expensive to purchase and to install because splices do not require termination in a junction box (see 590.4(G)) and the OSHA construction standard CFR 29 1926.405(a)(2)(ii)(H)].

OSHA standards require a junction box (and cover, 1926.405(b)(2)) for all splices and terminations where a raceway, metal-clad, or metal-sheathed cable is used in temporary wiring. This prohibits the use of metal-clad cables such as AC or MC from being used in temporary installations unless all terminations and splices are installed with fittings in junction boxes or enclosures. This requirement exists to prevent an ungrounded conductor from energizing the sheath of a metal-clad cable between two splice points, thereby creating a serious shock hazard.

Notes:

_____

_____

_____

Change Summary:
* The safety-driven provisions for GFCI protection for personnel in 590.6 have been clarified to apply to power derived from an electric utility company or from an onsite-generated power source.

Revised 2008 NEC Text:
590.6 GROUND-FAULT PROTECTION FOR PERSONNEL. Ground-fault protection for personnel for all temporary wiring installations shall be provided to comply with 590.6(A) and 590.6(B). This section shall apply only to temporary wiring installations used to supply temporary power to equipment used by personnel during construction, remodeling, maintenance, repair, or demolition of buildings, structures, equipment, or similar activities. This section shall apply to power derived from an electric utility company or from an onsite generated power source.

Change Significance:
This revision clarifies that all 125-volt single-phase 15-, 20-, and 30-ampere receptacle outlets that are not a part of the permanent wiring of the building or structure and that are in use by personnel shall have ground-fault circuit interrupter protection for personnel. This GFCI requirement exists for all temporary installations, regardless of whether a service is installed or a generator is used. The shock hazards are the same for personnel where temporary power is derived from a utility source (service) or an onsite generator.

It is important to note that in accordance with 590.6(A) where a receptacle is part of the permanent wiring and is in use by construction personnel GFCI shall be provided as follows: "If a receptacle(s) is installed or exists as part of the permanent wiring of the building or structure and is used for temporary electric power, ground-fault circuit-interrupter protection for personnel shall be provided. For the purposes of this section, cord sets or devices incorporating listed ground-fault circuit interrupter protection for personnel identified for portable use shall be permitted."

590.6 GROUND-FAULT PROTECTION FOR PERSONNEL

Chapter 5 Special Occupancies

Article 590 Temporary Installations

590.6 Ground-Fault Protection for Personnel

Change Type: Revision

Current Flow Through the Human Body	
Sensation in men	.4 mA
Muscle contraction and pain	3 mA
"Let Go" threshold	9 mA
Respiratory paralysis	30-75 mA
Heart fibrillation	100-200 mA
Tissue and organ burn	1500 mA

Class A ground-fault circuit interrupters trip when the current to ground has a value in the range of 4 mA to 6 mA. For further information, see UL 943, *Standard for Ground-Fault Circuit Interrupters.*

Three factors that determine the severity of injury from electrical shock are the path through the body, the amount of current, and the amount of time that current flows.

6

Special Equipment, Articles 600–695

Change Summary:

- The definition of *section sign* is modified to clarify that it is made up of subassemblies, which are either physically joined to form a single sign unit or installed as separate remote parts of an overall sign.
- A new first-level subdivision is added to 600.4 for the marking of section signs. The requirement includes markings to indicate that field wiring and installation instructions are required.

Revised 2008 NEC Text:

600.2 DEFINITIONS

SECTION SIGN.   A sign or outline lighting system, shipped as subassemblies, that requires field-installed wiring between the subassemblies to complete the overall sign. <u>The subassemblies are either physically joined to form a single sign unit or are installed as separate remote parts of an overall sign.</u>

600.4 MARKINGS

(C) SECTION SIGNS.   <u>Section signs shall be marked to indicate that field wiring and installation instructions are required.</u>

Change Significance:

This revision provides a more in-depth definition of *section sign*. The definition is revised to clarify that the subassemblies are either physically joined to form a single sign unit or installed as separate remote parts of an overall sign.

The installation of section signs is further clarified by a new first-level subdivision for required markings on the subassemblies. The required markings will inform the code user that field wiring is required and that installation instructions are provided.

600.2
DEFINITION OF
SECTION SIGN

600.4(C)
SECTION SIGNS

Chapter 6 Special Equipment

Article 600 Electric Signs and Outline Lighting

Part I General

600.2 Definition, Section Sign

600.4 Markings

(C) Section Signs

Change Type: Revision and New

Article 600 contains six definitions, including *section sign*, critical to the proper application of the article.

600.6
DISCONNECTS

Chapter 6 Special Equipment

Article 600 Electric Signs and Outline Lighting

Part I General

600.6 Disconnects

(A) Location

(1) Within Sight of the Sign

(2) Within Sight of the Controller

Change Type: Revision

A disconnecting means is required to allow persons to service and maintain signs.

Change Summary:
• The provisions for locking the disconnect required in 600.6 for signs and outline lighting systems in the open position must remain in place with or without the lock installed.

Revised 2008 NEC Text:
600.6 DISCONNECTS
(A) LOCATION
(1) WITHIN SIGHT OF THE SIGN.   The disconnecting means shall be within sight of the sign or outline lighting system that it controls. Where the disconnecting means is out of the line of sight from any section that is able to be energized, the disconnecting means shall be capable of being locked in the open position. The provision for locking or adding a lock to the disconnecting means must remain in place at the switch or circuit breaker whether the lock is installed or not. Portable means for adding a lock to the switch or circuit breaker shall not be permitted.

(2) WITHIN SIGHT OF THE CONTROLLER.   The following shall apply for signs or outline lighting systems operated by electronic or electromechanical controllers located external to the sign or outline lighting system:
(1) The disconnecting means shall be permitted to be located within sight of the controller or in the same enclosure with the controller.
(2) The disconnecting means shall disconnect the sign or outline lighting system and the controller from all ungrounded supply conductors.
(3) The disconnecting means shall be designed such that no pole can be operated independently and shall be capable of being locked in the open position. The provisions for locking or adding a lock to the disconnecting means must remain in place at the switch or circuit breaker whether the lock is installed or not. Portable means for adding a lock to the switch or circuit breaker shall not be permitted.

Change Significance:
This is another safety-driven revision requiring that the provisions for locking a disconnect means in the open position remain in place with or without the lock installed. This revision requires that all disconnects required in 600.6(A), where a disconnect is within sight of the sign, and 600.6(A)(2), where a disconnect is within sight of the controller, must be capable of being rendered inoperable with nothing more than a lock.

Where the NEC requires that a disconnecting means be capable of being locked in the open position it is for a single reason: the safety of persons. Installer/maintainers should not rely on having dozens of portable devices and hope that one fits the switch or circuit breaker in a given installation. Circuit breakers used as disconnects will require an identified accessory device, readily available from the manufacturer to be installed. A new last sentence clarifies that portable means of adding the lock is not permitted.

Change Summary:

- The requirements of 600.7 have been clarified by separation of grounding and bonding requirements into two first-level subdivisions. These requirements are further clarified by separating grounding and bonding requirements into second-level subdivisions.

The installation of signs mandates modifications and supplemental requirements in addition to the general rules of Chapters 1 through 4.

Revised 2008 NEC Text:

600.7 GROUNDING AND BONDING
(A) GROUNDING
(1) EQUIPMENT GROUNDING. (Revised to require connection to the equipment grounding conductor of a type specified in 250.118)

Exception: (A new exception permits portable cord-connected signs, where protected by a system of double insulation or its equivalent to not be connected to an equipment grounding conductor)

(2) SIZE OF EQUIPMENT GROUNDING CONDUCTOR. (Requires the EGC be sized in accordance with 250.122

(3) CONNECTIONS. (Requires connections in accordance with 250.130 & 250.8)

(4) SUPPLEMENTARY GROUNDING ELECTRODE. (Permits the use of a supplementary grounding electrode where installed in accordance with 250.54)

(continued)

600.7
GROUNDING AND BONDING

Chapter 6 Special Equipment

Article 600 Electric Signs and Outline Lighting

Part I General

600.7 Grounding and Bonding

(A) Grounding

(1) Equipment Grounding

(2) Size of Equipment Grounding Conductor

(3) Connections

(4) Supplementary Grounding Electrode

(5) Metal Building Parts

(B) Bonding

(1) Bonding of Metal Parts

(2) Bonding Connections

(3) Metal Building Parts

(4) Flexible Metal Conduit Length

(5) Small Metal Parts

(6) Nonmetallic Conduit

(7) Bonding Conductors

(8) Signs in Fountains

Change Type: Revision

(5) METAL BUILDING PARTS. (Prohibits building parts to be used as return conductor or EGC)

(B) BONDING
(1) BONDING OF METAL PARTS (Requires bonding of all system parts and power supply or EGC in accordance with 250.90)

(2) BONDING CONNECTIONS. (Requires connections be made in accordance with 250.8)

(3) METAL BUILDING PARTS. (Prohibits building parts as means for bonding)

(4) FLEXIBLE METAL CONDUIT LENGTH. (Permits listed FMC or LFMC as bonding means where total length does not exceed 100-feet)

(5) SMALL METAL PARTS. (Parts smaller than 2-inches do not need bonding)

(6) NONMETALLIC CONDUIT. (Requires bonding conductor to be outside of conduit)

(7) BONDING CONDUCTORS. (Requires bonding conductors be protected where subject to damage, and conductors made of copper not smaller than 14 AWG)

(8) SIGNS IN FOUNTAINS. (Requires all metal parts be bonded to EGC near the fountain and connected metal piping in accordance with 680.53)
    FPN: (References 600.32(J) for high-voltage secondary conductors)

Change Significance:

This revision provides the code user with a clear outline for grounding and bonding requirements of signs. By separating this section into two first-level subdivisions the requirements for grounding and bonding are clarified in a user-friendly format.

For clarity, five second-level subdivisions in 600.7(A), Grounding, provide the code user with an outline form similar to a list format. In addition, for clarity eight second-level subdivisions in 600.7(B), Bonding, provide the code user with an outline form similar to a list format.

Notes:

_____

_____

_____

_____

Change Summary:
- The requirement for a switched lighting outlet in attics and soffits where ballasts, transformers, or electronic power supplies are located is clarified with new text in 600.21(E).
- The required width of the access door is reduced from 24 inches to 22½ inches to coordinate with typical 2-foot spacing of trusses.

Revised 2008 NEC Text:
600.21 BALLASTS, TRANSFORMERS, AND ELECTRONIC POWER SUPPLIES (E) ATTIC AND SOFFIT LOCATIONS. Ballasts, transformers, and electronic power supplies shall be permitted to be located in attics and soffits, provided there is an access door at least 900 mm by ~~600~~ 562.5 mm (~~3 ft~~ 36 in. by 22 ½ in. ~~2 ft~~) and a passageway of at least 900 mm (3 ft) high by 600 mm (2 ft) wide with a suitable permanent walkway at least 300 mm (12 in.) wide extending from the point of entry to each component. <u>At least one lighting outlet containing a switch or controlled by a wall switch shall be installed in such spaces. At least one point of control shall be at the usual point of entry to these spaces. The lighting outlet shall be provided at or near the equipment requiring servicing.</u>

Change Significance:
The requirement for a switched lighting outlet to service ballasts, transformers, or electronic power supplies for signs is added in 600.21(E) for attic and soffit locations. The requirements of 210.70(C) apply wherever ballasts, transformers, or electronic power supplies are located in attics. However, the installation of signs often involves locating ballasts, transformers, and power supplies in soffits—which are spaces built in some cases for the sole purpose of housing this equipment to support an outdoor sign.

As required in 210.70(C), the lighting outlet must be located at or near the ballasts, transformers, or electronic power supplies and a switch must be located at the usual point of entry, which may be in the ceiling or soffit. Access to these spaces is in many cases through the ceiling. The required width of the access door is reduced from 24 inches to 22-½ inches to coordinate with typical 2-foot spacing of trusses.

600.21(E)
ATTIC AND SOFFIT LOCATIONS

Chapter 6 Special Equipment

Article 600 Electric Signs and Outline Lighting

Part I General

600.21 Ballasts, Transformers, and Electronic Power Supplies

(E) Attic and Soffit Locations

Change Type: Revision

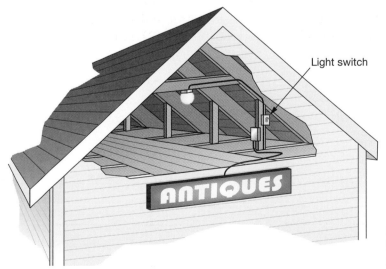

Light switch

ANTIQUES

Access to sign equipment and illumination in the area are necessary to maintain and repair signs.

600.32(K)
SPLICES

Chapter 6 Special Equipment

Article 600 Electric Signs and Outline Lighting

Part II Field-Installed Skeleton Tubing

600.32 Neon Secondary Circuit Conductors, Over 1000 Volts, Nominal

(K) Splices

Change Type: New

The installation of signs and conductors to supply signs requires equipment and enclosures identified for the purpose.

Change Summary:
- Splices in neon secondary circuit conductors are now required to be contained in an enclosure rated greater than 1000 volts. The splice enclosure must be accessible and listed for the location.

Revised 2008 NEC Text:
600.32 NEON SECONDARY CIRCUIT CONDUCTORS, OVER 1000 VOLTS, NOMINAL

(K) SPLICES.   Splices in high-voltage secondary circuit conductors shall be made in listed enclosures rated over 1000 volts. Splice enclosures shall be accessible after installation and listed for the location where they are installed.

Change Significance:
The 2005 NEC contains no requirements for splices of neon secondary circuit conductors. This new first-level subdivision requires that all splices be contained in an enclosure rated at greater than 1000 volts. The enclosure must also be rated for the location in which it is installed, which would include indoor, outdoor, damp, and wet locations. The splice enclosure is also required to be accessible after installation. Listed enclosures are readily available for all sign installations. Open splices are prohibited.

Change Summary:
* All installations of neon tubing are now required to be protected from physical damage. Where neon tubing is readily accessible to other than qualified persons, it is now required to be protected.

Revised 2008 NEC Text:
600.41 NEON TUBING
(D) PROTECTION. Field-installed skeleton tubing shall not be subject to physical damage. Where the tubing is readily accessible to other than qualified persons, field-installed skeleton tubing shall be provided with suitable guards or protected by other approved means.

Change Significance:
This revision provides protection of neon tubing in all installations from physical damage. The requirements for physical protection of neon tubing in 600.9 are limited to other than dry locations where "pedestrians" may be in close proximity.

All installations of neon tubing readily accessible to other than qualified persons are now required to be guarded or protected. This protection is required only where the neon tubing is readily accessible to other than qualified persons. Neon tubing installed in a restaurant at 10 feet above the floor would not be required to be protected. However, neon tubing installed in a restaurant window at 5 feet above the floor would be "readily accessible" and would be required to be guarded or protected in an approved manner.

600.41(D)
PROTECTION

Chapter 6 Special Equipment

Article 600 Electric Signs and Outline Lighting

Part II Field-Installed Skeleton Tubing

600.41 Neon Tubing

(D) Protection

Change Type: New

Neon tubing types of signs are installed indoors and outdoors and must be protected from physical damage.

600.42(A)
POINTS OF TRANSITION

Chapter 6 Special Equipment

Article 600 Electric Signs and Outline Lighting

Part II Field-Installed Skeleton Tubing

600.42 Electrode Connections

(A) Points of Transition

Change Type: New

Change Summary:
• Listed assemblies are now required for all points of transition from high-voltage secondary conductors to neon tubing.

Revised 2008 NEC Text:
600.42 ELECTRODE CONNECTIONS
(A) POINTS OF TRANSITION.   Where the high voltage secondary circuit conductors emerge from the wiring methods specified in 600.32(A), they shall be enclosed in a listed assembly.

Change Significance:
All points of transition from neon tubing to high-voltage secondary conductors are required to be enclosed in listed assemblies. These assemblies are rated for the voltage and protect the exposed splice from high-voltage conductors to the neon tubing.

These listed assemblies are now required by 600.42(H)(1) and (2) to be listed and identified as being suitable for dry, damp, or wet locations.

The point at which a neon tube is energized with high-voltage secondary conductors requires a listed assembly.

Change Summary:

* Where the NEC requires that a disconnect be "capable of being locked in the open position," revisions are modified to require that the provisions for adding a lock must remain in place with or without the lock in place.

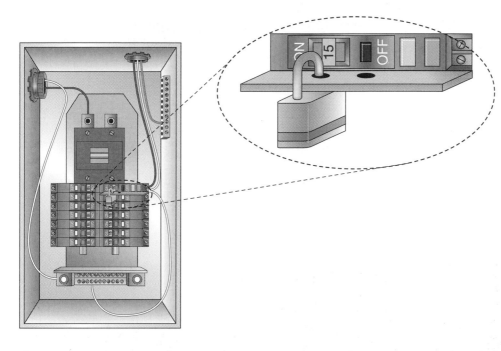

The 2008 NEC contains many prescriptive requirements designed to provide safe working conditions to those who will repair and maintain these systems.

Revised 2008 NEC Text:

The following requirements (marked with a checkmark) have been modified with new text to require that provisions for adding a lock remain in place with or without the lock installed and portable locking means are not permitted.

600.6 DISCONNECTS
   (A) LOCATION
     (1) WITHIN SIGHT OF THE SIGN ✔
     (2) WITHIN SIGHT OF THE CONTROLLER ✔
610.31 RUNWAY CONDUCTOR DISCONNECTING MEANS ✔
610.32 DISCONNECTING MEANS FOR CRANES AND MONORAIL HOISTS ✔
620.51 DISCONNECTING MEANS
   (A) TYPE ✔
   Exception ✔
   (C) LOCATION
     (1) ON ELEVATORS WITHOUT GENERATOR FIELD CONTROL ✔
620.53 CAR LIGHT, RECEPTACLE(S), AND VENTILATION DISCONNECTING MEANS ✔
620.54 HEATING AND AIR-CONDITIONING DISCONNECTING MEANS ✔
620.55 UTILIZATION EQUIPMENT DISCONNECTING MEANS ✔

(continued)

620.53 Car Light, Receptacle(s), and Ventilation Disconnecting Means

620.54 Heating and Air-Conditioning Disconnecting Means

620.55 Utilization Equipment Disconnecting Means

Article 625 Electric Vehicle Charging System

625.23 Disconnecting Means

Article 647 Sensitive Electronic Equipment

647.8 Lighting Equipment

(A) Disconnecting Means

Article 665 Induction and Dielectric Heating Equipment

665.12 Disconnecting Means

665.22 Access to Internal Equipment

Article 675 Electrically Driven or Controlled Irrigation Machines

Article 675.8(B) Main Disconnecting Means

675.8 Disconnecting Means

(B) Main Disconnecting Means

Change Type: New

625.23 DISCONNECTING MEANS ✔
647.8 LIGHTING EQUIPMENT
    (A) DISCONNECTING MEANS ✔
665.12 DISCONNECTING MEANS ✔
665.22 ACCESS TO INTERNAL EQUIPMENT ✔
675.8 DISCONNECTING MEANS
    (B) MAIN DISCONNECTING MEANS ✔

Change Significance:

Significant changes have occurred throughout the NEC with respect to required means of disconnect. This change references all sections in Chapter 6 regarding special equipment that have been modified to require that the provisions for adding a lock must remain in place with or without the lock installed. Additional text in most changes also prohibits the use of portable lockout devices. The safety-driven intent of these changes is to provide installations that can be locked out with the use of only a lock. Installers and maintenance personnel should not be required to carry dozens of portable lockout devices and hope that they have one to fit every switch or circuit breaker they need to lock in the open position.

The provisions for adding a lock to this disconnecting means will remain in place with or without the lock installed.

Change Summary:

* The definition of *manufactured wiring system* is expanded for clarity and to include busway. Permitted construction of manufactured wiring systems is expanded to include listed continuous plug-in types of busway.

Revised 2008 NEC Text:

604.2 DEFINITION

MANUFACTURED WIRING SYSTEM. A system containing component parts that are assembled in the process of manufacture and cannot be inspected at the building site without damage or destruction to the assembly <u>and used for the connection of luminaries, utilization equipment, continuous plug-in type busways and other devices.</u>

604.6 CONSTRUCTION

(A) CABLE OR CONDUIT TYPES

(4) BUSWAYS. <u>Busways shall be listed continuous plug-in type containing factory mounted, bare or insulated conductors, which shall be copper or aluminum bars, rods or tubes. The busway shall be grounded and provided with an equipment ground busbar equivalent in size to the ungrounded busbar. The busway shall be rated nominal 600 volts 20, 30, or 40 amperes. Busways shall be installed in accordance with 368.12, 368.17(D) and 368.30.</u>

Change Significance:

The definition of *manufactured wiring system* is expanded with explanatory text to include equipment utilizing energy and busway as a construction method. A new second-level subdivision is added to 604.6(A) clarifying that manufactured wiring systems are now clearly permitted to include busway. The new text specifically permits only "listed continuous plug-in type busway."

The proposal that generated this change sought to include the "trolley-type" busway. The technical committee chose to permit only listed continuous plug-in types of busway because the trolley type may be provided with accessible uninsulated live parts and is not intended to be placed within reach of individuals.

604.2 and 604.6(A)(4)
DEFINITION OF MANUFACTURED WIRING SYSTEM AND BUSWAYS

Chapter 6 Special Equipment

Article 604 Manufactured Wiring Systems

604.2 Definition, Manufactured Wiring System

604.6 Construction

(A) Cable or Conduit Types

(4) Busways

Change Type: Revision and New

Manufactured wiring systems may consist of cable, conduit, flexible cord, or busway in accordance with 604.6(A).

620.2

DEFINITIONS OF *REMOTE MACHINE ROOM AND CONTROL ROOM (FOR ELEVATOR, DUMBWAITER) AND REMOTE MACHINERY SPACE AND CONTROL SPACE (FOR ELEVATOR, DUMBWAITER)*

Chapter 6 Special Equipment

Article 620 Elevators, Dumbwaiters, Escalators, Moving Walks, Platform Lifts, and Stairway Chair Lifts

Part I General

620.2 Definitions

Change Type: New

Change Summary:
* Two new definitions are added in 620.2, Definitions: *remote, machine room and control room (for elevator, dumbwaiter) and remote machinery space* and *control space (for elevator, dumbwaiter).*

Revised 2008 NEC Text:
ARTICLE 620 ELEVATORS, DUMBWAITERS, ESCALATORS, MOVING WALKS, ~~WHEELCHAIR~~ PLATFORM LIFTS, AND STAIRWAY CHAIR LIFTS
620.2 DEFINITIONS
REMOTE MACHINE ROOM AND CONTROL ROOM (FOR ELEVATOR, DUMBWAITER).   A machine room or control room that is not attached to the outside perimeter or surface of the walls, ceiling or floor of the hoistway.

REMOTE MACHINERY SPACE AND CONTROL SPACE (FOR ELEVATOR, DUMBWAITER).   A machinery space or control space that is not within the hoistway, machine room or control room, and that is not attached to the outside perimeter or surface of the walls, ceiling or floor of the hoistway.

Change Significance:
The use of definitions, where necessary, is critical to proper application of the NEC. Two new terms are defined in 620.2. The style manual requires that a term defined in an individual article of the NEC be located in the second section of the article. These definitions apply only within the article in which they are defined.

The terms *remote machine room and control room (for elevator, dumbwaiter)* and *remote machinery space and control space (for elevator, dumbwaiter)* are used several times in Article 620. These definitions come from the Safety Code for Elevators and Escalators A17.1-2004. Inclusion of these definitions in the NEC also helps to coordinate electrical installation requirements with the Elevator Safety Code. The title of this article has been modified to replace the term *wheelchair* with the term *platform* to coordinate with the applicable standard, ASME A18.1 – 2003 Safety Standard for Platform Lifts and Stairway Lifts.

An elevator machine room will contain the drive mechanism and other machinery and control equipment to operate the elevator.

Change Summary:
- A new Article 626, Electrified Truck Parking Spaces, is added to the NEC to provide prescriptive requirements for installations designed to reduce the idling of trucks to comply with regulatory and environmental requirements.

Revised 2008 NEC Text:
ARTICLE 626 ELECTRIFIED TRUCK PARKING SPACE
I. GENERAL
626.1 SCOPE.   The provisions of the article cover the electrical conductors and equipment external to the truck or transport refrigerated unit that connect trucks or transport refrigerated units to a supply of electricity, and the installation of equipment and devices related to electrical installations within an electrified truck parking space

II. ELECTRIFIED TRUCK PARKING SPACE ELECTRICAL WIRING SYSTEMS
III. ELECTRIFIED TRUCK PARKING SPACE SUPPLY EQUIPMENT
IV. TRANSPORTATION REFRIGERATED UNITS (TRUS)

Change Significance:
This new article is intended to apply to all conductors, equipment, and devices external to trucks and refrigeration units in an "electrified truck parking space." The submitter of this proposal provided compelling substantiation for the need for electrified truck parking spaces and the void in the NEC to specifically address such an installation.

The substantiation pointed out that a typical class 8 truck (heavy-duty long-haul truck with a gross vehicle weight of 33,001 pounds plus) idles 3000 hours per year. There are approximately 1.4 million class 8 trucks on the road. A tremendous amount of pollution is created and more than 1.2 billion gallons of diesel fuel is used to idle at a cost of more than one trillion dollars.

This new article addresses wiring systems to supply theses electrified truck parking spaces along with all required supply equipment. This article does not cover the equipment on the trucks and refrigeration units in the electrified truck parking space.

Article 626
ELECTRIFIED TRUCK PARKING SPACE

Chapter 6 Special Equipment

Article 626 Electrified Truck Parking Space

Part I General

626.1 Scope

Part II Electrified Truck Parking Space Electrical Wiring Systems

Part III Electrified Truck Parking Space Supply Equipment

Part IV Transportation Refrigerated Units (TRUs)

Change Type: New

A typical class 8 truck (heavy-duty long-haul truck with a gross vehicle weight of 33,001 pounds plus) idles 3000 hours per year. The provisions of this article provide the electrical requirements for systems designed to eliminate the waste of fuel and buildup of pollution.

640.6
MECHANICAL EXECUTION OF WORK

Chapter 6 Special Equipment

Article 640 Audio Signal Processing, Amplification, and Reproduction Equipment

Part I General

640.6 Mechanical Execution of Work

(A) Neat and Workmanlike Manner

(B) Installation of Audio Distribution Cables

(C) Abandoned Audio Distribution Cables

(D) Installed Audio Distribution Cable Identified for Future Use

Change Type: Revision

Change Summary:
- 640.6 is revised to add clarity to requirements for the installation and removal of audio signal processing, amplification, and reproduction equipment and conductors.

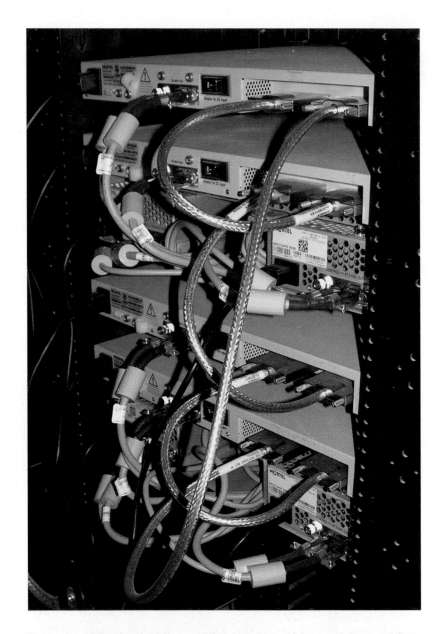

The NEC requires that the installation of audio signal processing, amplification, and reproduction equipment; cables; and circuits "be installed in a neat and workmanlike manner."

Revised 2008 NEC Text:
640.6 MECHANICAL EXECUTION OF WORK
(A) NEAT AND WORKMANLIKE MANNER. Audio signal processing, amplification, and reproduction ~~E~~equipment, ~~and~~ cables, and circuits shall be installed in a neat workmanlike manner.

(continued)

(B) INSTALLATION OF AUDIO DISTRIBUTION CABLES.   Cables installed exposed on the surface of ceilings and sidewalls shall be supported in such a manner that the <u>audio distribution</u> cables will not be damaged by normal building use. Such cables shall be supported straps, staples, hangers, or similar fittings designed and installed so as not to damage the cable. The installation shall conform to 300.4~~(D)~~ and 300.11<u>(A)</u>.

(C) ABANDONED AUDIO DISTRIBUTION CABLES.   <u>The accessible portion of abandoned audio distribution cables shall be removed.</u>

(D) INSTALLED AUDIO DISTRIBUTION CABLE IDENTIFIED FOR FUTURE USE.
(1) <u>Cables identified for future use shall be marked with a tag of sufficient durability to withstand the environment involved.</u>
(2) <u>Cable tags shall have the following information:</u>
   (a) <u>Date cable was identified for future use</u>
   (b) <u>Date of intended use</u>
   (c) <u>Information relating to the intended future use of cable</u>
   ~~FPN: Accepted industry practices are described in ANSI/NECA/BICSI 568-2001, Standard for Installing Commercial Building Telecommunications Cabling, and other ANSI-approved installation standards.~~

Change Significance:

This revision provides needed clarity for the installation and removal of audio signal processing, amplification, and reproduction equipment and conductors. This section is now separated into four first-level subdivisions for clarity. 640.6(A) provides clarity by requiring that all "audio signal processing, amplification, reproduction equipment, and cables, and circuits" be installed in a neat and workmanlike manner. 640.6(B) retains the previous installation requirements in a new first-level subdivision dedicated to installation.

The installation of audio distribution cables is now clarified to be in accordance with all of the requirements of 300.4 and 300.11(A). 640.6(C) relocates requirements for abandoned audio distribution cables from 640.3(A) into a new first-level subdivision in this section. Combining requirements for both the installation and removal of cables into the same section is user friendly. A new 640.6(D) is included to provide prescriptive requirements for cables that are not abandoned but are identified for future use. The fine print note (FPN) is deleted because it was in conflict with 90.1(C). The NEC is not intended as a training manual.

Notes:

_____

_____

_____

_____

645.2
DEFINITION OF ABANDONED SUPPLY CIRCUITS AND INTERCONNECTING CABLES

Chapter 6 Special Equipment

Article 645 Information Technology Equipment

645.2 Definitions

Change Type: New

Change Summary:
- A new definition is added to Article 645 defining *abandoned supply circuits and interconnecting cables*.

Revised 2008 NEC Text:
645.2 DEFINITIONS

ABANDONED SUPPLY CIRCUITS AND INTERCONNECTING CABLES. Installed supply circuits and interconnecting cables that are not terminated at equipment and not identified for future use with a tag.

Change Significance:
645.5(D) provides requirements for supply circuits and interconnecting cables under raised floors, where the installation is designed in accordance with Article 645, Information Technology Equipment. List item 645.5(D)(6) required in the 2005 NEC that all "abandoned cables" be removed unless installed in metal raceways. This requirement is relocated in two new first-level subdivisions 645.5(F) and (G). The buildup of abandoned cable in under-floor spaces is a serious concern. In many cases, as new equipment and technologies are installed the existing cables are abandoned in place. This creates serious problems, which include the inability to install new cable and the buildup of cables that in some cases may be combustible.

This revision identifies an abandoned cable as any cable not terminated or marked with a tag for future use. All abandoned cable in an information technology (IT) installation is required to be removed.

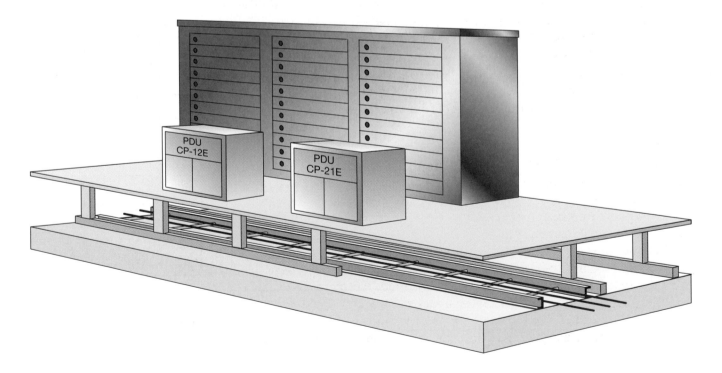

In general, all abandoned cable in an information technology installation must be removed.

Change Summary:
- A new table 645.5 is added to provide clarity and usability for the code user to determine the cable types permitted under raised floors in an Article 645 installation.
- 645.5(D) now recognizes supply cords of listed IT equipment for installation under raised floors.

Revised 2008 NEC Text:
645.5 SUPPLY CIRCUITS AND INTERCONNECTING CABLES
(D) UNDER RAISED FLOORS.   Power cables, communications cables, connecting cables, interconnecting cables, <u>cord-and-plug connections,</u> and receptacles associated with the information technology equipment shall be permitted under a raised floor, provided the following conditions are met: (List items (1), (2) are not changed, renumber list item (3) as (4), (4) as (5) and (5) as (6))

<u>(3) Supply cords of listed information technology equipment in accordance with 645.5(B).</u>

(Existing list item (6) is relocated to new first level subdivision 645.5(F))
<u>(6)</u>~~(5)~~ Cables, other than those covered in (D)(2) and those complying with (D)(5)(a), (D)(5)(b), or (D)(5)(c), shall be listed as Type DP cable having adequate fire-resistant characteristics suitable for use under raised floors of an information technology equipment room.

   a. Interconnecting cables enclosed in a raceway.

   b. Interconnecting cables listed with equipment manufactured prior to July 1, 1994, being installed with that equipment.

   c. Cable type designations <u>shown in Table 645.5 shall be permitted.</u> ~~Type TC (Article 336); Types CL2, CL3, and PLTC (Article 725); Type ITC (Article 727); Types NPLF and FPL (Article 760); Types OFC and OFN (Article 770); Type CM (Article 800); and Type CATV (Article 820). These designations shall be permitted to have an additional letter P or R or G.~~ Green, or green with one or more yellow stripes, insulated single conductor cables, 4 AWG and larger, marked "for use in cable trays" or "for CT use" shall be permitted for equipment grounding.

Change Significance:
The previous text of 645.5(D)(5)(c) has been revised to allow to code user to easily identify the types of cable permitted under raised floors in an Article 645 installation. It is extremely important to understand that the wiring methods permitted in this article may be used only when all of the requirements of this article are applied.

    The previous code text identified 10 cable types and stated that additional suffix letter P, R, or G could be applied to the permitted cable types. The new table, 645.5, identifies 28 types of cable permitted. All permitted cable types and their corresponding NEC articles are listed in a user-friendly table. New list item (3) permits listed cords under a raised floor where the length of the supply cord and attachment plug cap does not exceed 15 feet in length.

645.5(D)
UNDER RAISED FLOORS

Chapter 6 Special Equipment

645.5 Supply Circuits and Interconnecting Cables

(D) Under Raised Floors

Change Type: Revision

Table 645.5 Types Permitted Under Raised Floors			
Article	Plenum	Riser	General Purpose
336			TC
725	CL2P & CL3P	CL2R & CL3R	CL2, CL3 & PLTC
727			ITC
760	NPLFP & FPLP	NPLFR & FPLR	NPLF & FPL
770	OFNP & OFCP	OFNR & OFCR	OFN & OFC
800	CMP	CMR	CM & CMG
820	CATVP	CATVR	CATV

Compliance with the provisions of Article 645 is not mandatory.

645.5(F)
ABANDONED SUPPLY CIRCUITS AND INTERCONNECTING CABLES

645.5(G)
INSTALLED SUPPLY CIRCUITS AND INTERCONNECTING CABLES IDENTIFIED FOR FUTURE USE

Chapter 6 Special Equipment

Article 645 Information Technology Equipment

645.5 Supply Circuits and Interconnecting Cables

(D) Under Raised Floors

(F) Abandoned Supply Circuits and Interconnecting Cables

(G) Installed Supply Circuits and Interconnecting Cables Identified for Future Use

Change Type: Revision and New

Change Summary:
- 645.5(D)(6) is deleted and two new first-level subdivisions (F) and (G) are added to provide clarity to requirements for removal of abandoned cable throughout an Article 645 installation. The previous requirement addressed only abandoned cable under raised floors.

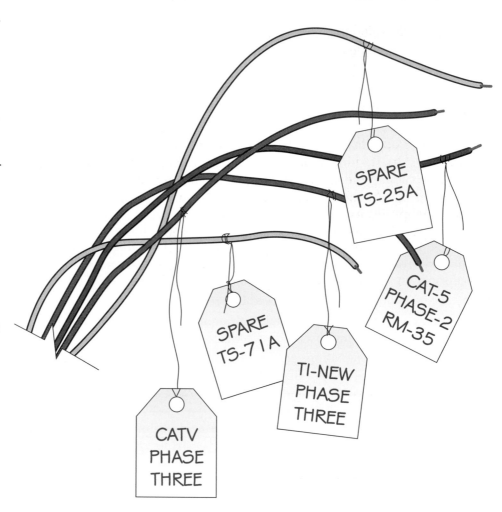

Circuits and cables in information technology installations which are not in use but are to be used in the future or are spares may remain as long as they are properly tagged.

Revised 2008 NEC Text:
645.5 SUPPLY CIRCUITS AND INTERCONNECTING CABLES
(D) UNDER RAISED FLOORS
~~(6) Abandoned cables shall be removed unless contained in metal raceways.~~

(F) ABANDONED SUPPLY CIRCUITS AND INTERCONNECTING CABLES.
<u>The accessible portion of A</u>~~a~~bandoned <u>supply circuits and interconnecting</u> cables shall be removed unless contained in <u>a</u> metal ~~raceways~~.

(continued)

(G) INSTALLED SUPPLY CIRCUITS AND INTERCONNECTING CABLES IDENTIFIED FOR FUTURE USE.

(1) Supply circuits and interconnecting cables identified for future use shall be marked with a tag of sufficient durability to withstand the environment involved.

(2) Supply circuit tags and interconnecting cable tags shall have the following information:

(a) Date identified for future use

(b) Date of intended use

(c) Information relating to the intended future use

Change Significance:

This revision provides needed clarity for the removal of abandoned cables in Article 645 applications. The previous requirement for abandoned cable in Article 645 existed in 645.5(D)(6), which was meant to be applied only to abandoned cable under a raised floor. This revision adds a new first-level subdivision (F), Abandoned Supply Circuits and Interconnecting Cables, which now applies to the entire installation—not just under a raised floor. This requirement mandates removal of abandoned cable unless it is installed in a metal raceway. A new 645.5(G), Installed Supply Circuits and Interconnecting Cables Identified for Future Use, is included to provide prescriptive requirements for cables that are not abandoned but are identified for future use.

Notes:

645.10
DISCONNECTING MEANS

Chapter 6 Special Equipment

645.10 Disconnecting Means

Change Type: Revision

Change Summary:
- The requirement for disconnecting all electronic equipment and the HVAC system is modified from a "whole room" approach to permit "designated zones within the room."

Revised 2008 NEC Text:
645.10 DISCONNECTING MEANS. ~~A~~ An approved means shall be provided to disconnect power to all electronic equipment in the information technology equipment room or in designated zones within the room. There shall also be a similar approved means to disconnect the power to all dedicated HVAC systems serving the room or designated zones and shall cause all required fire/smoke dampers to close. The control for these disconnecting means shall be grouped and identified and shall be readily accessible at the principal exit doors. A single means to control both the electronic equipment and HVAC systems in the room or in a zone shall be permitted. Where a pushbutton is used as a means to disconnect power, pushing the button in shall disconnect the power. Where multiple zones are created, each zone shall have an approved means to confine fire or products of combustion to within the zone.

Exception: Installations qualifying under the provisions of Article 685.

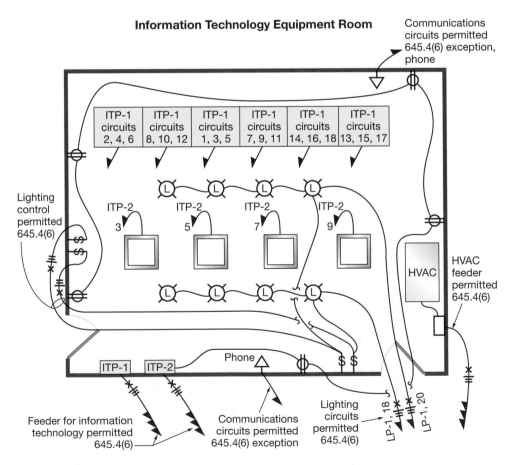

Information Technology Equipment Room

Article 645 requires that all equipment (including HVAC equipment) within a specified zone be capable of being disconnected at the principal exit doors.

(continued)

Change Significance:

This revision removes the long-standing requirement that a single disconnecting means remove power to all electronic equipment and HVAC systems. It is now permitted to create separate zones that in effect will become multiple IT equipment rooms.

Proponents of this change argued that disconnecting all of the IT equipment could create loss of communications and control and therefore impact life safety. In addition, the impact of de-energizing all electronic equipment could result in huge financial loss for business.

In some industries, such as in telecommunications facilities, a zoned shutdown approach is widely used. It is, however, unclear how the enforcement community will determine and enforce this new zone approach. The zones within an area or room will need to be clearly outlined in the design of the facility.

It is important to note that the requirements of Article 645 are optional. This revision may result in more Article 645 installations. A single point of failure, which is represented in the disconnecting means, may have driven owners to disregard the provisions of Article 645.

The disconnect requirement in 645.10 is now required for all electronic equipment in the information technology room *or in designated zones within the room.*

680.2
DEFINITIONS

Chapter 6 Special Equipment

Article 680 Swimming Pools, Fountains, and Similar Installations

Part I General

680.2 Definitions

Permanently Installed Swimming, Wading, Immersion, and Therapeutic Pools

Pool

Storable Swimming, Wading or Immersion Pool

Change Type: Revision

Change Summary:

• There are three definitions of "pool" in 680.2. Each of these definitions is modified to include the term *immersion.* A pool of water used to immerse people such as a baptistery now meets one of the definitions of Article 680.

Revised 2008 NEC Text:

680.2 DEFINITIONS

PERMANENTLY INSTALLED SWIMMING, WADING, <u>IMMERSION,</u> AND THERAPEUTIC POOLS.   Those that are constructed in the ground or partially in the ground, and all others capable of holding water in a depth greater than 1.0 m (42 in.), and all pools installed inside of a building, regardless of water depth, whether or not served by electrical circuits of any nature.

POOL.   Manufactured or field-constructed equipment designed to contain water on a permanent or semipermanent basis and used for swimming, wading, <u>immersion,</u> or other purposes.

STORABLE SWIMMING, WADING OR <u>IMMERSION</u> POOL.   Those that are constructed on or above the ground and are capable of holding water to a maximum depth of 1.0 m (42 in.), or a pool with nonmetallic, molded polymeric walls or inflatable fabric walls regardless of dimension.

Change Significance:

The three definitions in Article 680 that define types of pools are revised to include "immersion." The substantiation provided with this change was directed toward pools of water used as a baptistery to immerse persons in water. Substantiation provided included a fatality involving immersion in a baptistery. This revision now includes all pools used strictly for immersion of persons into the scope of all requirements of Article 680.

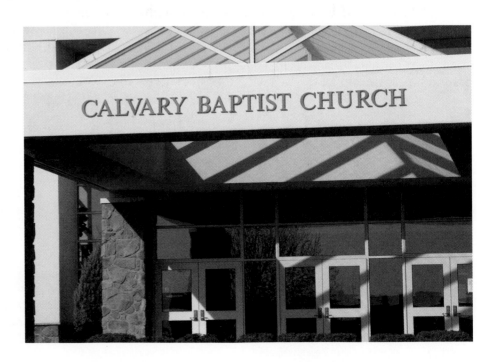

By definition, a pool may be an installation that is designed for swimming, wading, immersion, therapeutic reasons, or for any combination of these purposes.

Change Summary:

- The requirements of 680.10 are clarified to require a complete raceway system. A cable assembly is not permitted to be "sleeved."
- The depth requirements are now clarified as "cover" and not "burial" depth. The distance given in Table 680.10 is for "cover" to the top of the raceway.

Revised 2008 NEC Text:

680.10 UNDERGROUND WIRING LOCATION. Underground wiring shall not be permitted under the pool or within the area extending 1.5 m (5 ft) horizontally from the inside wall of the pool unless this wiring is necessary to supply pool equipment permitted by this article. Where space limitations prevent wiring from being routed a distance 1.5 m (5 ft) or more from the pool, such wiring shall be permitted where installed in <u>complete raceway systems of</u> rigid metal conduit, intermediate metal conduit, or a nonmetallic raceway system. All metal conduit shall be corrosion resistant and suitable for the location. The minimum ~~burial~~ <u>cover</u> depth shall be as given in Table 680.10.

Change Significance:

The general rule of 680.10 prohibits underground wiring of any type under a pool or within the area extending 5 feet horizontally from the inside wall of the pool unless the wiring is necessary to supply pool equipment permitted by Article 680. However, where space limitations do not allow for a distance of 5 feet, other wiring is permitted as specified in 680.10. This revision now clarifies that wiring installed within 5 feet of the pool must be in a "complete raceway system" of rigid metal conduit, intermediate metal conduit, or nonmetallic raceway. A cable assembly permitted for direct burial specified in 300.5 is now clearly not permitted to be sleeved with a raceway where it is within 5 feet of the pool.

The term *burial* is deleted and replaced with the term *cover* for the depth of the raceway. This provides for consistency with the general rules for underground installations in 300.5 and provides clarity for the implementation of this requirement. The measurement is not to the bottom of the trench but to the top of the raceway after placement in the trench.

680.10
UNDERGROUND
WIRING LOCATION

Chapter 6 Special Equipment

Article 680 Swimming Pools, Fountains, and Similar Installations

Part I General

680.10 Underground Wiring Location

Change Type: Revision

Underground wiring not associated with a pool is not permitted within 5 feet measured horizontally from the inside wall of the pool.

680.12
MAINTENANCE DISCONNECTING MEANS

Chapter 6 Special Equipment

680 Swimming Pools, Fountains, and Similar Installations

Part I General

680.12 Maintenance Disconnecting Means

Change Type: Revision

Change Summary:
- The location of the maintenance disconnecting means required by 680.12 for all utilization equipment other than lighting is now clearly required to be at least 5 feet from the inside wall of a pool, spa, or hot tub.

Revised 2008 NEC Text:
680.12 MAINTENANCE DISCONNECTING MEANS.   One or more means to <u>simultaneously</u> disconnect all ungrounded conductors shall be provided for all utilization equipment other than lighting. Each means shall be readily accessible and within sight from its equipment <u>and shall be located at least 1.5 m (5 ft) horizontally from the inside walls of a pool, spa, or hot tub unless separated from the open water by a permanently installed barrier that provides a 1.5 m (5 ft) reach path or greater. This horizontal distance is to be measured from the water's edge along the shortest path required to reach the disconnect.</u>

Change Significance:
This revision replaces the requirement for the maintenance disconnecting means to be at least 5 feet from a pool, spa, or hot tub as it existed in 680.12 in the 1999 NEC. The replacement of this text is significant. 680.12 is located in Part I of Article 680, which means that it applies to all installations under the scope of Article 680.

Part II of Article 680 addresses "permanently installed pools," and 680.22(C) requires "switching devices" to be located at least 5 feet from the pool. Not all maintenance disconnects will be considered switching devices. Part IV of Article 680 addresses "spas and hot tubs," and 680.43(C) requires that "wall switches" must be located at least 5 feet from the spa or hot tub. Not all maintenance disconnects will be considered wall switches.

The relocation of this requirement into Part I, General of Article 680 provides the code user with clarity and usability in the application of this requirement. Provisions for placement of the disconnect closer than 5 feet are included. This would include, for example, a disconnect to be mounted behind a glass partition. It would be within sight, but the operator would have to walk around the partition to open the disconnect.

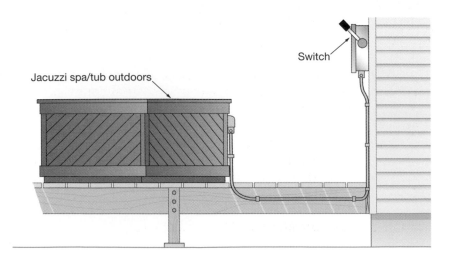

Maintenance and repair of pools, spas, and hot tubs requires a disconnecting means readily accessible and within sight of the equipment.

Change Summary:
- The permission to use any of the wiring methods recognized in Chapter 3 in the interior of a dwelling unit or accessory building is qualified as an exception to the general rule in 680.21(A)(1).

680.21(A)
WIRING METHODS

Chapter 6 Special Equipment

Article 680 Swimming Pools, Fountains, and Similar Installations

Part II Permanently Installed Pools

680.21 Motors

(A) Wiring Methods

(1) General

(2) On or Within Buildings

(3) Flexible Connections

(4) One Family Dwellings

(5) Cord and Plug Connections

Change Type: Revision

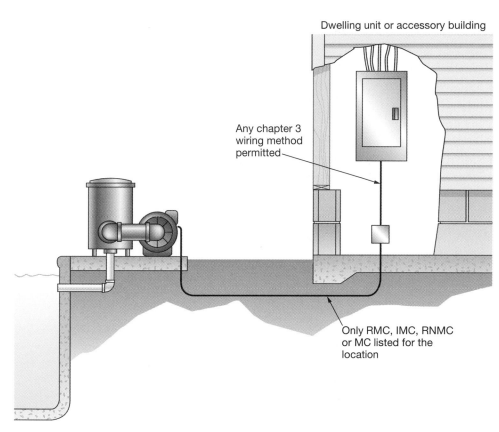

Dwelling unit or accessory building

Any chapter 3 wiring method permitted

Only RMC, IMC, RNMC or MC listed for the location

Wiring methods for pools must comply with the special requirements of Article 680.

Revised 2008 NEC Text:
680.21 MOTORS
(A) WIRING METHODS. <u>The wiring to a pool motor shall comply with (A)(1) unless modified for specific circumstances by (A)(2), (A) (3), (A) (4), or (A) (5).</u>

(1) GENERAL. The branch circuits for pool-associated motors shall be installed in rigid metal conduit, intermediate metal conduit, rigid <u>polyvinyl chloride</u> ~~nonmetallic~~ conduit, <u>reinforced thermosetting resin conduit</u> or Type MC cable listed for the location. Other wiring methods and materials shall be permitted in specific locations or applications as covered in this section. Any wiring method employed shall contain an insulated copper equipment grounding conductor sized in accordance with 250.122 but not smaller than 12 AWG.

(2) ON OR WITHIN BUILDINGS. (No change)

(continued)

(3) FLEXIBLE CONNECTIONS.   Where necessary to employ flexible connections at or adjacent to the motor, liquidtight flexible metal or <u>liquidtight flexible</u> nonmetallic conduit with approved fittings shall be permitted.

(4) ONE-FAMILY DWELLINGS.   (No change)

(5) CORD AND PLUG CONNECTIONS.   (No change)

Change Significance:

680.21(A)(4) in the 2005 NEC was essentially an exception to the provisions of 680.21(A)(1). The location of this permissive text as a second-level subdivision was confusing to the user of the NEC.

680.21(A)(1) requires branch circuits for pool-associated motors to be installed in rigid metal conduit, intermediate metal conduit, rigid polyvinyl chloride conduit, or type MC cable listed for the location. 680.21(A)(1) now recognizes type reinforced thermosetting resin conduit (RTRC) as a permitted raceway. Second-level subdivision (4) for one-family dwelling units was not an expansion of the rule in 680.21(A). Rather, it was an exception for dwelling units. The additional parent text added in 680.21(A) now provides clear guidance that the general rule is in 680.21(A)(1) and second-level subdivisions 680.21(A)(2), (3), (4), and (5) are essentially exceptions to the general rule. 680.21(A)(3) is editorially revised to permit "liquid-tight flexible" nonmetallic conduit.

Notes:

_____

_____

_____

_____

_____

_____

_____

_____

_____

Change Summary:
- All pool pump motors cord-and-plug connected or hard wired and rated at 15 or 20 amps at 125 or 240 volts are now required to be ground-fault circuit interrupter (GFCI) protected.

Revised 2008 NEC Text:
680.22 AREA LIGHTING, RECEPTACLES, AND EQUIPMENT
(A) RECEPTACLES
(4)~~(5)~~ GFCI PROTECTION.   All 15- and 20-ampere, single-phase, 125-volt receptacles located within 6.0 m (20 ft) of the inside walls of a pool shall be protected by a ground-fault circuit interrupter. ~~Receptacles that supply pool pump motors and that are rated 15 or 20 amperes, 125 volts through 250 volts, single phase, shall be provided with GFCI protection.~~

(B) GFCI PROTECTION. <u>Outlets supplying pool pump motors from branch circuits with short-circuit and ground-fault protection rated 15 or 20 amperes, 125 volt or 240 volt, single phase, whether by receptacle or direct connection, shall be provided with ground-fault circuit-interrupter protection for personnel.</u>

Change Significance:
This revision reinstates the 1999 NEC requirement for GFCI protection on single-phase hard-wired pump motors. This change will dramatically improve the safety of pool installations.

This revision impacts 680.22(A), which addresses only "receptacles." The last sentence of 680.22(A)(4) is deleted. This second-level subdivision requires all 15- or 20-amp 125-volt receptacles within 20 feet of the pool to be GFCI protected. The deleted text required all receptacles at 15 or 20 amps at 125 or 240 volts that supplied pool pump motors to be GFCI protected. This sentence is no longer necessary because a new first-level subdivision is added. A new 680.22(B), GFCI Protection, requires that all "outlets" supplying pool pump motors at 15 or 20 amps and 125 or 240 volts be GFCI protected.

680.22
AREA LIGHTING, RECEPTACLES, AND EQUIPMENT

Chapter 6 Special Equipment

Article 680 Swimming Pools, Fountains, and Similar Installations

Part II Permanently Installed Pools

680.22 Area Lighting, Receptacles, and Equipment

(A) Receptacles

(4) GFCI Protection

(B) GFCI Protection

Change Type: Revision and New

The pool pump motor represents a source of electrical energy that could energize the pool water or other conductive material in the pool area.

680.22(E)
OTHER OUTLETS

Chapter 6 Special Equipment

Article 680 Swimming Pools, Fountains, and Similar Installations

Part II Permanently Installed Pools

680.22 Area Lighting, Receptacles, and Equipment

(E) Other Outlets

Change Type: New

Change Summary:

- New subdivision 680.22(E) requires that other outlets be located a minimum of 10 feet from the pool. This requirement is directed at "other systems" and an informational FPN is added to aid the code user.

Revised 2008 NEC Text:

680.22 AREA LIGHTING, RECEPTACLES, AND EQUIPMENT
(E) OTHER OUTLETS. Other outlets shall be not less than 3.0 m (10 ft) from the inside walls of the pool. Measurements shall be determined in accordance with 680.22(A)(5).

FPN: Other outlets may include, but are not limited to, remote-control, signaling, fire alarm, and communications circuits.

Change Significance:

This revision now clearly prohibits outlets from other systems to be located closer than 10 feet from the inside wall of the pool. This new requirement is directed at "other systems." These other systems are noted in an informational FPN that explains to the code user that the "Other outlets may include, but are not limited to, remote-control, signaling, fire alarm, and communications circuits."

Other receptacle outlets are addressed by 680.22(A)(2). Lighting outlets are addressed in 680.22(B). Switching devices are addressed by 680.22(C).

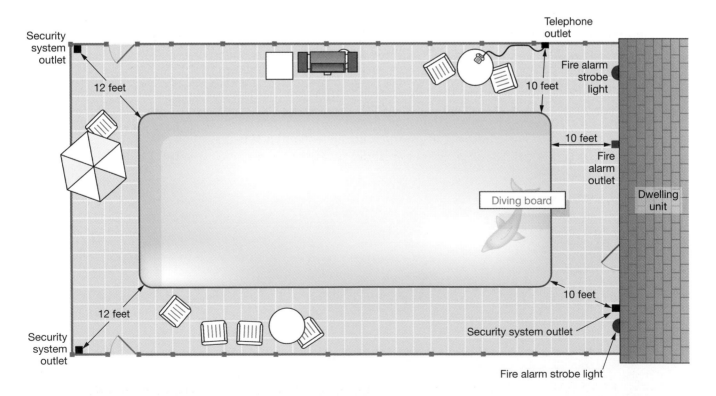

Requiring electrical outlets and outlets of other systems to be installed away from the pool area is essential to the safety of persons.

Change Summary:
- Upward-facing underwater luminaires in permanently installed pools and fountains are now permitted to be adequately guarded or listed for use without a guard.

Revised 2008 NEC Text:
II. PERMANENTLY INSTALLED POOLS
680.23 UNDERWATER LUMINAIRES
(A) GENERAL
(6) BOTTOM-MOUNTED LUMINAIRES.   A luminaire facing upward shall comply with either (1) or (2): ~~have the lens adequately guarded to prevent contact by any person.~~

(1) Have the lens adequately guarded to prevent contact by any person.

(2) Be listed for use without a guard.

V. FOUNTAINS
680.51 LUMINAIRES, SUBMERSIBLE PUMPS, AND OTHER SUBMERSIBLE EQUIPMENT
(C) LUMINAIRE LENSES.   Luminaires shall be installed with the top of the luminaire lens below the normal water level of the fountain unless listed for above-water locations. A luminaire facing upward shall comply with either (1) or (2): ~~have the lens adequately guarded to prevent contact by any person.~~
(1) Have the lens adequately guarded to prevent contact by any person.

(2) Be listed for use without a guard.

Change Significance:
This revision will now permit an upward-facing lighting fixture/luminaire to be installed in a permanently installed pool or fountain where it is adequately guarded or listed for use without a guard. Luminaires with high-strength plastic lenses and those with small-diameter and particularly thick glass lenses can be found suitable as well. The Standard for Underwater Luminaires and Junction Box, UL 676, requires submersible luminaires with an integral lens guard to withstand a 250-pound loading test and a 100-foot-pound impact (a 50-pound 9-inch-diameter cylinder dropped 2 feet).

680.23(A)(6) and 680.51(C)
BOTTOM-LINED LUMINAIRES AND LUMINAIRE LENSES

Chapter 6 Special Equipment

Article 680 Swimming Pools, Fountains, and Similar Installations

Part II Permanently Installed Pools

680.23 Underwater Luminaires

(A) General

(6) Bottom-Lined Luminaires

Part V Fountains

680.51 Luminaires, Submersible Pumps, and Other Submersible Equipment

(C) Luminaire Lenses

Change Type: Revision

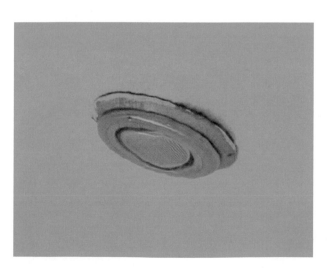

Underwater luminaires represent a source of electrical energy that could energize the pool water and must be installed in accordance with the requirements of Article 680.

680.23(B)
WET-NICHE LUMINAIRES

Chapter 6 Special Equipment

Article 680 Swimming Pools, Fountains, and Similar Installations

Part II Permanently Installed Pools

680.23 Underwater Luminaires

(B) Wet-Niche Luminaires

(6) Servicing

Change Type: Revision

Change Summary:

Requirements for servicing of all wet-niche luminaires in permanently installed pools is revised to require that:

- All wet-niche fixtures be removable for inspection, relamping, or other maintenance.
- The forming shell location and cord length must allow for the luminaire to be placed on the deck or other dry location for maintenance.
- The luminaire maintenance location must be accessible without entering the pool.

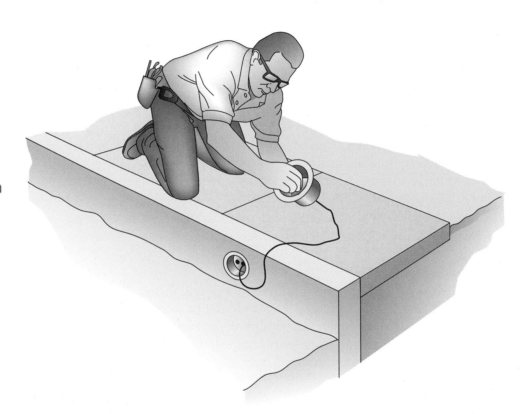

Maintenance of wet-niche luminaires must be performed in a safe and dry location.

Revised 2008 NEC Text:

680.23 UNDERWATER LUMINAIRES
(B) WET-NICHE LUMINAIRES
(1) (through)   **(5)** (No change)

(6) SERVICING.   All <u>wet-niche</u> luminaires shall be removable from the water for <u>inspection,</u> relamping, or <u>other</u> ~~normal~~ maintenance. <u>The forming shell location and length of cord in the forming shall permit personnel to place the removed luminaire on the deck or other dry location for such maintenance. The luminaire maintenance location shall be accessible with-</u>

(continued)

out entering or going in the pool water. ~~Luminaires shall be installed in such a manner that personnel can reach the luminaire for relamping, maintenance, or inspection while on the deck or equivalently dry location.~~

Change Significance:

The previous text of 680.23(B)(6) required that the wet-niche luminaire be removable and that relamping, maintenance, and inspection could be performed. As written in the 2005 NEC, it could be interpreted that as long as a person could lie down on the deck and reach the luminaire the installation was code compliant. This safety-driven revision now clearly requires that:

(1) All wet-niche luminaires shall be removable from the water for inspection, relamping, or other maintenance.

(2) The forming shell location and length of cord in the forming shell shall permit personnel to place the removed luminaire on the deck or other dry location for such maintenance.

(3) The luminaire maintenance location shall be accessible without entering or going in the pool water.

The intent of this requirement has been clarified: the installation of the luminaire must allow service to be safely performed with the fixture accessible on the pool deck or other surface. No one should need to enter or go into the water to perform maintenance of any type. The cord in the forming shell must be long enough to allow the luminaire to be placed on the pool deck for servicing.

Chapters 1 through 4 of the NEC apply to all pool installations unless modified in Article 680. Working space about equipment is addressed in 110.26. The new requirement for servicing wet-niche luminaires is special (limited to pools) and belongs in Article 680 of Chapter 6.

680.25(A)
WIRING METHODS

Chapter 6 Special Equipment

Article 680 Swimming Pools, Fountains, and Similar Installations

Part II Permanently Installed Pools

680.25 Feeders

(A) Wiring Methods

Change Type: Revision

Change Summary:
- A new last sentence in 680.25(A) prohibits aluminum conduit in the pool area where subject to corrosion.
- Type RTRC reinforced thermosetting resin conduit is now permitted.

Revised 2008 NEC Text:
680.25 FEEDERS

(A) WIRING METHODS. Feeders shall be installed in rigid metal conduit, intermediate metal conduit, liquidtight flexible nonmetallic conduit, or rigid polyvinyl chloride ~~nonmetallic~~ conduit <u>or reinforced thermosetting resin conduit</u>. Electrical metallic tubing shall be permitted where installed on or within a building, and electrical nonmetallic tubing shall be permitted where installed within a building. <u>Aluminum conduits shall not be permitted in the pool area where subject to corrosion.</u>

Exception: (No change)

Change Significance:
Aluminum conduit is now clearly prohibited in the pool area where subject to corrosion. 680.25(A) permits rigid metal conduit, which includes aluminum. The UL guide information for "Rigid Nonferrous Metallic Conduit (DYWV)" requires that where aluminum conduit is used in concrete or in contact with soil supplementary corrosion protection is required. Aluminum conduit would be subject to corrosion where exposed to pool chemicals or treated pool water. The term *nonmetallic* is deleted and type PVC is properly referenced as rigid polyvinyl chloride conduit. Type RTRC reinforced thermosetting resin conduit is now permitted.

Feeders are commonly installed in pool areas to supply a panelboard, which will control the pool pump, lighting, water heaters, and other equipment.

Change Summary:

- 680.26 has been extensively revised for clarity and usability.
- The requirement for perimeter surface bonding with an alternate means has been revised to allow a single, solid, 8-AWG copper wire conductor in the center of the 3-foot horizontal plane.

680.26
EQUIPOTENTIAL BONDING

Chapter 6 Special Equipment

Article 680 Swimming Pools, Fountains, and Similar Installations

Part II Permanently Installed Pools

680.26 Equipotential Bonding

(A) Performance

(B) Bonded Parts

(1) Conductive Pool Shells

(2) Perimeter Surfaces

(3) Metallic Components

(4) Underwater Lighting

(5) Metal Fittings

(6) Electrical Equipment

(7) Metal Wiring Methods and Equipment

Change Type: Revision

Equipotential Bonding Perimeter Surfaces

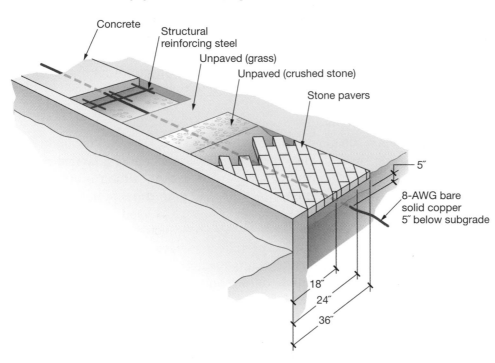

Concrete
Structural reinforcing steel
Unpaved (grass)
Unpaved (crushed stone)
Stone pavers
5″
8-AWG bare solid copper 5″ below subgrade
18″
24″
36″

The bonding of perimeter surfaces surrounding a pool is designed to reduce voltage gradients in the pool area.

Revised 2008 NEC Text:

Note: The use of <u>underscore</u> and ~~strikethrough~~ is omitted due to space.

680.26 EQUIPOTENTIAL BONDING

(A) PERFORMANCE.   The equipotential bonding required by this section shall be installed to reduce voltage gradients in the pool area.

(B) BONDED PARTS.   The parts specified in 680.26(B)(1) through (B)(7) shall be bonded together using solid copper conductors, insulated covered, or bare, not smaller than 8 AWG or with rigid metal conduit of brass or other identified corrosion-resistant metal. Connections to bonded parts shall be made in accordance with 250.8. An 8 AWG or larger solid copper bonding conductor provided to reduce voltage gradients in the pool area shall not be required to be extended or attached to any remote panelboard, to service equipment, or electrodes.

(continued)

(1) CONDUCTIVE POOL SHELLS. Bonding to conductive pool shells shall be provided as specified in 680.26(B)(1)(a) or 680.26(B)(1)(b). Poured concrete, pneumatically applied or sprayed concrete, and concrete block with painted or plastered coatings shall all be considered conductive materials due to water permeability and porosity. Vinyl liners and fiberglass composite shells shall be considered to be non-conductive materials.

> **(a)** *Structural Reinforcing Steel.* Unencapsulated structural reinforcing steel shall be bonded together by steel tie wires or the equivalent. Where structural reinforcing steel is encapsulated in a nonconductive compound, a copper conductor grid shall be installed in accordance with 680.26(B)(1)(b).
>
> **(b)** *Copper Conductor Grid.* A copper conductor grid shall be provided and shall comply with (b)(1) through (b)(4).

(1) Be constructed of minimum 8 AWG bare solid copper conductors bonded to each other at all points of crossing.

(2) Conform to the contour of the pool and the pool deck.

(3) Be arranged in a 300 mm (12 in.) by 300 mm (12 in.) network of conductors in a uniformly spaced perpendicular grid pattern with a tolerance of 100 mm (4 in.).

(4) Be secured within or under the pool no more than 150 mm (6 in.) from the outer contour of the pool shell.

(2) PERIMETER SURFACES. The perimeter surface shall extend for 1 m (3 ft) horizontally beyond the inside walls of the pool and shall include unpaved surfaces as well as poured concrete and other types of paving. Bonding to perimeter surfaces shall be provided as specified in 680.26(B)(2)(a) or 680.26(B)(2)(b), and shall be attached to the pool reinforcing steel or copper conductor grid at a minimum of four (4) points uniformly spaced around the perimeter of the pool. For non-conductive pool shells, bonding at four points shall not be required.

> **(a)** *Structural Reinforcing Steel.* Structural reinforcing steel shall be bonded In accordance with 680.26 (B)(1)(a).
>
> **(b)** *Alternate Means.* Where structural reinforcing steel is not available or is encapsulated in a nonconductive compound, a copper conductor(s) shall be utilized where the following conditions are met:

(1) At least one minimum 8 AWG bare solid copper conductor shall be provided.

(2) The conductors shall follow the contour of the perimeter surface.

(3) Only listed splices shall be permitted.

(4) The required conductor shall be 450 to 600 mm (18 to 24 in.) from the inside walls of the pool.

(5) The required conductor shall be secured within or under the perimeter surface 100 to 150 mm (4 to 6 in.) below the subgrade.

(3) METALLIC COMPONENTS. All metallic parts of the pool structure, including reinforcing metal not addressed in 680.26(1)(a), shall be bonded. Where reinforcing steel is encapsulated with a nonconductive compound, the reinforcing steel shall not be required to be bonded.

(continued)

(4) UNDERWATER LIGHTING. All metal forming shells and mounting brackets of no-niche luminaires shall be bonded.

Exception: Listed low-voltage lighting systems with nonmetallic forming shells shall not require bonding.

(5) METAL FITTINGS. All metal fittings within or attached to the pool structure shall be bonded. Isolated parts that are not over 100 mm (4 in.) in any dimension and do not penetrate into the pool structure more than 25 mm (1 in.) shall not require bonding.

(6) ELECTRICAL EQUIPMENT. Metal parts of electrical equipment associated with the pool water circulating system, including pump motors and metal parts of equipment associated with pool covers, including electric motors, shall be bonded.

Exception: Metal parts of listed equipment incorporating an approved system of double insulation shall not be bonded.

(a) *Double-Insulated Water Pump Motors.* Where a double-insulated water-pump motor is installed under the provisions of this rule, a solid 8 AWG copper conductor of sufficient length to make a bonding connection to a replacement motor shall be extended from the bonding grid to an accessible point in the vicinity of the pool pump motor. Where there is no connection between the swimming pool bonding grid and the equipment grounding system for the premises, this bonding conductor shall be connected to the equipment grounding conductor of the motor circuit.

(b) *Pool Water Heaters.* For pool water heaters rated at more than 50 amperes and having specific instructions regarding bonding and grounding, only those parts designated to be bonded shall be bonded and only those parts designated to be grounded shall be grounded.

(7) METAL WIRING METHODS AND EQUIPMENT. Metal-sheathed cables and raceways, metal piping, and all fixed metal parts shall be bonded.

Exception No. 1: Those separated from the pool by a permanent barrier shall not be required to be bonded.

Exception No. 2: Those greater than 1.5 m (5 ft) horizontally of the inside walls of the pool shall not be required to be bonded.

Exception No. 3: Those greater than 3.7 m (12 ft) measured vertically above the maximum water level of the pool, or as measured vertically above any observation stands, towers, or platforms, or any diving structures shall not be required to be bonded.

Change Significance:

680.26 has been completely revised. The changes are as follows:

680.26: EQUIPOTENTIAL BONDING

(A) PERFORMANCE. The term *eliminate* has been deleted and the performance requirement is clarified to *reduce* voltage gradients. It is impossible to eliminate them. The information in the FPN has been deleted.

(continued)

(B) BONDED PARTS.   This parent text now includes (1) the requirements for the bonding conductor that previously existed in (C) and (2) the connection requirements that previously existed in (D). The FPN is rewritten to positive text from the previous (A).

(B)(1) CONDUCTIVE POOL SHELLS.   This new second-level subdivision includes requirements for the bonding of the pool shell. Information on nonconductive pool shells that existed previously as an exception in TIA 05-2 is also included. Text is included to explain that poured concrete, pneumatically applied or sprayed concrete, and concrete block with painted or plastered coatings are all considered conductive materials due to water permeability and porosity—whereas vinyl liners and fiberglass composite shells are considered nonconductive materials. Requirements for how to bond the pool shell include prescriptive requirements found previously in (C).

(B)(2) PERIMETER SURFACES.   This requirement addresses the 3-foot horizontal plane that makes up the walking surface surrounding the pool, which previously existed in (C). Unpaved surfaces as well as poured concrete and other types of paving are now clearly included. Two methods are permitted: structural reinforcing steel and alternate means. It is significant to note that the alternate means consists of a single solid 8-AWG copper conductor run 18 to 24 inches from the inside pool wall.

(B)(3) METALLIC COMPONENTS.   This requirement for bonding all metallic components existed previously in (B)(1).

(B)(4) UNDERWATER LIGHTING.   This requirement, which existed previously in (B)(2), is for clarity separated into positive text and an Exception.

(B)(5) METAL FITTINGS.   This is relocated from (B)(3).

(B)(6) ELECTRICAL EQUIPMENT.   This requirement contains text that existed previously in (B)(4) and (E). Requirements for bonding metal parts of pump motors and circulating systems are for clarity separated into positive text and an Exception. The text is further separated into third-level subdivisions addressing (a) double insulated pump motors and (b) pool water heaters.

(B)(7) METAL WIRING METHODS AND EQUIPMENT.   This is relocated from previous (B)(5). Previous subdivisions (C), (D), are (E) are deleted—with their requirements moved into the new text.

Notes:

Change Summary:
- 680.26 now includes a requirement to "bond" the pool water.

Revised 2008 NEC Text:
680.26 EQUIPOTENTIAL BONDING
(C) POOL WATER.   An intentional bond of a minimum conductive surface area of 5806 mm² (9 in²) shall be installed in contact with the pool water. This bond shall be permitted to consist of parts that are required to be bonded in 680.26(B).

Change Significance:
This new requirement will essentially guarantee that all permanently installed pools bond the pool water to all other required elements addressed in 680.26. Bonding of metal parts in and around a swimming pool to an equipotential bonding grid is extensively covered in 680.26. The intent of this bonding is to equalize the voltages between the pool water and the deck, including any attached metal structures or parts.

680.26 requires various metal parts and equipment to be bonded. It is to some degree assumed that one or more of these parts are in contact with the pool water. This may not always be the case. Some pools do not have any bonded metal parts in contact with the water. In such a case, this new requirement for intentional bonding of the water is necessary to equalize the water-to-deck voltages.

This bonding may be accomplished through a metal ladder, in the pool, which is bonded. Where no metal parts exist in contact with the pool, the pool manufacturer, installer, and electrical contractor must ensure that 9 square inches of bonded metal is placed in contact with the water.

680.26(C)
POOL WATER

Chapter 6 Special Equipment

Article 680 Swimming Pools, Fountains, and Similar Installations

Part II Permanently Installed Pools

680.26 Equipotential Bonding

(C) Pool Water

Change Type: New

Nine square inches of conductive surface area are required in the pool to bond the pool water.

680.31

PUMPS

Chapter 6 Special Equipment

Article 680 Swimming Pools, Fountains, and Similar Installations

Part III Storable Pools

680.31 Pumps

Change Type: Revision

Change Summary:

- Cord-connected pool filter pumps for storable pools are now required to have a ground-fault circuit interrupter as an integral part of the attachment plug-or located in the power supply cord within 12 inches of the attachment plug.

Revised 2008 NEC Text:

680.31 PUMPS.   A cord-connected pool filter pump shall incorporate an approved system of double insulation or its equivalent and shall be provided with means for grounding only the internal and nonaccessible non-current-carrying metal parts of the appliance.

The means for grounding shall be an equipment grounding conductor run with the power-supply conductors in the flexible cord that is properly terminated in a grounding-type attachment plug having a fixed grounding contact member.

<u>Cord-connected pool filter pumps shall be provided with a ground-fault circuit interrupter that is an integral part of the attachment plug or located in the power supply cord within 300 mm (12 in.) of the attachment plug.</u>

Change Significance:

This revision now requires that GFCI protection be "built in" to a cord-connected pool pump used with a storable pool. The technical committee substantiation for this addition is to ensure that a storable pool pump is GFCI protected regardless of whether or not the pump is supplied from a GFCI-protected source.

The requirement is for the GFCI protective device to be an integral part of the attachment plug or located in the power supply cord 12 inches of the attachment plug.

680.32 requires all 125-volt receptacles within 20 feet of a storable pool to be GFCI protected. Where a storable pool is installed, the nearest receptacle outlet for supplying the pool pump motor may be more than 20 feet away and may not be GFCI protected. This safety-driven revision will ensure that all pool pump motors for storable pools are GFCI protected.

The safety driven provisions of Article 680 for GFCI protection are enhanced with a new requirement for "built-in" GFCI protection of pump motors for storable pools.

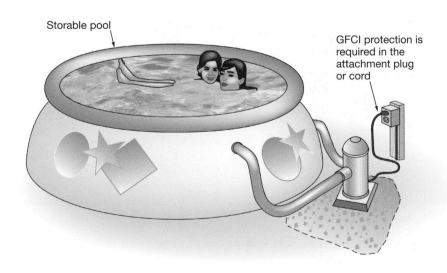

Storable pool

GFCI protection is required in the attachment plug or cord

Change Summary:

- All metal "raceways" within 5 feet of a spa or hot tub must be bonded to metal fittings, metal parts of the electrical/circulation system, metal surfaces, and other electrical devices/controls.
- Metal parts of electrical equipment associated with the water circulating system, including pump motors that are part of a listed self-contained spa or hot tub, are not required to be bonded in the field.

Revised 2008 NEC Text:

680.43 INDOOR INSTALLATIONS

(D) BONDING.   The following parts shall be bonded together:

(1) (No change)

(2) (No change)

(3) Metal <u>raceway</u> ~~conduit~~ and metal piping that are within 1.5 m (5 ft) of the inside walls of the spa or hot tub and that are not separated from the spa or hot tub by a permanent barrier

(4) (No change)

  Exception <u>No 1</u>: (No change)
  <u>*Exception No. 2: Metal parts of electrical equipment associated with the water circulating system, including pump motors that are part of a listed self-contained spa or hot tub.*</u>

(5) (No change)

Change Significance:

The term *conduit* is deleted in 680.43(D)(4) and replaced with the term *raceway.* The term *raceway* is more appropriate. It is defined and applies to all wiring methods noted in the Article 100 definition, not just conduit. The bonding requirements of 680.43(D) now apply to all raceways.

A new exception exempts listed self-contained spas or hot tubs. Grounding and bonding in listed self-contained spas is evaluated and controlled as part of the listing. Field-assembled spas would require field bonding because they are not "self-contained."

680.43(D)
BONDING

Chapter 6 Special Equipment

Article 680 Swimming Pools, Fountains, and Similar Installations

Part IV Spas and Hot Tubs

680.43 Indoor Installations

(D) Bonding

Change Type: Revision

Spas and hot tubs must be installed in accordance with Article 680.

680.50
GENERAL

Chapter 6 Special Equipment

Article 680 Swimming Pools, Fountains, and Similar Installations

Part V Fountains

680.50 General

Change Type: Revision

Change Summary:
- It is clarified that portable fountains of any size are covered in Article 422, Appliances, and not in Article 680.

Revised 2008 NEC Text:
680.50 GENERAL.   The provisions of Part I and Part V of this article shall apply to all permanently installed fountains as defined in 680.2. Fountains that have water common to a pool shall additionally comply with the requirements in Part II of this article. Part V does not cover self-contained, portable fountains ~~not larger than 1.5 m (5 ft) in any dimension.~~ Portable fountains shall comply with Parts II and III of Article 422.

Change Significance:
680.50 explains how this article covers fountains. Part I, General and Part V, Fountains apply to all permanently installed fountains. Where a fountain has water common to a pool, Part II, Permanently Installed Pools also applies. This revision now clarifies that all portable fountains, regardless of size, are covered by Article 422, Appliances and not by Article 680.

Permanent fountains are covered in Article 680. Portable fountains are considered appliances and are installed in accordance with Article 422.

Change Summary:
- Hydromassage bathtubs are now required to be supplied by an individual branch circuit.
- The GFCI required to protect a hydromassage bathtub is now required to be "readily accessible."

Revised 2008 NEC Text:
680.71 PROTECTION. Hydromassage bathtubs and their associated electrical components shall be <u>on an individual branch circuit(s) and</u> protected by a <u>readily accessible</u> ground-fault circuit interrupter. All 125-volt, single-phase receptacles not exceeding 30 amperes and located within 1.5 m (~~56~~ ft) measured horizontally of the inside walls of a hydromassage tub shall be protected by a ground-fault circuit interrupter(s).

Change Significance:
Hydromassage bathtub installations in existing dwelling units and other occupancies are quite common. Installers in some cases supply the hydromassage bathtub from an existing circuit, which may be supplying outlets in bedrooms, living rooms, or other areas. The requirement for an "individual" branch circuit ensures that no other outlets will be served.

The location of the required GFCI device for the hydromassage bathtub is in many cases behind an access door within the assembly. Homeowners and others may be unaware of the location of the device for regular testing and to reset the device after a ground fault. This revision requires this GFCI device to be "readily accessible."

680.71
PROTECTION

Chapter 6 Special Equipment

Article 680 Swimming Pools, Fountains, and Similar Installations

Part VII Hydromassage Bathtubs

680.71 Protection

Change Type: Revision

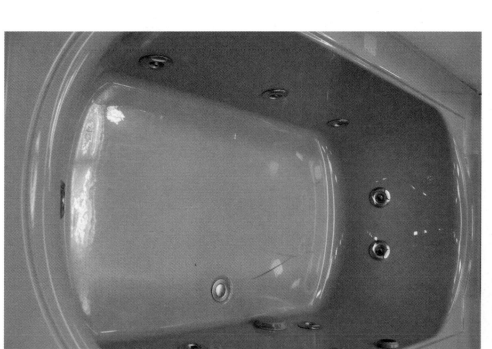

A hydromassage bathtub is a permanently installed bathtub equipped with a recirculating piping system, pump, and associated equipment. It is designed so it can accept, circulate, and discharge water upon each use.

680.74
BONDING

Chapter 6 Special Equipment

Article 680 Swimming Pools, Fountains, and Similar Installations

Part VII Hydromassage Bathtubs

680.74 Bonding

Change Type: Revision

Change Summary:

- Bonding of hydromassage tubs is clarified to require that the solid 8 AWG be connected to all metal piping systems and that all grounded metal parts in contact with the water be bonded to the circulating pump motor unless it is double insulated.
- It is clarified that the bonding jumper is not required to be connected to or extended to any remote panelboard, service equipment, or electrode.

Revised 2008 NEC Text:

680.74 BONDING.   All metal piping systems and all grounded metal parts in contact with the circulating water shall be bonded together using a <u>solid</u> copper bonding jumper, insulated, covered, or bare, not smaller than 8 AWG ~~solid.~~ <u>The bonding jumper shall be connected to the terminal on the circulating pump motor that is intended for this purpose. The bonding jumper shall not be required to be connected to a double insulated circulating pump motor. The 8 AWG or larger solid copper bonding jumper shall be required for equipotential bonding in the area of the hydromasage bathtub and shall not be required to be extended or attached to any remote panelboard, service equipment, or any electrode.</u>

Change Significance:

This revision provides clarity for the code user for bonding in a hydromassage bathtub. 680.74 is clarified to require that the solid 8 AWG be connected to all metal piping systems and that all grounded metal parts in contact with the water be bonded to the circulating pump motor unless the circulating pump is double insulated. The bonding of metal parts in contact with the water, metal water piping, and the circulating motor completes this requirement. Additional text is provided to inform the code user that it is not required to connect to or extend the bonding conductor to any remote panelboard, service equipment, or electrode.

A hydromassage bathtub is designed to accept, circulate, and discharge water upon each use.

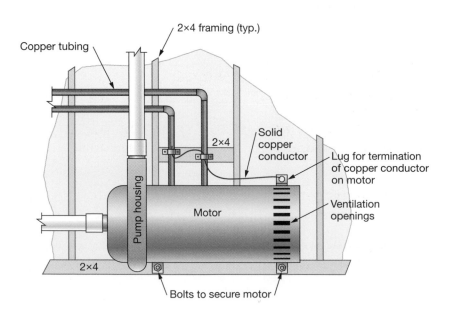

Change Summary:
- Minimum receptacle locations from pools, spas, and hot tubs have been increased from 5 feet to 6 feet.
- Receptacle requirements for permanently installed pools have been decreased from a minimum of 10 feet to 6 feet.
- GFCI requirements for receptacles near pools and tubs for therapeutic use and hydromassage bathtubs has been increased from 5 feet to 6 feet.

Specifications regarding the location of receptacles around pools, spas, and hot tubs are modified and supplemented in Article 680.

Revised 2008 NEC Text:
II. PERMANENTLY INSTALLED POOLS
680.22 AREA LIGHTING, RECEPTACLES, AND EQUIPMENT
(A) RECEPTACLES
(1) CIRCULATION AND SANITATION SYSTEM, LOCATION. (Minimum receptacle location increases from 5 feet to 6 feet from the inside walls of the pool)

(2) OTHER RECEPTACLES, LOCATION. (Minimum location for other receptacles decreases from 10 feet to 6 feet from the inside walls of a pool)

(3) DWELLING UNIT(S). (Minimum location of the required receptacle decreases from 10 feet to 6 feet from the inside walls of a pool)

(continued)

Article 680
RECEPTACLE LOCATIONS, PARTS II, III, IV, VI, AND VII

Chapter 6 Special Equipment

Article 680 Swimming Pools, Fountains, and Similar Installations

Part II Permanently Installed Pools

680.22 Area Lighting, Receptacles, and Equipment

(A) Receptacles

(1) Circulation and Sanitation System, Location

(2) Other Receptacles, Location

(3) Dwelling Unit(s)

Part III Storable Pools

680.34 Receptacle Locations

Part IV Spas and Hot Tubs

680.43 Indoor Installations

(A) Receptacles

(1) Location

Part VI Pools and Tubs for Therapeutic Use

680.62 Therapeutic Tubs (Hydrotherapeutic Tanks)

(E) Receptacles

Part VII Hydromassage Bathtubs

680.71 Protection

Change Type: Revision

(4) RESTRICTED SPACE. (This subdivision which permitted a relaxation of the minimum receptacle location from 10 feet to 5 feet is deleted. It is no longer necessary)

III. STORABLE POOLS
680.34 RECEPTACLE LOCATIONS. (Minimum location of the receptacles decreases from 10 feet to 6 feet from the inside walls of a pool)

IV. SPAS AND HOT TUBS
680.43 INDOOR INSTALLATIONS
(A) RECEPTACLES. (Minimum location of the required receptacle increases from 5 feet to 6 feet from the inside wall of the spa or hot tub)

(1) LOCATION. (Minimum receptacle locations increases from 5 feet to 6 feet from the inside wall of the spa or hot tub)

VI. POOLS AND TUBS FOR THERAPEUTIC USE
680.62 THERAPEUTIC TUBS (HYDROTHERAPEUTIC TANKS)
(E) RECEPTACLES. (GFCI requirements for receptacles increases from 5 feet to 6 feet)

VII. HYDROMASSAGE BATHTUBS
680.71 PROTECTION. (GFCI requirements for receptacles increases from 5 feet to 6 feet)

Change Significance:

Location requirements for receptacles near pools, spas, hot tubs (including those for therapeutic use), and hydromassage bathtubs have been revised. Minimum receptacle location requirements have been expanded from 5 feet to 6 feet. Standard cord length on radios and other appliances generally used in the vicinity of pools and the like is 6 feet. The expansion is designed to make the minimum distance far enough away to prevent a radio or other appliance from falling into the water.

Minimum locations of required and other receptacles have been decreased from 10 feet to 6 feet. The technical committee recognizes that the 10-foot distance was created before GFCI protection was readily available and mandated in Article 680. The minimum 6-foot distance now provides consistency throughout Article 680.

Receptacles near pools and tubs for therapeutic use and hydromassage bathtubs are required to be GFCI protected. This requirement has been increased from 5 feet to 6 feet.

Notes:

_____

_____

_____

Change Summary:

- Wiring methods permitted for use in or adjacent to natural and artificially made bodies of water are clarified. Permitted raceways and cable assemblies are listed, and references to Chapter 3 and other articles are removed.

Revised 2008 NEC Text:

682.13 WIRING METHODS AND INSTALLATION. <u>Liquidtight flexible metal conduit or liquidtight flexible nonmetallic conduit with approved fittings shall be permitted for feeders and where flexible connections are required for services. Extra-hard usage portable power cable listed for both wet locations and sunlight resistance shall be permitted for a feeder or a branch circuit where flexibility is required. Other wiring methods, suitable for the location shall be permitted to be installed where flexibility is not required. Temporary wiring in accordance with 590.4 shall be permitted.</u>

~~Wiring methods and installations of Chapter 3 and Articles 553, 555, and 590 shall be permitted where identified for use in wet locations.~~

Change Significance:

The previous text of 682.13 was confusing to the code user and not in conformance with the editorial guidelines of the NEC style manual. The general references to all of Chapter 3 and to Articles 553, 550, and 590 are deleted. Specific text is added permitting liquid-tight flexible metal and nonmetallic conduits with approved fittings for feeders and where flexible connections are required for services.

Additional text permits extra-hard-usage portable power cable listed for both wet locations and sunlight resistance for a feeder or a branch circuit where flexibility is required. Other wiring methods suitable for the location are also permitted where flexibility is not required. A new last sentence specifically permits temporary wiring methods, in accordance with 590.4, where a temporary electrical installation occurs near a natural or artificially made body of water.

682.13
WIRING METHODS AND INSTALLATION

Chapter 6 Special Equipment

Article 682 Natural and Artificially Made Bodies of Water

Part II Installation

682.13 Wiring Methods and Installation

Change Type: Revision

Electrical installations in and around natural and artificially made bodies of water must comply with Article 682.

682.30
GROUNDING

Chapter 6 Special Equipment

Article 682 Natural and Artificially Made Bodies of Water

Part III Grounding and Bonding

682.30 Grounding

Change Type: Revision

Change Summary:

- The requirements for grounding electrical installations near or in a natural or artificially made body of water are clarified. References to entire articles are moved and replaced with clear references.

Revised 2008 NEC Text:

682.30 GROUNDING. Wiring and equipment within the scope of this article shall be grounded as specified in <u>Part III of</u> ~~Articles 250,~~ 553, ~~and~~ 555.<u>15</u> and with the requirements in Part III <u>of this Article.</u>

Change Significance:

The previous text of 682.30 referenced all of Articles 250, 553, and 555. Referencing an entire article is not user friendly and is not in conformance with the editorial guidelines of the NEC style manual.

The general references to all of Articles 250, 553, and 555 are deleted. This revision clarifies the grounding requirements by specifically referencing only Part III of Article 553, Section 555.15, and Part III of Article 682.

A fish pond with a filtering pump installed near a dwelling unit is an artificially made body of water.

Change Summary:

- Numerous significant changes occurred throughout Article 690. Some of the most significant changes are listed below with brief explanations of revisions for solar PV installations.

The use of solar energy and wind energy systems provides clean renewable energy sources for the future.

Article 690
SOLAR PHOTOVOLTAIC SYSTEMS

Chapter 6 Special Equipment

Article 690 Solar Photovoltaic Systems

Part I General

Part II Circuit Requirements

Part III Disconnecting Means

Part IV Wiring Methods

Part V Grounding

Part VI Marking

Part VII Connection to Other Sources

Change Type: Revision

Revised 2008 NEC Text:
ARTICLE 690 SOLAR PHOTOVOLTAIC SYSTEMS
I. GENERAL
690.4 INSTALLATION
(D) EQUIPMENT.   *All equipment intended for use in photovoltaic power systems shall be identified and listed for the application*

II. CIRCUIT REQUIREMENTS
690.7 MAXIMUM VOLTAGE
(A) MAXIMUM PHOTOVOLTAIC SYSTEM VOLTAGE.   *Use manufacturers instructions when supplied*

III. DISCONNECTING MEANS
690.13 ALL CONDUCTORS.   *Disconnection required for <u>ungrounded</u> conductors, disconnection of grounded conductor is clarified*

IV. WIRING METHODS
690.35 UNGROUNDED PHOTOVOLTAIC POWER SYSTEMS
(D)   *Listed, identified Photovoltaic (PV) Wire now permitted*

(F)   *Revised marking requirements*

(continued)

V. GROUNDING
690.42 POINT OF SYSTEM GROUNDING CONNECTION. *Ground fault protective device may contain required connection*

690.45 SIZE OF EQUIPMENT GROUNDING CONDUCTOR. *Size of EGC clarified*

690.47 GROUNDING ELECTRODE SYSTEM
(D) GROUNDING ELECTRODES FOR ARRAY GROUNDING. *New requirement for grounding electrodes*

690.50 EQUIPMENT BONDING JUMPERS. *New section, requires compliance with 250.120(C)*

VI. MARKING
690.53 DIRECT-CURRENT PHOTOVOLTAIC POWER SOURCE. *"Permanent" marking required*

690.57 LOAD DISCONNECT. *Disconnect shall open all sources when in the off position.*

VII. CONNECTION TO OTHER SOURCES
690.62 AMPACITY OF NEUTRAL CONDUCTOR. *Minimum size grounded conductor clarified*

690.64 POINT OF CONNECTION. *Output of <u>utility-interactive inverter</u>*

(A) SUPPLY SIDE. *Output of <u>utility-interactive inverter</u> ~~photovoltaic power source~~*

(B) LOAD SIDE. *Output of <u>utility-interactive inverter</u> ~~photovoltaic power source~~*

Change Significance:
Article 690 continues to be modified as the solar PV industry becomes larger and larger. Modifications have been made throughout this article to improve clarity and usability. Modifications to 690 include safety-driven revisions for grounding the system, ground-fault protection, equipment grounding conductors, and requirements that part of these systems be listed/labeled or identified for such use.

Notes:

_____

_____

_____

Change Summary:

- 690.5 has been revised to clarify the following: (1) where ground-fault protection of grounded DC photovoltaic (PV) arrays is required, (2) opening of the grounded conductor, (3) isolating faulted circuits, and (4) labels and markings.

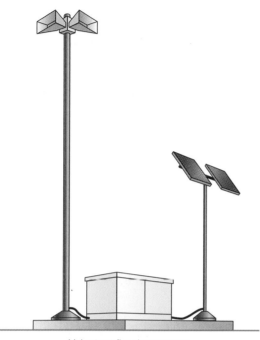

Volunteer fire department
siren system

Ground-mounted or pole-mounted photovoltaic arrays with not more than two paralleled source circuits and with all DC source and DC output circuits isolated from buildings are not required to have ground-fault protection.

690.5
GROUND-FAULT PROTECTION

Chapter 6 Special Equipment

Article 690 Solar Photovoltaic Systems

Part I General

690.5 Ground-Fault Protection

(A) Ground-Fault Detection and Interruption

(B) Isolating Faulted Circuits

(C) Labels and Markings

Change Type: Revision

Revised 2008 NEC Text:

690.5 GROUND-FAULT PROTECTION. ~~Roof-mounted~~ <u>Grounded</u> dc photovoltaic arrays ~~located on dwellings~~ shall be provided with dc ground-fault protection <u>meeting the requirements of 690.5 (A) through (C)</u> to reduce fire hazards. <u>Ungrounded dc photovoltaic arrays shall comply with 690.35.</u>

<u>Exception 1: Ground-mounted or pole-mounted photovoltaic arrays with not more than two paralleled source circuits and with all dc source and dc output circuits isolated from buildings shall be permitted without ground-fault protection.</u>

<u>Exception 2: PV arrays installed at other than dwelling units shall be permitted without ground-fault protection where the equipment-grounding conductors are sized in accordance with 690.45.</u>

(A) GROUND-FAULT DETECTION AND INTERRUPTION. The ground-fault protection device or system shall be capable of detecting a ground-fault current, interrupting the flow of fault current, and providing an indication of the fault.

<u>Automatically opening the grounded conductor of the faulted circuit to interrupt the ground-fault current path shall be permitted. If a grounded conductor is opened to interrupt the ground-fault current path, all conductors of the faulted circuit shall be automatically and simultaneously opened.</u>

(continued)

Manual operation of the main PV dc disconnect shall not activate the ground-fault protection device or result in grounded conductors becoming ungrounded.

(B) ISOLATING FAULTED CIRCUITS ~~DISCONNECTION OF CONDUCTORS~~. The faulted circuits shall be isolated by one of the two following methods.

(1) The ungrounded conductors of the faulted ~~source~~ circuit shall be automatically disconnected.

(2) The inverter or charge controller fed by the faulted circuit shall automatically cease to supply power to output circuits.

~~If the grounded conductors of the faulted source circuit are disconnected to comply with the requirements of 690.5(A), all conductors of the faulted source circuit shall be opened automatically and simultaneously. Opening the grounded conductor of the array or opening the faulted sections of the array shall be permitted to interrupt the ground-fault current path.~~

(C) LABELS AND MARKINGS.   A warning label ~~Labels and markings~~ shall appear on the utility-interactive inverter or be applied by the installer near the ground-fault indicator at a visible location, stating ~~that, if a ground fault is indicated, the normally grounded conductors may be energized and ungrounded~~ the following:

WARNING
ELECTRIC SHOCK HAZARD
IF A GROUND FAULT IS INDICATED,
NORMALLY GROUNDED CONDUCTORS MAY
BE UNGROUNDED AND ENERGIZED

When the photovoltaic system also has batteries, the same warning shall also be applied by the installer in a visible location at the batteries.

Change Significance:

The parent text of 690.5 is revised to clarify that this section applies to "grounded" DC photovaltic (PV) arrays. Two new exceptions are added to 690.5. The first exception clarifies that ground or pole-mounted arrays with not more than two paralleled source circuits (DC source and DC output) isolated from buildings do not require ground-fault protection (GFP). The second allows PV arrays on other than dwelling units to be permitted without GFP where the EGC is sized in accordance with 690.45.

690.5(A) is revised, permitting the GFP to open the grounded conductor provided all conductors of the faulted circuit are opened automatically and simultaneously. Additional new text requires that manual operation of the PV disconnect shall not operate the GFP or result in grounded conductors becoming ungrounded. 690.5(B) is revised to more clearly address how faulted circuits are to be isolated. The deleted text from 690.5(B) is addressed in revisions to 690.5(A).

695.5(C) is revised to require a warning label (1) on the utility-interactive inverter or one applied by the installer near the ground-fault indicator at a visible location and (2) that the same warning shall also be applied by the installer in a visible location at the batteries if present in the system. The warning explains the possibility of grounded conductors becoming energized.

Change Summary:

- The requirements for stand-alone solar PV systems are revised to clarify the size of the inverter output. A new first-level subdivision is added to clarify that both energy storage and backup power supplies are not required.

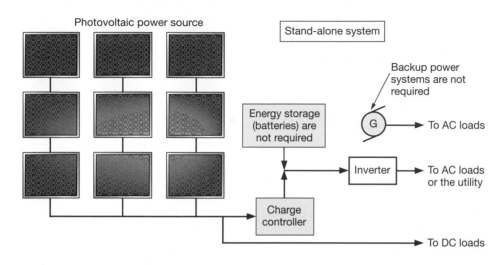

A solar photovoltaic system may be installed with a direct grid tie to sell power back to the utility grid.

690.10
STAND-ALONE SYSTEMS

Chapter 6 Special Equipment

Article 690 Solar Photovoltaic Systems

Part II Circuit Requirements

690.10 Stand-Alone Systems

(A) Inverter Output

(D) Energy Storage or Backup Power System Requirements

Change Type: Revision

Revised 2008 NEC Text:

690.10 STAND-ALONE SYSTEMS   The premises wiring system shall be adequate to meet the requirements of this Code for a similar installation connected to a service. The wiring on the supply side of the building or structure disconnecting means shall comply with this Code except as modified by 690.10(A), <u>through (D).</u><s>(B), and (C).</s>

(A) INVERTER OUTPUT.   The ac <s>inverter</s> output from a stand-alone <u>inverter(s)</u> <s>system</s> shall be permitted to supply ac power to the building or structure disconnecting means at current levels <s>below the rating of that disconnecting means</s> <u>less than the calculated load connected to that disconnect. The inverter output rating or the rating of an alternate energy source shall be equal to or greater than the load posed by the largest single utilization equipment connected to the system. Calculated, general lighting loads shall not be considered as a single load.</u>

(B)   (No change)

(C)   (No change)

(D) ENERGY STORAGE OR BACKUP POWER SYSTEM REQUIREMENTS. <u>Energy storage or backup power supplies are not required.</u>

Change Significance:

As previously written, this requirement provided no minimum or maximum size for the inverter output. This revision clarifies that the inverter

(continued)

output need only be sized equal to or greater than the largest single load connected to the system. Many stand-alone solar PV systems are supplemented with other alternative energy sources such as generators of wind systems. Owners of stand-alone systems facilitate the use of these systems through conservation of energy. Sizing the inverter for the largest single load provides the system with the ability to handle that, whereas larger loads may be supplied by other alternative energy sources such as a generator. Lighting is not permitted as the largest load because it is owner controlled and may be turned off completely.

A new first-level subdivision (D) clarifies that a stand-alone system does not require a system of batteries for storage and no backup power system is required. A stand-alone system may be installed to pump water only during daylight hours to feed livestock. Each system is designed to supply the needs of the individual installation.

This installation consists of multiple inverters with a grid tie to supply power to the local utility.

Change Summary:

- New text in 690.31(A) requires that where PV source and output circuits operating at greater than 30 volts are installed in a readily accessible location they must be enclosed in a metallic raceway.

- Single-conductor cables used in exposed outdoor locations for PV source circuits for PV module interconnections within the PV array are now required to be type USE-2 or be listed and labeled as PV wire.

- Flexible, fine-stranded cables are now required to be terminated with terminals, lugs, devices, or connectors that are identified and listed for such use.

690.31
METHODS
PERMITTED

Chapter 6 Special Equipment

Article 690 Solar Photovoltaic Systems

Part IV Wiring Methods

690.31 Methods Permitted

(A) Wiring Systems

(B) Single-Conductor Cable

(F) Flexible, Fine-Stranded Cables

Change Type: Revision

Wiring methods for solar photovoltaic systems are supplemented and modified in Article 690.

Revised 2008 NEC Text:
690.31 METHODS PERMITTED
(A) WIRING SYSTEMS. All raceway and cable wiring methods included in this Code and other wiring systems and fittings specifically intended and

(continued)

identified for use on photovoltaic arrays shall be permitted. Where wiring devices with integral enclosures are used, sufficient length of cable shall be provided to facilitate replacement.

<u>Where photovoltaic source and output circuits operating at maximum system voltages greater than 30 volts are installed in readily accessible locations, circuit conductors shall be installed in a raceway.</u>

FPN: (No change)

(B) SINGLE-CONDUCTOR CABLE.   <u>Single-conductor cable</u> ~~T~~types ~~SE, UF, USE, and~~ USE-2 <u>and</u> single-conductor cable <u>listed and labeled as Photovoltaic (PV) wire</u> shall be permitted <u>in exposed outdoor locations</u> in photovoltaic source circuits <u>for photovoltaic module interconnections within the photovoltaic array.</u> ~~where installed in the same manner as a Type UF multiconductor cable in accordance with Part II of Article 340. Where exposed to sunlight, Type UF cable identified as sunlight-resistant shall be used.~~

Exception: Raceways shall be used when required by 690.31(A).

(C) & (D)   (No change), **(E)**   (<u>metallic</u> enclosures)

(F) FLEXIBLE, FINE-STRANDED CABLES.   <u>Flexible, fine-stranded cables shall only be terminated with terminals, lugs, devices, or connectors that are identified and listed for such use.</u>

Change Significance:

690.31(A) is revised to require that all PV source and output circuits operating at system voltages greater than 30 volts, which are installed in readily accessible locations, shall have circuit conductors installed in a metallic raceway. These systems may operate at 600 volts. Physical protection of these conductors is now required where they are readily accessible. A roof mounted installation on a dwelling unit, would require a ladder to access the wiring and would not be considered readily accessible.

Single-conductor cables permitted are now limited to type USE-2 or cables which are listed and labeled as PV wire. As noted in the existing FPN, temperatures of conductors in direct sunlight with the heat of the modules, may routinely exceed 70°C. Conductors must be rated for 90°C and be rated for wet locations. It is for these reasons that cable types SE, UF, and USE are deleted. A new standard has been developed for listed and labeled PV wire. UL 4703 requires a 90°C, wet-rated insulation that is more durable than SE and USE cable insulation and it has passed the long-duration 720-hour accelerated sunlight/UV exposure tests allowing wire to be marked as sunlight resistant.

Where flexible, fine-stranded cables are used in a solar PV system they are now required to be terminated with terminals, lugs, devices, or connectors that are identified and listed for such use. The lack of proper termination has led to multiple documented cable failures. A similar requirement is added to 690.74 for battery terminations.

Change Summary:

* Connectors in solar PV systems operating at a system or nominal voltage of over 30 volts now require a tool to open if installed in a readily accessible location. Connectors must also be rated for safely interrupting current or marked to warn the installer or maintainer.

Revised 2008 NEC Text:

690.33 CONNECTORS.   The connectors permitted by Article 690 shall comply with 690.33(A) through 690.33(E).

(A) & (B)   (No change)

(C) TYPE.   The connectors shall be of the latching or locking type. <u>Connectors that are readily accessible and that are used in circuits operating at over 30 volts, nominal, maximum system voltage for dc circuits, or 30 volts for ac circuits, shall require a tool to open.</u>

(D)   (No change)

(E) INTERRUPTION OF CIRCUIT.   ~~The~~ ~~c~~Connectors shall be <u>either (1) or (2):</u> ~~capable of~~
(1) <u>Be rated for</u> interrupting ~~the circuit~~ current without hazard to the operator.
(2) <u>Be a type that requires the use of a tool to open and marked "Do Not Disconnect Under Load" or "Not for Current Interrupting."</u>

Change Significance:

690.33(C) requires that connectors be of the latching or locking type. New text now requires that where connectors are installed in a readily accessible location, circuits operating at over 30 volts (maximum system voltage for DC circuits or nominal voltage for ac circuits) shall require a tool to open. This revision mirrors the new requirement in 690.31(A) for conductors to be installed in metallic raceways where in a readily accessible location. This revision is intended to prevent unqualified persons from opening circuits which could be operating at up to 600 volts.

690.33(E) now permits connectors to interrupt the circuit when they are "rated" for the current or they can only be opened with a tool, in a de-energized state, as marked.

Connectors are now required in 690.33(E)(1) to be "rated" for interrupting current without hazard to the operator. This requirement is more stringent than the previous requirement that the connector be "capable" of interrupting current. 690.33(E)(2) permits connector types which are not rated for interrupting current, provided a tool is required to open the connector and they are marked, "Do Not Disconnect Under Load" or "Not for Current Interrupting."

690.33
CONNECTORS

Chapter 6 Special Equipment

Article 690 Solar Photovoltaic Systems

Part IV Wiring Methods

690.33 Connectors

(C) Type

(E) Interruption of Circuit

Change Type: Revision

Readily accessible connectors in systems operating more than 30 volts now require a tool to open

Part IV of Article 690 provides requirements for wiring methods that modify and supplement the provisions of Chapter 3 of the NEC.

690.43
EQUIPMENT GROUNDING

Chapter 6 Special Equipment

Article 690 Solar Photovoltaic Systems

Part V. Grounding

690.43 Equipment Grounding

Change Type: Revision

690.43 supplements and modifies the provisions of Article 250 with respect to equipment grounding conductors. 250. An EGC 6 AWG and smaller may be green or green with one or more yellow stripes.

Change Summary:
* Revisions to 690.43 include: (1) a new last sentence requiring an EGC in accordance with 250.110; (2) a new second paragraph is added to address devices which are listed and identified for grounding and/or bonding the metallic frames of PV modules; and (3) a new third paragraph is added to address installation of the EGC.

Revised 2008 NEC Text:
690.43 EQUIPMENT GROUNDING.   Exposed non-current-carrying metal parts of module frames, equipment, and conductor enclosures shall be grounded in accordance with 250.134 or 250.136(A) regardless of voltage. An equipment grounding conductor between a PV array and other equipment shall be required in accordance with 250.110.

Devices listed and identified for grounding the metallic frames of PV modules shall be permitted to ground the exposed metallic frames of PV modules to grounded mounting structures. Devices identified and listed for bonding the metallic frames of PV modules shall be permitted to bond the exposed metallic frames of PV modules to the metallic frames of adjacent PV modules.

Equipment-grounding conductors for the PV array and structure (where installed) shall be contained within the same raceway or cable, or otherwise run with the PV array circuit conductors when those circuit conductors leave the vicinity of the PV array.

Change Significance:
A new second sentence is added to the first paragraph of 690.43 requiring an EGC in accordance with 250.110. Problems occur when installers do not install an EGC. In many cases a supplementary grounding electrode system is installed for the installation and installers do not install an EGC. This can lead to very dangerous situations, placing voltages on electrodes, arrays, and misapplication of ground fault devices.

A new second paragraph is added to recognize module grounding clips and other devices which are being developed and listed that will effectively penetrate the oxide or anodizing on aluminum framed PV modules. These module grounding clips and other devices will be listed to ground the modules to grounded PV array mounting structures or to effectively bond them to adjacent PV modules which, in turn, may be grounded. This permission in Article 690 supplements and modifies the requirements of 250.134 and 250.136.

A new third paragraph is added to require that EGCs are installed in the same raceway or cable, or are otherwise run with the circuit conductors when those circuit conductors leave the vicinity of the PV array. When EGCs are not run in close proximity to circuit conductors in a DC system, overcurrent protective devices may not operate properly. By eliminating spacing and separation of circuit conductors from the EGC, time constants are kept low, ensuring proper operation of overcurrent protective devices.

Change Summary:
- Requirements for sizing equipment grounding conductors for PV source and PV output circuits are completely revised.

Revised 2008 NEC Text:
(This is a complete revision, strikethrough and underline are omitted)

690.45 SIZE OF EQUIPMENT-GROUNDING CONDUCTOR. Equipment grounding conductors for photovoltaic source and photovoltaic output circuits shall be sized in accordance with 690.45(A) or (B).

(A) Equipment-grounding conductors in photovoltaic source and photovoltaic output circuits shall be sized in accordance with Table 250.122. When no overcurrent protective device is used in the circuit, an assumed overcurrent device rated at the photovoltaic rated short-circuit current shall be used in Table 250.122. Increases in equipment-grounding conductor size to address voltage drop considerations shall not be required. The equipment-grounding conductors shall be no smaller than 14 AWG.

(B) For other than dwelling units where ground-fault protection is not provided in accordance with 690.5(A) through (C), each equipment grounding conductor shall have an ampacity of at least two (2) times the temperature and conduit fill corrected circuit conductor ampacity.

FPN: The short circuit current of photovoltaic modules and photovoltaic sources is just slightly above the full load normal output rating. In ground fault conditions, these sources are not able to supply the high levels of short-circuit or ground-fault currents necessary to quickly activate overcurrent devices as in typical ac systems. Protection for equipment grounding conductors in photovoltaic systems that are not provided with ground-fault protection is related to size and withstand capability of the equipment grounding conductor, rather than overcurrent device operation.

Change Significance:
This revision inserts a general rule for all equipment grounding conductors installed in PV source and PV output circuits. The general rule is to size equipment grounding conductors according to Table 250.122 in accordance with the upstream OCPD or an assumed upstream OCPD. The requirement for sizing equipment grounding conductors at 125% where no ground fault protection exists is removed. The requirement for larger equipment grounding conductors is now limited to "other than dwelling units where ground-fault protection is not provided in accordance with 690.5(A) through (C)." These installations will now require the equipment grounding conductors to be sized at twice the required ampacity of the circuit conductors.

The reason for sizing the EGC at twice the ampacity of the circuit conductors is to ensure the survivability of the EGC. A new informational FPN is added to explain the reason for larger EGCs.

690.45
SIZE OF EQUIPMENT-GROUNDING CONDUCTOR

Chapter 6 Special Equipment

Article 690 Solar Photovoltaic Systems

Part V. Grounding

690.45 Size of Equipment-Grounding Conductor

Change Type: Revision

Solar photovoltaic systems have unique characteristics and must have equipment grounding conductors sized in accordance with 690.45.

692.41
SYSTEM GROUNDING

Chapter 6 Special Equipment

Article 692 Fuel Cell Systems

Part V Grounding

692.41 System Grounding

(A) AC Systems

(B) DC Systems

(C) Systems With Alternating-Current and Direct-Current Grounding Requirements

Change Type: Revision and New

Change Summary:
- Fuel cell system grounding requirements have been clarified in three first-level subdivisions as follows: 692.41(A) AC systems, (B) DC systems, and (C) systems with AC and DC.

Revised 2008 NEC Text:
692.41 SYSTEM GROUNDING. (Existing text completely deleted and revised)

(A) AC SYSTEMS. Grounding of ac systems shall be in accordance with 250.20, and 250.30 for stand-alone systems.

(B) DC SYSTEMS. Grounding of dc systems shall be in accordance with 250.160.

(C) SYSTEMS WITH ALTERNATING-CURRENT AND DIRECT-CURRENT GROUNDING REQUIREMENTS. When fuel cell power systems have both alternating-current (ac) and direct-current (dc) grounding requirements, the dc grounding system shall be bonded to the ac grounding system. The bonding conductor shall be sized according to 692.45. A single common grounding electrode and grounding bar may be used for both systems, in which case the common grounding electrode conductor shall be sized to meet the requirements of both 250.66(ac) and 250.166(dc).

Change Significance:
The existing requirements of 692.41 are deleted and replaced for clarity with new first-level subdivisions (A) for AC systems and (B) for DC systems. AC systems must be installed in accordance with 250.20 and 250.30 for stand-alone systems, and DC systems in accordance with 250.160.

A new first-level subdivision is added to address systems with both AC and DC grounding requirements. This requires a bonding jumper sized in accordance with 250.122 to bond the AC and DC grounding systems. A single common grounding electrode system is permitted. The common grounding electrode conductor must be installed in accordance with both 250.66 for AC systems and 250.166 for DC systems. This revision ensures that bonding will occur between the AC and DC grounding systems.

Fuel cell systems may be designed to deliver DC current, AC current, or both.

Change Summary:
- Fire pump supply conductors are now required to be encased in a minimum of two inches of concrete or a 2-hour fire rating.

Revised 2008 NEC Text:
695.6 POWER WIRING

(B) CIRCUIT CONDUCTORS.   Fire pump supply conductors on the load side of the final disconnecting means and overcurrent device(s) permitted by 695.4(B) shall be kept entirely independent of all other wiring. They shall supply only loads that are directly associated with the fire pump system, and they shall be protected to resist potential damage by fire, structural failure or operational accident. They shall be permitted to be routed through a building(s) using one of the following methods:

(1) Be encased in a minimum 50 mm (2 in.) of concrete

(2) Be ~~within an enclosed construction~~ <u>protected by a fire-rated assembly listed to achieve a minimum fire rating of 2-hour and</u> dedicated to the fire pump circuit(s)<u>.</u> ~~and having a minimum of a 1-hour fire resistive rating~~

(3) Be a listed electrical circuit protective system with a minimum ~~1~~<u>2</u>-hour fire rating

<u>FPN: UL guide information for electrical circuit protective systems (FHIT) contains information on proper installation requirements to maintain the fire rating.</u>

Exception: The supply conductors located in the electrical equipment room where they originate and in the fire pump room shall not be required to have the minimum 1-hour fire separation or fire resistance rating, unless otherwise required by 700.9(D) of this code.

Change Significance:
This revision will now require 2-hour fire-rated enclosures and/or listed 2-hour fire-rated electrical circuit protective systems. Previous code text permitted a 1-hour rating. There are three options listed for the installation of the fire pump supply conductors on the load side of the permitted disconnecting means and overcurrent protection in 695.4(B).

(1) Encase a raceway in 2 inches of concrete. This mirrors the requirements of service conductors in 230.6. These conductors could be considered outside the building.

(2) Protect the conductors with a fire-rated assembly, such as building an enclosure dedicated to the fire pump supply conductors—which could include layering core-board or drywall to achieve a 2-hour rating.

(3) Install a listed electrical circuit protective system rated at 2 hours, such as type MI cable or conductors—which are installed in metal raceways or metal cable armor and have a 2-hour rating.

It is important to note that purchasing a 2-hour-rated electrical circuit protective system is does not automatically achieve the 2-hour rating. The system must be installed and supported in accordance with the manufacturer's instructions. A new FPN is added to provide the code user with a UL reference for information on the installation requirements to achieve a 2-hour rating.

695.6(B)
CIRCUIT CONDUCTORS

Chapter 6 Special Equipment

Article 695 Fire Pumps

695.6 Power Wiring

(B) Circuit Conductors

Change Type: Revision

A 2-hour fire rating is now required for fire pump circuit conductors.

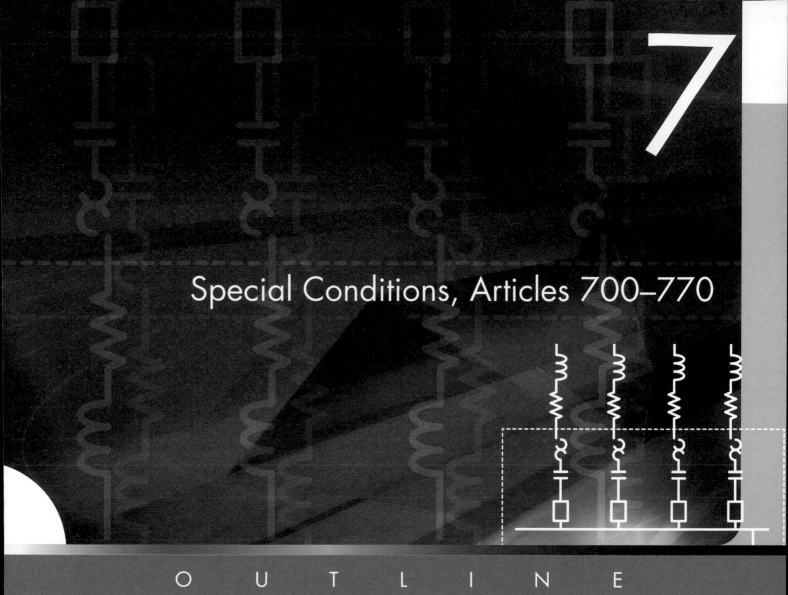

7

Special Conditions, Articles 700–770

OUTLINE

Change Summary:

- Transfer switches used in emergency or legally required standby systems are now required to be listed for that purpose.

Revised 2008 NEC Text:

700.6 TRANSFER EQUIPMENT

(C) AUTOMATIC TRANSFER SWITCHES. Automatic transfer switches shall be electrically operated and mechanically held. <u>Automatic transfer switches, rated 600 VAC and below, shall be listed for emergency system use.</u>

701.7 TRANSFER EQUIPMENT

(C) AUTOMATIC TRANSFER SWITCHES. Automatic transfer switches shall be electrically operated and mechanically held. <u>Automatic transfer switches, rated 600 VAC and below, shall be listed for legally required standby system use.</u>

Change Significance:

This revision now requires that transfer switches used for emergency systems or legally required standby systems be listed for the purpose. Substantiation provided for this revision points out serious problems that can occur if the transfer equipment is not listed for the purpose of emergency or legally required standby systems.

Problems with equipment not listed for the purpose can occur during testing or generator exercise, when both sources are available in the transfer switch. The system voltages will be out of phase with each other and the switch must be suitable for switching between out-of-phase power sources. Where a short circuit occurs downstream of the transfer switch, a loss of power brings in the alternate source. The transfer switch must be suitable to close in a short-circuit situation.

700.6(C) and 701.7(C)
AUTOMATIC TRANSFER SWITCHES

Chapter 7 Special Conditions

Article 700 Emergency Systems

Part I General

700.6 Transfer Equipment

(C) Automatic Transfer Switches

Article 701 Legally Required Standby Systems

Part I General

701.7 Transfer Equipment

(C) Automatic Transfer Switches

Change Type: Revision

A transfer switch is used to supply power to a load from both normal sources and alternative sources in an emergency, legally required, or optional standby system.

700.9(B)
WIRING

Chapter 7 Special Conditions

Article 700 Emergency Systems

Part II Circuit Wiring

700.9 Wiring, Emergency System

(B) Wiring

Change Type: Revision and New

Change Summary:
- A new list item 700.9(B)(5) is added to clarify separation of emergency circuits from other wiring and equipment at the source of supply.

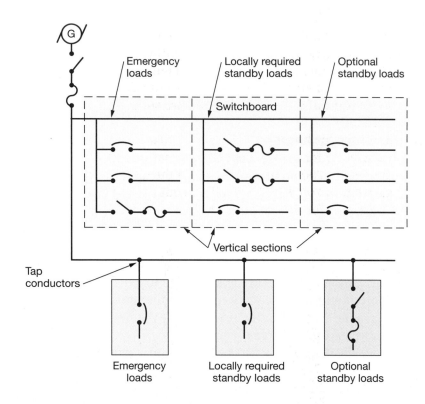

In general, all wiring for emergency loads must be kept entirely independent of other wiring.

Revised 2008 NEC Text:
700.9 WIRING, EMERGENCY SYSTEM

(B) WIRING.   Wiring of two or more emergency circuits supplied from the same source shall be permitted in the same raceway, cable, box, or cabinet. Wiring from an emergency source or emergency source distribution overcurrent protection to emergency loads shall be kept entirely independent of all other wiring and equipment, unless otherwise permitted in (1) through ~~(4)~~ (5): (Note: (1) through (4) No change)

(5) Wiring from an emergency source shall be permitted to supply any combination of emergency, legally required, or optional loads in accordance with (a), (b) and (c).

(a) From separate vertical switchboard sections, with or without a common bus, or from individual disconnects mounted in separate enclosures.

(b) The common bus or separate sections of the switchboard or the individual enclosures shall be permitted to be supplied by single or multiple feeders without overcurrent protection at the source.

Exception to (5) (b). Overcurrent protection shall be permitted at the source or for the equipment, provided the overcurrent protection is selectively coordinated with the down stream overcurrent protection.

(continued)

(c) Legally required and optional standby circuits shall not originate from the same vertical switchboard section, panelboard enclosure or individual disconnect enclosure as emergency circuits.

Change Significance:

This new list item 700.9(B)(5) clarifies the separation of emergency circuits at the source of supply. The technical committee has clearly conveyed that it is permitted to supply any combination of emergency, legally required, or optional loads from a single feeder or from multiple feeders or from separate vertical sections of a switchboard that are supplied by either a common bus or individually.

The committee further stated that the use of an overcurrent protective device at the source or for the equipment is a matter of reliability and design. The requirements of 700.9(B)(5)(b) maintain the highest degree of reliability, and the exception to 700.9(B)(5)(b) will permit the use of an overcurrent device at the source or for the equipment.

700.9(B)(5) now clearly permits an emergency source to supply a combination of emergency, legally required, or optional loads.

700.9(D)
FIRE PROTECTION

Chapter 7 Special Conditions

Article 700 Emergency Systems

Part II Circuit Wiring

700.9 Wiring, Emergency System

(D) Fire Protection

(1) Feeder-Circuit Wiring

(2) Feeder Circuit Equipment

(3) Generator Control Wiring

Change Type: Revision

Change Summary:
- 700.9(D)(1)(4) is revised to prohibit emergency feeder-circuit wiring from being enclosed in fire-rated assemblies with other wiring circuits.
- A new 700.9(D)(3) requires control conductors installed between the transfer equipment and the emergency generator to be installed in spaces fully protected by suppression systems and entirely independent of other wiring.

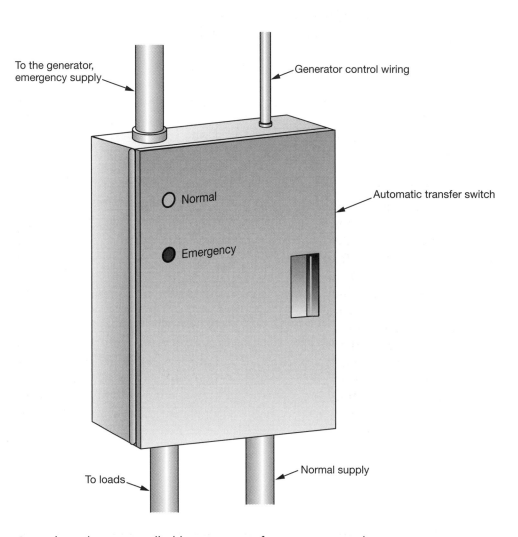

Control conductors installed between transfer equipment and an emergency generator must be kept entirely independent of *all* other wiring.

Revised 2008 NEC Text:
700.9 WIRING, EMERGENCY SYSTEM
(D) FIRE PROTECTION.   Emergency systems shall meet the additional requirements in 700.9(D)(1) and (D)(2) assembly occupancies for not less than 1000 persons or in buildings above 23 m (75 ft) in height with any of the following occupancy classes: assembly, educational, residential, detention and correctional, business, and mercantile.

(continued)

(1) FEEDER-CIRCUIT WIRING.   Feeder-circuit wiring shall meet one of the following conditions:

(1), & (3) (No change)

(2) Be a listed electrical protective system with a minimum 1-hour fire rating.

FPN: UL guide information for electrical circuit protective systems (FHIT) contains information on proper installation requirements to maintain the fire rating.

(4) Be protected by a <u>listed</u> fire-rated assembly ~~listed to achieve~~ <u>that has</u> a minimum fire rating of 1 hour <u>and contains only emergency wiring circuits.</u>

(2) FEEDER CIRCUIT EQUIPMENT.   (No change)

(3) <u>GENERATOR CONTROL WIRING.</u>   <u>Control conductors installed between the transfer equipment and the emergency generator shall be kept entirely independent of all other wiring and shall meet the conditions of 700.9(D)(1).</u>

Change Significance:

This revision adds a new fine print note (FPN) to 700.9(D)(1)(2) to provide the code user with a reference to information regarding the installation of a fire-rated electrical protective system. These systems must be installed in accordance with manufacturers' instructions to maintain a 1- or 2-hour fire rating. List item 700.9(D)(1)(4) has been revised to clearly prohibit emergency circuits from being enclosed in a fire-rated assembly, such as a drywall shaft, with other wiring.

Such an installation must now consist of only emergency circuits wrapped in a fire-rated assembly such as drywall. The generator control circuit is a critical element in an emergency system. A new second-level subdivision (700.9(D)(3), Generator Control Wiring) requires control conductors installed between the transfer equipment and the emergency generator to be installed spaces fully protected by suppression systems and entirely independent of other wiring.

Notes:

_____

_____

_____

_____

_____

700.12(B)(6), 701.11(B)(5), and 702.11

GENERATOR DISCONNECTING MEANS LOCATED OUTDOORS

Change Type: Revision

Change Summary:
- An outdoor generator disconnecting means that serves a building or structure in an emergency, legally required standby, or optional standby system is required to be suitable for use as service equipment.

Revised 2008 NEC Text:
700.12(B)(6), 701.11(B)(5), 702.11 OUTDOOR GENERATOR SETS. Where an outdoor housed generator set is equipped with a readily accessible disconnecting means located within sight of the building or structure supplied, an additional disconnecting means shall not be required where ungrounded conductors serve or pass through the building or structure. The disconnecting means shall meet the requirements of 225.36.

Change Significance:
A new last sentence has been added to 700.12(B)(6), 701.11(B)(5), and 702.11—which contain requirements for outdoor generator sets. This new text requires that the disconnecting means meet the requirements of 225.36, which requires that the disconnecting means be "suitable for use as service equipment."

In most cases, a generator used in an Article 700, 701, or 702 application is used as a backup or standby source to resupply all or part of the electrical distribution system in the building or structure through transfer switches. These conductors from a generator to the building or structure are feeders. They are also "outdoor feeders," which means that the provisions of Article 225 apply unless supplemented or modified in 700, 701, or 702.

Optional standby systems installed as an alternate supply system for dwelling units must meet the requirements of Chapter 1 through 4 as supplemented or modified by Article 702.

Change Summary:
- Dimmer systems listed for use in emergency systems are now permitted to be used as a control device for energizing emergency lighting circuits.

Revised 2008 NEC Text:
700.23 DIMMER SYSTEMS. A dimmer system containing more than one dimmer and listed for use in emergency systems shall be permitted to be used as a control device for energizing emergency lighting circuits. Upon failure of normal power, the dimmer system shall be permitted to selectively energize only those branch circuits required to provide minimum emergency illumination. All the branch circuits supplied by the dimmer system cabinet shall comply with the wiring methods of Article 700.

Change Significance:
This new section clarifies that a dimmer system listed for emergency use may be applied to energize only those circuits required for minimum emergency illumination levels, rather than all circuits in the dimmer cabinet. Dimmer systems listed for emergency use under UL 924 and containing more than one dimmer are now common. These systems include a method to sense failure of normal power and selectively energize branch circuits fed from the dimmer cabinet using a reliable method, regardless of the setting of control switches or panels normally used to control the dimmer system.

700.23
DIMMER SYSTEMS

Chapter 7 Special Conditions

Article 700 Emergency Systems

Part V Control— Emergency Lighting Circuits

700.23 Dimmer Systems

Change Type: New

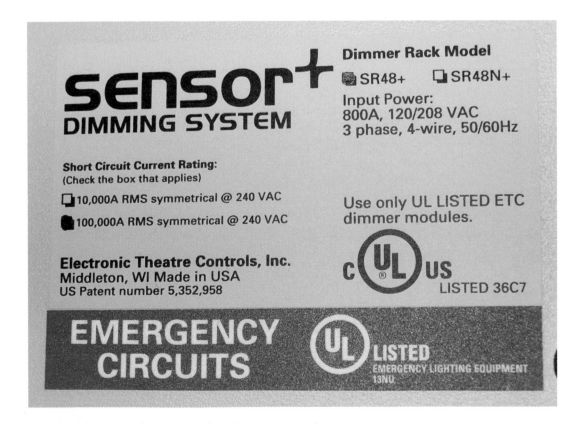

A dimmer system is sometimes used to control all lighting in a given area, including lighting served from both normal and emergency power sources. Courtesy of Electronic Theater Controls Inc.

700.27 and 701.18

COORDINATION

Chapter 7 Special Conditions

Article 700 Emergency Systems

Part VI Overcurrent Protection

700.27 Coordination

Article 701 Legally Required Standby Systems

Part IV Overcurrent Protection

701.18 Coordination

Change Type: New

Change Summary:
* An exception is added to 700.27 and 701.18 relaxing the selectivity requirement for two overcurrent devices in series under given conditions.

Revised 2008 NEC Text:
700.27 COORDINATION. Emergency system(s) . . .

701.18 COORDINATION. Legally required standby systems . . . overcurrent devices shall be selectively coordinated with all supply side overcurrent protective devices.

Exception: Selective coordination shall not be required in (1) or (2):
(1) Between transformer primary and secondary overcurrent protective devices, where only one overcurrent protective device or set of overcurrent protective devices exist(s) on the transformer secondary.
(2) Between overcurrent protective devices of the same size (ampere rating) in series.

Change Significance:
This new exception recognizes two situations in an emergency or legally required standby distribution system that if not selectively coordinated will not impact the safety or reliability of the remainder of the system. The first situation is where a transformer is installed with a single circuit breaker or single set of fuses on both the transformer primary and secondary. These two overcurrent devices (primary and secondary) are not required to be selectively coordinated because the activation of one or both overcurrent devices results in the same loss of power in the emergency or legally required standby system.

The second part of the exception relaxes the selective coordination requirement for two overcurrent devices of the same size or amp rating installed in series. An example would be a circuit breaker supplying a 200-amp feeder that supplies a 200-amp panelboard with a 200-amp main circuit breaker. These two overcurrent devices (feeder and panelboard main) are not required to be selectively coordinated because the activation of one or both overcurrent devices results in the same loss of power in the emergency and/or legally required standby system. This revision does not decrease the safety or reliability of the emergency system.

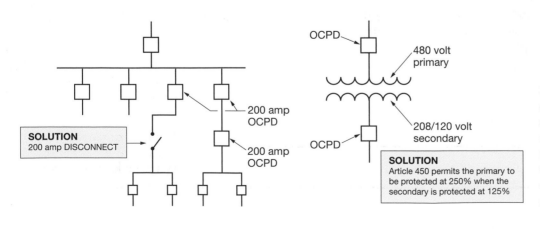

SOLUTION
200 amp DISCONNECT

200 amp OCPD

200 amp OCPD

OCPD

480 volt primary

208/120 volt secondary

OCPD

SOLUTION
Article 450 permits the primary to be protected at 250% when the secondary is protected at 125%

Selective Coordination is essential in emergency and legally required standby systems to restrict the loss of power from an overcurrent to only the affected equipment.

Change Summary:
*   702.5 is clarified, providing the code user with clear text for optional standby systems operated with manual or automatic transfer.

Revised 2008 NEC Text:
702.5 CAPACITY AND RATING
(A) AVAILABLE SHORT CIRCUIT CURRENT. Optional standby system equipment shall be suitable for the maximum available ~~fault~~ <u>short-circuit</u> current at its terminals.

(B) SYSTEM CAPACITY. <u>The calculations of load on the standby source shall be made in accordance with Article 220 or by another approved method.</u>

(1) MANUAL TRANSFER EQUIPMENT. <u>Where manual transfer equipment is used</u> an optional standby system shall have adequate capacity and rating for the supply of all equipment intended to be operated at one time. The user of the optional standby system shall be permitted to select the load connected to the system.

(2) AUTOMATIC TRANSFER EQUIPMENT. <u>Where automatic transfer equipment is used, an optional standby system shall comply with (2)(a) or (2)(b).</u>
    (a) *Full Load.* <u>The standby source shall be capable of supplying the full load that is transferred by the automatic transfer equipment.</u>
    (b) *Load Management.* <u>Where a system is employed that will automatically manage the connected load, the standby source shall have a capacity sufficient to supply the maximum load that will be connected by the load management system.</u>

Change Significance:
This revision separates the requirements of 702.5 in two first-level subdivisions. 702.5(A), Available Short Circuit Current, is user friendly and clarifies the requirement that optional standby equipment must be rated for the short-circuit current at its terminals.

702.5(B), System Capacity, requires compliance with Article 220 or another approved method. Some areas of the country may allow sizing based on recorded load measurements. 702.5(B) is further separated in two second-level subdivisions. The first, 702.5(B)(1), addresses systems with manual transfer equipment and allows the user to connect the load to the system. The second, 702.5(B)(2), addresses systems with automatic transfer equipment and two options are provided: the source handles the full load or a load management system is employed.

Notes:

_____

_____

702.5
CAPACITY AND RATING

Chapter 7 Special Conditions

Article 702 Optional Standby Systems

Part I General

702.5 Capacity and Rating

(A) Available Short-Circuit Current

(B) System Capacity

(1) Manual Transfer Equipment

(2) Automatic Transfer Equipment

(a) Full Load

(b) Load Management

Change Type: Revision and New

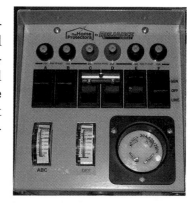

Manual transfer equipment allows the user/operator to select the loads connected to the system.

Article 705
INTERCONNECTED ELECTRIC POWER PRODUCTION SOURCES

Chapter 7 Special Conditions

Article 705 Interconnected Electric Power Production Sources

Part I General

Part II Utility Interactive Inverters

Part III Generators

Change Type: Revision

Change Summary:
- Article 705 is significantly revised for clarity and usability and to coordinate with Article 690, Solar Photovoltaic Systems and Article 692, Fuel Cell Systems.

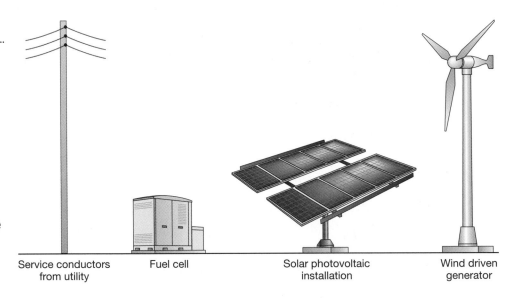

Service conductors from utility   Fuel cell   Solar photovoltaic installation   Wind driven generator

Article 705 covers the installation of one or more electric power production sources operating in parallel with a primary/utility source.

Revised 2008 NEC Text:
The following is an outline of the significant changes in Article 705

705 INTERCONNECTED ELECTRIC POWER PRODUCTION SOURCES
I. GENERAL
705.2 DEFINITIONS
~~INTERACTIVE SYSTEM.~~   *Moved to Article 100*

HYBRID SYSTEM.   *New definition to coordinate with multiple sources*

POINT OF COMMON COUPLING.   *New to define point at which power production and distribution network and the customer interface meet in an interactive system*

UTILITY-INTERACTIVE INVERTER OUTPUT CIRCUIT.   *New definition to define conductors between the utility interactive inverter and the service equipment*

705.4 EQUIPMENT APPROVAL.   *New requirement for approved equipment and liste interconnection systems*

705.12 POINT OF CONNECTION
(A) SUPPLY SIDE.   *Permits power production source on supply side of service disconnecting means, 230.82(6)*

(continued)

(B) INTEGRATED ELECTRIC SYSTEMS.   *Relocated from 705.12(A)*

(C) GREATER THAN 100KW ~~GENERAL~~.   *Rename for clarity. Documented procedures now required*

(D) UTILITY INTERACTIVE INVERTERS.   *Addresses the output of the inverter*

(1) DEDICATED OVERCURRENT AND DISCONNECT.   *Each source protected interconnection*

(2) BUS OR CONDUCTOR RATING.   *Sum of OCPD not over 120%*

(3) GROUND-FAULT PROTECTION.   *Interconnection on line side of GFP*

(4) MARKING.   *Presence of all sources identified*

(5) SUITABLE FOR BACKFEED.   *Circuit breakers must be suitable*

(6) FASTENING.   *Listed devices not required to be fastened per 408.36(D)*

(7) INVERTER OUTPUT CONNECTION.   *Size and marking requirement*

705.30 OVERCURRENT PROTECTION
~~(A) GENERATORS~~   *Relocated to new 705.130 in new Part III, Generators*

(D) UTILITY INTERACTIVE INVERTERS.   *Protection as per 705.65*

705.40 LOSS OF PRIMARY SOURCE
Exception. New exception requires the utility interactive inverter cease exporting power while not required to disconnect

705.42 ~~UNBALANCED CONNECTIONS~~ **LOSS OF THREE-PHASE PRIMARY SOURCE.**   *Rename for clarity*
Exception. New exception requires the utility interactive inverter cease exporting power while not required to disconnect

~~**705.43 SYNCHRONOUS GENERATORS.**~~   *Relocated to new 705.143 in new Part III, Generators*

II. UTILITY INTERACTIVE INVERTERS
705.60 CIRCUIT SIZING AND CURRENT.   *New section addresses calculation of maximum circuit current along with ampacity and overcurrent device ratings*

705.65 OVERCURRENT PROTECTION.   *New section addresses overcurrent protection for circuits, equipment, power transformers, inverter source circuits, DC devices, and series overcurrent protection*

705.70 UTILITY-INTERACTIVE INVERTERS MOUNTED IN NOT-READILY-ACCESSIBLE LOCATIONS.   *New section outlines requirements for inverters on roofs, etc.*

(continued)

705.80 UTILITY INTERACTIVE POWER SYSTEMS EMPLOYING ENERGY STORAGE.   *Requires systems with energy storage be marked with maximum operating voltage, equalization voltage and grounded conductor polarity*

705.82 HYBRID SYSTEMS.   *New section, permits interconnection of hybrid systems with utility interactive inverter*

705.95 AMPACITY OF NEUTRAL CONDUCTOR.   *New section, sizing neutral conductor*

705.100 UNBALANCED INTERCONNECTIONS.   *New section, provides requirements to eliminate unbalanced voltages in single-phase and three-phase systems*

III. GENERATORS
705.130 OVERCURRENT PROTECTION.   *Relocated from 705.30(A) to new Part III, Generators*

705.143 SYNCHRONOUS GENERATORS.   *Relocated from 705.43 to new Part III, Generators*

Change Significance

Article 705 covers the installation of one or more electric power production sources operating in parallel with a primary source such as a utility service connection. As more solar, fuel cell, and wind energy systems are installed, this article will play a more prominent role in these installations. Where multiple sources are installed for private consumption and for the purpose of selling energy back to the utility through a grid tie, the requirements of Article 705 and the use of equipment listed for the purpose are necessary for safe installations.

Notes:

_____

_____

_____

_____

_____

_____

Change Summary:

- A new Article 708, Critical Operations Power Systems (COPS), is added to provide prescriptive and performance-based requirements for systems designated as critical. This article will apply only where required by a municipal, state, federal, or governmental agency having jurisdiction or by facility engineering documentation establishing the necessity for such a system.

Article 708 will only apply when required in design or by municipal, state, federal, or other governmental agencies.

Revised 2008 NEC Text:

I. GENERAL

708.1 SCOPE.   The provisions of this article apply to the installation, operation, monitoring, control, and maintenance of the portions of the premises wiring system intended to supply, distribute and control electricity to designated critical operations areas (DCOA) in the event of disruption to elements of the normal system.

Critical operations power systems are those systems so classed by municipal, state, federal, or other codes, by any governmental agency having jurisdiction, or by facility engineering documentation establishing the necessity for such a system. These systems include but are not limited to power systems, HVAC, fire alarm, security, communications and signaling for designated critical operations areas.

FPN No. 1: Critical Operations Power Systems are generally installed in vital infrastructure facilities that, if destroyed or incapacitated, would disrupt national security, the economy, public health or safety; and where enhanced electrical infrastructure for continuity of operation has been deemed necessary by governmental authority.

FPN No. 2: For further information on disaster and emergency management see NFPA 1600-2004 edition, *Standard on Disaster/Emergency Management and Business Continuity Programs.*

FPN No. 3: For further information regarding performance of emergency and standby power systems, see NFPA 110-2005, *Standard for Emergency and Standby Power Systems.*

(continued)

Article 708
CRITICAL OPERATIONS POWER SYSTEMS (COPS)

Chapter 5 Special Occupancies

Article 708 Critical Operations Power Systems (COPS)

Part I General

Part II Circuit Wiring and Equipment

Part III Power Sources and Connection

Part IV Overcurrent Protection

Part V System Performance and Analysis

Change Type: New

FPN No. 4: For further information regarding performance and maintenance of emergency systems in health care facilities, see NFPA 99-2005, *Standard for Health Care Facilities.*

FPN No. 5: For specification of locations where emergency lighting is considered essential to life safety, see NFPA 101®-2006, *Life Safety Code®.*

FPN No.6: For further information regarding physical security see NFPA 730-2006, *Guide for Premises Security.*

FPN No. 7: Threats to facilities that may require transfer of operation to the critical systems include both naturally occurring hazards and human caused events.

See also Section A.5.3.2, NFPA 1600-2004.

708.2 DEFINITIONS
COMMISSIONING. *Acceptance testing*

CRITICAL OPERATIONS POWER SYSTEMS (COPS). Power systems for facilities or parts of facilities that require continuous operation for the reasons of public safety, emergency management, national security, or business continuity.

DESIGNATED CRITICAL OPERATIONS AREAS (DCOA). *Critical areas*

SUPERVISORY CONTROL AND DATA ACQUISITION (SCADA). *Electronic monitoring/control*

708.3 APPLICATION OF OTHER ARTICLES. *90.3 applies*

708.4 RISK ASSESSMENT. *Document hazards and mitigation*

708.5 PHYSICAL SECURITY. *Assess risk and restrict access*

708.6 TESTING AND MAINTENANCE. *Test, maintain and document*

708.8 COMMISSIONING. *Test, document baseline, develop performance test program*

II. CIRCUIT WIRING AND EQUIPMENT
708.10 FEEDER AND BRANCH CIRCUIT WIRING. *Installation requirements*

708.11 BRANCH CIRCUIT AND FEEDER DISTRIBUTION EQUIPMENT. *Location of equipment*

708.12 FEEDERS AND BRANCH CIRCUITS SUPPLIED BY COPS. *Critical service only*

708.14 WIRING OF HVAC, FIRE ALARM, SECURITY, EMERGENCY COMMUNICATIONS, AND SIGNALING SYSTEMS. *Installation requirements*

III. POWER SOURCES AND CONNECTION
708.20 SOURCES OF POWER. *Requirements for power sources*

(continued)

708.22 CAPACITY OF POWER SOURCES.   *Rating, capability and duration of sources*

708.24 TRANSFER EQUIPMENT.   *General requirements, COPS loads only*

708.30 BRANCH CIRCUITS SUPPLIED BY COPS.   *Critical service only*

IV. OVERCURRENT PROTECTION
708.50 ACCESSIBILITY.   *Access to authorized persons only*

708.52 GROUND-FAULT PROTECTION OF EQUIPMENT.   *Selectivity requirements*

708.54 COORDINATION.   *Selective coordination of overcurrent devices is required*

V. SYSTEM PERFORMANCE AND ANALYSIS
708.64 EMERGENCY OPERATIONS PLAN.   *Documented emergency plan*

ANNEX F
AVAILABILITY AND RELIABILITY FOR CRITICAL OPERATIONS POWER SYSTEMS; AND DEVELOPMENT AND IMPLEMENTATION OF FUNCTIONAL PERFORMANCE TESTS (FPT'S) FOR CRITICAL OPERATIONS POWER SYSTEMS

ANNEX G
SUPERVISORY CONTROL AND DATA ACQUISITION (SCADA)

Change Significance:

This new article provides requirements to achieve a superior level of survivability and performance for power systems considered critical. This article is crafted to provide the code user with prescriptive requirements for the installation of an electrical infrastructure that can survive natural disasters, terrorist events, or any other scenario that could render the system inoperable. This new article goes far beyond basic installation requirements.

Many requirements included in this article would not be appropriate elsewhere in the NEC because they would be considered design issues. These new requirements in Article 708 include design, performance, planning, and documentation necessary for the survivability of an electrical system considered necessary for the safety of persons. Venues that may utilize the requirements of Article 708 include military, police, rescue, fire department, municipal, state, federal, utility, communications, and health care institutions. The following two new informational annexes are created in support of this new article:

- Annex F: Availability and Reliability for Critical Operations Power Systems, and Development and Implementation of Functional Performance Tests (FPTs) for Critical Operations Power Systems
- Annex G: Supervisory Control and Data Acquisition (SCADA)

725.3(B)
SPREAD OF FIRE OR PRODUCTS OF COMBUSTION

Chapter 7 Special Conditions

Article 725 Class 1, Class 2, and Class 3 Remote-Control, Signaling, and Power-Limited Circuits

Part I General

725.3 Other Articles

(B) Spread of Fire or Products of Combustion

Change Type: Revision

Change Summary:
- 725.3(B) has been revised to clarify that the provisions of 300.21 shall apply to all Article 725 installations.
- Abandoned cable requirements are moved to 725.25.

Revised 2008 NEC Text:
725.3 OTHER ARTICLES.   Circuits and equipment shall comply with the articles or sections listed in 725.3(A) through 725.3(G). Only those sections of Article 300 referenced in this article shall apply to Class 1, Class 2, and Class 3 circuits.

(B) SPREAD OF FIRE OR PRODUCTS OF COMBUSTION.   Section 300.21 <u>shall provide the requirements for installations concerning the spread of fire or products of combustion.</u> ~~The accessible portion of abandoned Class 2, Class 3, and PLTC cables shall be removed.~~

Change Significance:
The arrangement of the NEC is outlined in Section 90.3. Chapters 1 through 4 apply generally. Chapters 5, 6, and 7 are special and modify or supplement Chapters 1 through 4. Chapter 8 stands alone.

The parent text of 725.3 mandates that only those sections of Article 300 referenced in Article 725 will apply to Class 1, Class 2, and Class 3 circuits. 725.3(B) is revised to provide clarity. This change now clearly requires that all Article 725 installations comply with 300.21. The last sentence of 725.3(B), which required the accessible portions of abandoned cable to be removed, is relocated to a new Section 725.25 addressing only abandoned cable.

Approved methods must be implemented to fire-stop openings in fire resistant rated walls, floors, or ceilings. Products which are listed and labeled for the purpose will be acceptable to the authority having jurisdiction.

Change Summary:
- The requirements for removing abandoned cable have been logically located in the XXX.25 section of Articles 725, 760, 770, 800, 820, and 830. Only cables identified for future use with a durable tag may remain.

Revised 2008 NEC Text:
725.25 ABANDONED CABLES.   Class 2, Class 3, and PLTC cables

760.25 ABANDONED CABLES.   fire alarm cables

770.25 ABANDONED CABLES.   optical fiber cables

800.25 ABANDONED CABLES.   communications cables

820.25 ABANDONED CABLES.   coaxial cables

830.25 ABANDONED CABLES.   network-powered broadband cables
   The accessible portion of . . . . shall be removed. Where cables are identified for future use with a tag, the tag shall be of sufficient durability to withstand the environment involved.

Change Significance:
This revision logically locates the requirements for removing abandoned cable in a user-friendly format in Articles 725, 760, 770, 800, 820, and 830. The NEC requires that abandoned cable be removed. The installer as well as the enforcement community must recognize that as upgrades are made to an existing system or an entire area is renovated all abandoned cable must be removed. This issue of abandoned cable has received a tremendous amount of attention over the past 10 years.

   Cables that are not terminated are not permitted to remain unless they are tagged for future use. This revision now requires that the tag be durable for the environment in which it is installed.

Chapters 7 and 8
ABANDONED CABLES

Chapter 7 Special Conditions

725.25 Abandoned Cables

760.25 Abandoned Cables

Chapter 8 Communications Systems

800.25 Abandoned Cables

820.25 Abandoned Cables

830.25 Abandoned Cables

Change Type: Revision and New

Renovations or modifications of data and communications systems in commercial occupancies require that installers and inspectors ensure that abandoned cables that are not identified for future use are removed.

725.48(B)
CLASS 1, CIRCUITS WITH POWER SUPPLY CIRCUITS

Chapter 7 Special Conditions

Article 725 Class 1, Class 2, and Class 3 Remote-Control, Signaling, and Power-Limited Circuits

Part II Class 1 Circuits

725.48 Conductors of Different Circuits in the Same Cable, Cable Tray, Enclosure, or Raceway

(B) Class 1 Circuits With Power Supply Circuits

(4) In Cable Trays

Change Type: Revision

The installation of Class 1 circuit conductors in a cable tray with other circuit conductors is addressed in 725.48.

Change Summary:
- 725.48(B)(4) now permits the installation of Class 1 circuit conductors and power-supply conductors not functionally associated with each other in the same cable tray without providing a barrier in the cable tray. This is only permitted where both the Class 1 circuit conductors and the power supply conductors are within multiconductor type AC, MC, MI, or TC cables and all cables are insulated at 600 volts.

Revised 2008 NEC Text:
725.~~26~~48 CONDUCTORS OF DIFFERENT CIRCUITS IN THE SAME CABLE, CABLE TRAY, ENCLOSURE, OR RACEWAY

(B) CLASS 1 CIRCUITS WITH POWER SUPPLY CIRCUITS. Class 1 circuits shall be permitted to be installed with power supply conductors as specified in 725.48~~26~~(B)(1) through (B)(4).

(4) IN CABLE TRAYS. Installations in cable trays shall comply with 725.48(B)(4)(1) or 725.48(B)(4)(2).

(1) ~~In cable trays where the~~ Class 1 circuit conductors and power-supply conductors not functionally associated with ~~them are~~ the Class 1 circuit conductors shall be separated by a solid fixed barrier of a material compatible with the cable tray.

(2) Class 1 circuit conductors and power-supply conductors not functionally associated with the Class 1 circuit conductors shall be permitted to be installed in a cable tray without barriers where all of the conductors are installed within separate multiconductor Type AC, Type MC, Type MI, or Type TC cables and all of the conductors in the cables are insulated at 600 volts.

Change Significance:
This revision now permits the installation of Class 1 circuit conductors and power-supply conductors not functionally associated with each other in the same cable tray without providing a barrier in the cable tray. This is only permitted where both the Class 1 circuit conductors and the power supply conductors are within multiconductor type AC, MC, MI, or TC cables and all cables are insulated at 600 volts.

The previous requirement for a barrier was intended to eliminate potential damage to power and Class 1 circuit conductors in a cable tray where the Class 1 circuit conductors were not functionally associated with other power circuits, thus affecting the operation and safety of the other nonrelated circuits. This revision provides for separation through limiting the wiring methods permitted.

The installation of single conductors in cable tray as permitted in 392.3(B) would require a barrier where Class 1 circuits were not functionally associated with the single conductors. This revised text in 725.48(B)(4)(2) will apply only where the wiring methods are type AC, MC, MI, or TC cables and all cables are insulated at 600 volts.

Change Summary:

- Revisions in 725.154 provide clarity by identifying permitted types of raceways for listed Class 2, Class 3, and PLTC cables installed in risers per 725.61(B) and cable trays per 725.61(C).

Revised 2008 NEC Text:

725.~~61~~154 APPLICATIONS OF LISTED CLASS 2, CLASS 3, AND PLTC CABLES

(B) RISER. Cables installed in risers shall be as described in any of (B)(1), (B)(2), or (B)(3):

(1) Cables installed in vertical runs and penetrating more than one floor, or cables installed in vertical runs in a shaft, shall be Type CL2R or CL3R. Floor penetrations requiring Type CL2R or CL3R shall contain only cables suitable for riser or plenum use. Listed riser signaling raceways <u>and listed plenum signaling raceways</u> shall be permitted to be installed in vertical riser runs in a shaft from floor to floor. Only Type CL2R, CL3R, CL2P, or CL3P cables shall be permitted to be installed in these raceways.

(2) and (3) (No change)

(C) CABLE TRAYS. Cables installed in cable trays outdoors shall be Type PLTC. Cables installed in cable trays indoors shall be Types PLTC, CL3P, CL3R, CL3, CL2P, CL2R, and CL2.

Listed <u>general-purpose</u> signaling raceways<u>, listed riser signaling raceways and listed plenum signaling raceways</u> shall be permitted for use with cable trays.

FPN: See 800.1<u>5</u>~~33~~(D)~~(B)~~ for cables permitted in cable trays.

Change Significance:

This revision in 725.154(B)(1) provides clarification that plenum raceways are permitted to substitute for riser and general-purpose raceways just as plenum cable is permitted to substitute for riser and general-purpose cables. The revision in 725.154(C) provides clarity by noting all signaling raceways explicitly so that it is clear that all of these raceways are permitted in cable trays.

725.154
APPLICATIONS OF LISTED CLASS 2, CLASS 3, AND PLTC CABLES

Chapter 7 Special Conditions

Article 725 Class 1, Class 2, and Class 3 Remote-control, Signaling, and Power-limited Circuits

Part III Class 2 and Class 3 Circuits

725.154 Applications of Listed Class 2, Class 3, and PLTC Cables

(B) Riser

(C) Cable Trays

Change Type: Revision

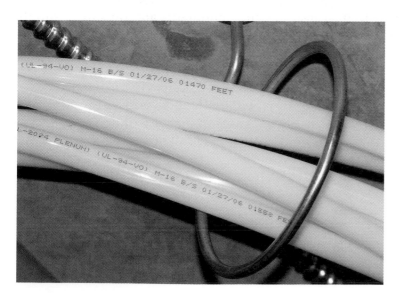

Listed general-purpose, riser, and plenum signaling raceways are permitted in cable tray.

725.154(D)
HAZARDOUS (CLASSIFIED) LOCATIONS

Chapter 7 Special Conditions

Article 725 Class 1, Class 2, and Class 3 Remote-Control, Signaling, and Power-Limited Circuits

Part III Class 2 and Class 3 Circuits

725.154 Applications of Listed Class 2, Class 3, and PLTC Cables

(D) Hazardous (Classified) Locations

(4) In Industrial Establishments

Change Type: Revision

Power-limited tray cable with a metallic sheath or armor is permitted to be installed exposed provided the requirements of 725.154(D)(4) are met.

Change Summary:

- The installation of type PLTC cable in hazardous locations in industrial establishments has been clarified by recognizing PLTC cable in two separate list items. List item (1) addresses PLTC cable with a metallic sheath and list item (2) addresses PLTC cable without a metallic sheath.

Revised 2008 NEC Text:

725.~~61~~154 APPLICATIONS OF LISTED CLASS 2, CLASS 3, AND PLTC CABLES

(D) HAZARDOUS (CLASSIFIED) LOCATIONS

(4) IN INDUSTRIAL ESTABLISHMENTS. In industrial establishments where the conditions of maintenance and supervision ensure that only qualified persons service the installation, <u>Type PLTC cable shall be permitted in accordance with either (1) or (2).</u>

<u>(1) Type PLTC cable, with a metallic sheath or armor in accordance with 725.179~~82~~(E), shall be permitted to be installed exposed. The cable shall be continuously supported and protected against physical damage using mechanical protection such as dedicated struts, angles, or channels. The cable shall be secured at intervals not exceeding 1.8 m (6ft).</u>

<u>(2)</u> Type PLTC cable, <u>without a metallic sheath or armor</u>, that complies with the crush and impact requirements of Type MC cable and is identified for such use <u>with the marking PLTC-ER,</u> shall be permitted to be <u>installed</u> exposed ~~between the cable tray and utilization equipment or device~~. The cable shall be continuously supported and protected against physical damage using mechanical protection such as dedicated struts, angles, or channels. The cable shall be secured at intervals not exceeding 1.8 m (6 ft).

Change Significance:

This revision recognizes that cable types TC, PLTC, and ITC are very similar and should have similar if not identical installation methods. Per the listing of these cable types, they are permitted to be installed as exposed routing (ER) when the cable is listed with the suffix ER. Cables listed for ER installations are a stronger cable capable of withstanding more abuse than cables not listed for use in ER installations.

Type PLTC cable, with a metallic sheath or armor, is now recognized as suitable where installed in accordance with list item (1). Type PLTC cable, without a metallic sheath or armor, that complies with the crush and impact requirements of type MC cable and is identified for such use with the marking PLTC-ER shall be permitted to be installed exposed—as permitted in list item (2). The revised text restructures the requirement in two list items, separating the types of PLTC for clarity.

Notes:

_____

_____

Change Summary:

*   Where type PLTC cable is used in a wet location for circuits installed in accordance with Article 725, the cable shall be listed for use in wet locations or have a moisture-impervious metal sheath.

Revised 2008 NEC Text:

725.~~82~~179 LISTING AND MARKING OF CLASS 2, CLASS 3, AND TYPE PLTC CABLES

(E) TYPE PLTC.   Type PLTC nonmetallic-sheathed, power-limited tray cable shall be listed as being suitable for cable trays and shall consist of a factory assembly of two or more insulated conductors under a nonmetallic jacket. The insulated conductors shall be 22 AWG through 12 AWG. The conductor material shall be copper (solid or stranded). Insulation on conductors shall be suitable for 300 volts. The cable core shall be either (1) two or more parallel conductors, (2) one or more group assemblies of twisted or parallel conductors, or (3) a combination thereof.

A metallic shield or a metallized foil shield with drain wire(s) shall be permitted to be applied either over the cable core, over groups of conductors, or both. The cable shall be listed as being resistant to the spread of fire. The outer jacket shall be a sunlight- and moisture-resistant nonmetallic material. <u>Type PLTC cable used in a wet location shall be listed for use in wet locations, or have a moisture impervious metal sheath.</u>

Change Significance:

Where type PLTC cable is used in a wet location for circuits installed in accordance with Article 725, the cable shall be listed for use in wet locations or have a moisture-impervious metal sheath. This revision now requires that PLTC cable be listed for use in wet locations or have a moisture-impervious metal sheath when the cable is installed in a wet location, including installation outdoors or directly buried—as permitted in 725.154.

725.179(E)
TYPE PLTC CABLES

Chapter 7 Special Conditions

Article 725 Class 1, Class 2, and Class 3 Remote-Control, Signaling, and Power-Limited Circuits

Part IV Listing Requirements

725.179 Listing and Marking of Class 2, Class 3, and Type PLTC Cables

(E) Type PLTC

Change Type: Revision

Equipment installed outdoors in wet locations (such as a personnel hoist) may utilize type PLTC cable for controls.

Articles 725 and 760

ARTICLE 725 CLASS 1, CLASS 2, AND CLASS 3 REMOTE-CONTROL, SIGNALING, AND POWER-LIMITED CIRCUITS

ARTICLE 760 FIRE ALARM SYSTEMS

Change Type: Revision

Change Summary:

- Sections within Article 725 have been renumbered in an effort to parallel the requirements for "Class 1, Class 2, and Class 3 Remote-Control, Signaling, and Power-Limited Circuits" with the requirements for "Fire Alarm Systems" in Article 760.

Article 725 provides installation requirements for systems such as thermostat controls for heating.

Article 725	Article 760
I. General	I. General
~~725.8~~	~~760.8~~
725.24	760.24
Mechanical Execution of Work	Mechanical Execution of Work
~~725.10~~	~~760.10~~
725.30 Class 1, Class 2, and Class 3 Circuit Identification	760.30 Fire Alarm Circuit Identification
II. Class 1 Circuits	II. Non–Power-Limited Fire Alarm (NPLFA) Circuits
~~725.21~~	~~760.21~~
725.41 Class 1 Circuit Classifications and Power Source Requirements	760.41 NPLFA Circuit Power Source Requirements
~~725.23~~	~~760.23~~
725.43 Class 1 Circuit Overcurrent Protection	760.43 NPLFA Circuit Overcurrent Protection

(continued)

Revised 2008 NEC Text:

Examples of this parallel reorganization of section numbers of Articles 725 and 760 are as follows:

III. Class 2 and Class 3 Circuits	III. Power-Limited Fire Alarm (PLFA) Circuits
~~725.42~~ 725.124 Circuit Marking	~~760.42~~ 760.124 Circuit Marking
IV. Listing Requirements	IV. Listing Requirements
~~725.82~~ 725.179 Listing and Marking of Class 2, Class 3, and Type PLTC Cables	~~760.82~~ 760.179 Listing and Marking of PLFA Cables and Insulated Continuous Line-Type Fire Detectors

Change Significance:

In an effort to enhance the usability of the NEC, Articles 725 and 760 have undergone a revision of section numbering to parallel and coordinate requirements. An effort was also made to parallel and coordinate section renumbering of Articles 725 and 760 with similar requirements located in Chapter 8.

The NEC has undergone several such reorganizations in an effort to increase clarity and make the entire document easier to use and apply. The reorganization of Article 250 and the renumbering of sections for cable assemblies and raceways are examples of how this type of revision increases clarity and usability for the code user.

Notes:

_____

_____

_____

_____

_____

_____

_____

_____

727.4
USES PERMITTED

Chapter 7 Special Conditions

Article 727 Instrumentation Tray Cable, Type ITC

727.4 Uses Permitted

Change Type: Revision

Change Summary:

- The permitted use of type ITC cable has been clarified by removing the 50-foot limitation on exposed cable. In addition, exposed runs of ITC must be marked with the suffix ER and be continuously supported with "dedicated" struts, angles, or channels.

Revised 2008 NEC Text:

(Delete list items (5) and (6) from the 2005 NEC, add list item (5) below and renumber remaining list items)

727.4 USES PERMITTED

(5) <u>Cable,</u> without a metallic sheath or armor, that complies with the crush and impact requirements of Type MC cable <u>and is identified for such use with the marking ITC-ER, shall be permitted to be installed exposed.</u> The cable shall be continuously supported and protected against physical damage using mechanical protection such as <u>dedicated </u>struts, angles, or channels. The cable shall be secured at intervals not exceeding 1.8 m (6ft).

Change Significance:

Note that the previously cited 2008 NEC text is only marked with substantive new text. Other revisions are purely editorial. This revision removes the 50-foot limitation on the installation of type ITC cable from cable tray to equipment. ITC cable is now permitted to be exposed provided it is marked with the suffix ER. Exposed runs of type ITC-ER cable are now required to be continuously supported and protected against physical damage using mechanical protection such as "dedicated" struts, angles, or channels.

Cable types TC, PLTC, and ITC are very similar and the technical committee has taken steps this cycle to provide similar if not identical installation methods. Per the listing of ITC cable, it is permitted to be installed as ER when the cable is listed with the suffix ER. Cables listed for ER installations are a stronger cable capable of withstanding more abuse than cables not listed for use in ER installations.

Article 727 provides requirements for instrumentation tray cable (ITC) which is permitted only for instrumentation and control circuits that operate at not more than 150 volts and 5 amps.

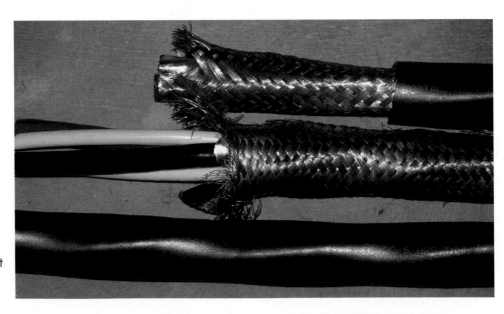

Change Summary:
- Conductors installed within the scope of Article 760 shall not be installed in raceways or cable tray that contain pipe, tube, or the equivalent for steam, water, air, gas, drainage, or any service other than electrical.

Revised 2008 NEC Text:
760.3 OTHER ARTICLES
(G) INSTALLATION OF CONDUCTORS WITH OTHER SYSTEMS. Installations shall comply with 300.8.

Change Significance:
The parent text of 760.3 limits the application of Article 300 to fire alarm installations with the following sentence: "Only those sections of Article 300 referenced in this article shall apply to fire alarm systems." 760.3, titled "Other Articles," references requirements in other areas of the NEC as applicable to fire alarm systems—including requirements in Articles 110, 300, 310, 517, 725, and 770, and Articles 500 through 516. The inclusion of a new first-level subdivision (G), which references 300.8, will now clearly prohibit the installation of fire alarm conductors in raceways or cable trays that contain systems other than electrical—as required in 300.8 as follows:

300.8 INSTALLATION OF CONDUCTORS WITH OTHER SYSTEMS. Raceways or cable trays containing electric conductors shall not contain any pipe, tube, or equal for steam, water, air, gas, drainage, or any service other than electrical.

760.3(G) INSTALLATION OF CONDUCTORS WITH OTHER SYSTEMS

Chapter 7 Special Conditions

Article 760 Fire Alarm Systems

Part I General

760.3 Other Articles

(G) Installation of Conductors With Other Systems

Change Type: New

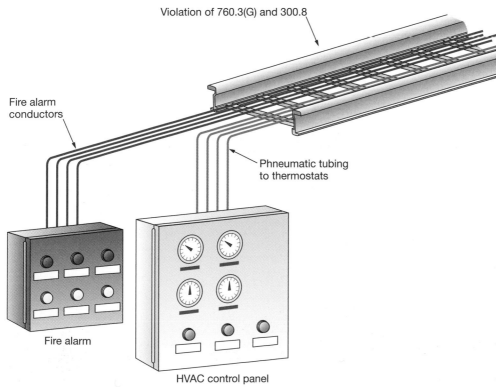

Violation of 760.3(G) and 300.8

Fire alarm conductors

Phneumatic tubing to thermostats

Fire alarm

HVAC control panel

The provisions of 300.8 do not permit raceways or cable tray to contain any pipe, tube, or equivalent for steam, water, air, gas, drainage, or any service other than electrical.

760.41
NPLFA CIRCUIT POWER SOURCE REQUIREMENTS

Chapter 7 Special Conditions

Article 760 Fire Alarm Systems

Part II Non–Power-Limited Fire Alarm (NPLFA) Circuits

760.41 NPLFA Circuit Power Source Requirements

(A) Power Source

(B) Branch Circuit

Change Type: Revision

Change Summary:
* 760.41 is editorially separated into two first-level subdivisions for clarity and usability. Branch circuits for NPLFA circuits are now required to be individual branch circuits, as defined in Article 100.

Revised 2008 NEC Text:
760.~~21~~41 NPLFA CIRCUIT POWER SOURCE REQUIREMENTS
(A) POWER SOURCE.   The power source of non–power-limited fire alarm circuits shall comply with Chapters 1 through 4, and the output voltage shall not be more than 600 volts, nominal.

(B) BRANCH CIRCUIT.   <u>An individual branch circuit shall be required for the supply of the power source. This branch circuit</u> ~~These circuits~~ shall not be supplied through ground-fault circuit interrupters or arc-fault circuit interrupters.

   FPN: See 210.8(A)(5), Exception ~~No. 3~~, for receptacles in dwelling-unit unfinished basements that supply power for fire alarm systems.

Change Significance:
This revision provides clarity and usability by separating the requirements of 760.41 into two first-level subdivisions. The requirements are now logically separated for NFPLA circuits into "(A) Power Source" and "(B) Branch Circuit."

   In addition to requiring that NFPLA circuits not be supplied through ground-fault circuit interrupters or arc-fault circuit interrupters, the requirement is expanded by mandating an "individual branch circuit." The Article 100 definition of *individual branch circuit* is as follows:

BRANCH CIRCUIT, INDIVIDUAL.   A branch circuit that supplies only one utilization equipment.

   This requirement for an individual branch circuit is made to bring the NEC requirements into alignment with NFPA 72, 4.4.1.4.1, which requires a dedicated branch circuit to supply fire alarm power.

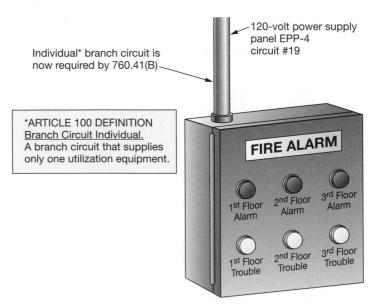

120-volt power supply panel EPP-4 circuit #19

Individual* branch circuit is now required by 760.41(B)

*ARTICLE 100 DEFINITION
Branch Circuit Individual.
A branch circuit that supplies only one utilization equipment.

FIRE ALARM

1st Floor Alarm   2nd Floor Alarm   3rd Floor Alarm
1st Floor Trouble   2nd Floor Trouble   3rd Floor Trouble

The power source for an NPLFA circuit must comply with Chapters 1 through 4 and be not more than 600 volts.

Change Summary:

- A reference to 300.7 in 760.46 for NFPLA circuit wiring methods and in 760.130 for wiring methods on the load side of a PLFA power source requires that seals and expansion fittings in raceways be installed where required.

Revised 2008 NEC Text:

760.~~25~~46 NPLFA CIRCUIT WIRING METHODS. Installation of non–power-limited fire alarm circuits shall be in accordance with 110.3(B), <u>300.7,</u> 300.11, 300.15, 300.17, and other appropriate articles of Chapter 3.

760.~~52~~130 WIRING METHODS AND MATERIALS ON LOAD SIDE OF THE PLFA POWER SOURCE
(B) PLFA WIRING METHODS AND MATERIALS. Power-limited fire alarm conductors and cables described in 760.<u>179</u>~~82~~ shall be installed as detailed in 760.<u>130</u>~~52~~(B)(1), (B)(2), or (B)(3) of this section <u>and 300.7</u>. Devices shall be installed in accordance with 110.3(B), 300.11(A), and 300.15.

Change Significance:

The parent text of 760.3 limits the application of Article 300 to fire alarm installations with the following sentence: "Only those sections of Article 300 referenced in this article shall apply to fire alarm systems." 760.3, titled "Other Articles," references requirements in other areas of the NEC as applicable to fire alarm systems—including requirements in Articles 110, 300, 310, 517, 725, and 770, and Articles 500 through 516.

The inclusion of a reference to 300.7 in both "760.46, NPLFA Circuit Wiring Methods" and "760.130(B), PLFA Wiring Methods and Materials" requires that expansion fittings and seals be installed where required. 300.7 requires that where portions of a cable raceway or sleeve are known to be subjected to different temperatures and where condensation is known to be a problem (as in cold storage areas of buildings or where passing from the interior to the exterior of a building), the raceway or sleeve shall be filled with an approved material to prevent the circulation of warm air to a colder section of the raceway or sleeve. 300.7 also requires that where necessary to compensate for thermal expansion and contraction raceways shall be provided with expansion fittings.

760.46 and 760.130(B)

NPLFA CIRCUIT WIRING METHODS AND PLFA WIRING METHODS AND MATERIALS

Chapter 7 Special Conditions

Article 760 Fire Alarm Systems

Part II Non–Power-Limited Fire Alarm (NPLFA) Circuits

760.46 NPLFA Circuit Wiring Methods

Part III Power-Limited Fire Alarm (PLFA) Circuits

760.130 Wiring Methods and Materials on Load Side of the PLFA Power Source

(B) PLFA Wiring Methods and Materials

Change Type: Revision

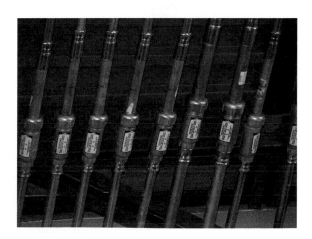

The provisions of 300.7(B) require expansion fittings where necessary for thermal expansion and contraction.

760.121
POWER SOURCES FOR PLFA CIRCUITS

Chapter 7 Special Conditions

Article 760 Fire Alarm Systems

Part III Power-Limited Fire Alarm (PLFA) Circuits

760.121 Power Sources for PLFA Circuits

(A) Power Source

(B) Branch Circuit

Change Type: Revision

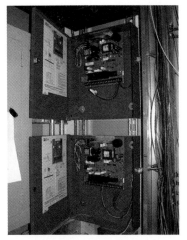

The power source for a PLFA circuit is not permitted to be protected with an AFCI or GFCI protective device.

Change Summary:
- 760.121 is editorially reorganized and separated into two first-level subdivisions for clarity and usability. A branch circuit supplying a PLFA power source is now required to be an individual branch circuit, as defined in Article 100.

Revised 2008 NEC Text:
760.4~~41~~**121** POWER SOURCES FOR PLFA CIRCUITS
(A) POWER SOURCE.   The power source for a power-limited fire alarm circuit shall be as specified in <u>760.121(A)(1), 760.121(A)(2) or 760.121(A)(3)</u>. ~~760.41(A), (B), or (C). These circuits shall not be supplied through ground-fault circuit interrupters or arc-fault circuit interrupters.~~

FPN No. 1: Tables 12(A) and 12(B) in Chapter 9 provide the listing requirements for power-limited fire alarm circuit sources.

FPN No. 2: See 210.8(A)(5), Exception ~~No. 3~~, for receptacles in dwelling-unit unfinished basements that supply power for fire alarm systems.

~~(A)~~<u>(1)</u> ~~Transformers.~~ A listed PLFA or Class 3 transformer.

~~(B)~~<u>(2)</u> ~~Power Supplies.~~ A listed PLFA or Class 3 power supply.

~~(C)~~<u>(3)</u> ~~Listed Equipment.~~ Listed equipment marked to identify the PLFA power source.

FPN: Examples of listed equipment are a fire alarm control panel with integral power source; a circuit card listed for use as a PLFA source, where used as part of a listed assembly; a current-limiting impedance, listed for the purpose or part of a listed product, used in conjunction with a non–power-limited transformer or a stored energy source, for example, storage battery, to limit the output current.

(B) BRANCH CIRCUIT.   <u>An individual branch circuit shall be required for the supply of the power source. This branch circuit shall not be supplied through ground-fault circuit interrupters or arc-fault circuit interrupters.</u>

Change Significance:
This revision provides clarity and usability by reorganizing and separating the requirements of 760.121 into two first-level subdivisions. The requirements are now logically separated for PLFA circuits into "(A) Power Source" and "(B) Branch Circuit." Permitted power sources are editorially revised into list items.

In addition to requiring that PLFA power sources not be supplied through ground-fault circuit interrupters or arc-fault circuit interrupters, the requirement is expanded by mandating that an "individual branch circuit" supply the PLFA power source. The Article 100 definition of *individual branch circuit* is as follows:

BRANCH CIRCUIT, INDIVIDUAL.   A branch circuit that supplies only one utilization equipment.

This requirement for an individual branch circuit is made to bring the NEC requirements into alignment with NFPA 72, 4.4.1.4.1, which requires a dedicated branch circuit to supply fire alarm power.

Change Summary:

- NPLFA and PLFA cables used in a wet location shall be listed for use in wet locations or have a moisture-impervious metal sheath.

Revised 2008 NEC Text:

760.~~81~~176 LISTING AND MARKING OF NPLFA CABLES. Non–power-limited fire alarm cables installed as wiring within buildings shall be listed in accordance with 760.~~176~~81 (A) and 760.~~176~~81 (B) and as being resistant to the spread of fire in accordance with 760.~~176~~81 (C) through 760.~~176~~81 (F), and shall be marked in accordance with 760.~~176~~81 (G). <u>Cable used in a wet location shall be listed for use in wet locations or have a moisture-impervious metal sheath.</u>

760.~~82~~179 LISTING AND MARKING OF PLFA CABLES AND INSULATED CONTINUOUS LINE-TYPE FIRE DETECTORS. Type FPL cables installed as wiring within buildings shall be listed as being resistant to the spread of fire and other criteria in accordance with 760.~~179~~82 (A) through 760.~~179~~82—(II) and shall be marked in accordance with 760.~~179~~82 (I). Insulated continuous line-type fire detectors shall be listed in accordance with 760.~~179~~82 (J). <u>Cable used in a wet location shall be listed for use in wet locations or have a moisture impervious metal sheath.</u>

Table 760.176(G) NPLFA Cable Markings		
Cable Marking	**Type**	**Reference**
NPLFP	Non–power-limited fire alarm circuit cable for use in "other space used for environmental air"	760.176(C) and (G)
NPLFR	Non–power-limited fire alarm circuit riser cable	760.176(D) and (G)
NPLF	Non–power-limited fire alarm circuit cable	760.176(E) and (G)

Note: Cables identified in 760.176(C), (D), and (E) and meeting the requirements for circuit integrity shall have the additional classification using the suffix "CI" (for example, NPLFP-CI, NPLFR-CI, and NPLF-CI).

FPN: Cable types are listed in descending order of fire resistance rating.

NPLFA cable markings include NPLFP, NPLFR, and NPLF.

Change Significance:

This revision mandates that where PLFA or NPLFA cable is used in a wet location the cable shall be listed for use in wet locations or have a moisture-impervious metal sheath.

760.176 and 760.179

LISTING AND MARKING OF NPLFA CABLES AND LISTING AND MARKING OF PLFA CABLES AND INSULATED CONTINUOUS LINE-TYPE FIRE DETECTORS

Chapter 7 Special Conditions

Article 760 Fire Alarm Systems

Part IV Listing Requirements

760.176 Listing and Marking of NPLFA Cables

760.179 Listing and Marking of PLFA Cables and Insulated Continuous Line-Type Fire Detectors

Change Type: Revision

Article 770
OPTICAL FIBER CABLE AND RACEWAYS

Chapter 7 Special Conditions

Article 770 Optical Fiber Cable and Raceways

Part I General

Part II Cables Outside and Entering Buildings

Part III Protection

Part IV Grounding Methods

Part V Installation Methods Within Buildings

Part VI Listing Requirements

Change Type: Revision

Change Summary:

- Article 770 has been editorially revised to organize requirements into parts parallel with the organization of Articles 800, 820, and 830.

Revised 2008 NEC Text:

ARTICLE: 770 OPTICAL FIBER CABLE AND RACEWAYS
  I. GENERAL
  II. CABLES OUTSIDE AND ENTERING BUILDINGS ~~PROTECTION~~
  III. PROTECTION ~~CABLES WITHIN BUILDINGS~~
  IV. GROUNDING METHODS ~~LISTING REQUIREMENTS~~
  V. INSTALLATION METHODS WITHIN BUILDINGS
  VI. LISTING REQUIREMENTS

Change Significance:

This revision reorganizes the requirements of Article 770 into parts that parallel the format of Articles 800, 820, and 830. Where articles within the NEC are used in a similar fashion, it is user friendly to organize those articles in the same manner to allow the user to move from one article to the next within a similar format. Section numbering within Article 770 has also been revised to accommodate the restructuring of parts within this article.

The best example of parallel formatting within the NEC is the common numbering system for cable assemblies and raceways. Once the code user finds a desired requirement in one article, it is easy to cross reference that requirement with other cable assemblies and/or raceways.

Article 770, Optical Fiber Cable and Raceways, is found in Chapter 7, Special Conditions. This article is sometimes confused as being strictly for communications. Fiber optic systems are used in a wide array of installations and are not limited to communications.

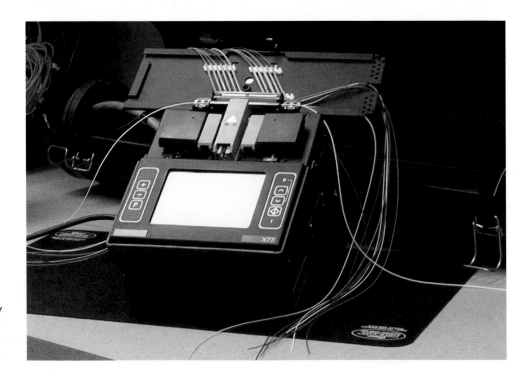

Article 770 is logically separated into six parts to aid the code user in quickly accessing necessary information and requirements.

Change Summary:

- Five new definitions are added to 770.2 for fiber optic cable, increasing the clarity and usability of this article.

Revised 2008 NEC Text:

770.2 DEFINITIONS

CABLE SHEATH. A covering over the optical fiber assembly that includes one or more jackets and may include one or more metallic members or strength members.

COMPOSITE OPTICAL FIBER CABLE. These cables contain optical fibers and current-carrying electrical conductors.

CONDUCTIVE OPTICAL FIBER CABLE. These optical fiber cables contain non–current-carrying conductive members such as metallic strength members, metallic vapor barriers, and metallic armor or sheath.

NONCONDUCTIVE OPTICAL FIBER CABLE. These optical fiber cables contain no metallic members and no other electrically conductive materials.

OPTICAL FIBER CABLE. A factory assembly of one or more optical fibers having an overall covering.

Change Significance:

The term *optical fiber cable* is used throughout Article 770 and is now defined in 770.2 for clarity. The three types of optical fiber cable (including composite, conductive, and nonconductive) are now defined in 770.2. These types of cable were described in 770.9 in previous editions of the NEC. The NEC style manual requires that definitions used within a single article be placed in the second section of the article for usability. The term *cable sheath* is also now defined in 770.2. There are many different types of cable sheaths for optical fiber cable, which are dependent on the individual application.

The definition of *exposed* has been revised to clarify that contact is between conductive members of the cable and an electrical circuit. The definition of *point of entrance* is revised to clarify that the point is within a building.

770.2
DEFINITIONS

Chapter 7 Special Conditions

Article 770 Optical Fiber Cable and Raceways

Part I General

770.2 Definitions

Change Type: New

Optical fiber cable is a factory assembly of one or more optical fibers. The cable may contain conductive or nonconductive materials and current-carrying conductors.

8

Communication Systems, Articles 800–820

800.1 and 800.2
SCOPE AND DEFINITIONS

Chapter 8
Communications
Systems

Article 800
Communications
Circuits

Part I General

800.1 Scope

800.2 Definition,
Communications Circuit

Change Type: Revision
and New

Change Summary:

- The scope of Article 800, Communications Circuits, is revised to clarify that it only applies to "communications circuits and equipment." A new definition of *communications circuit* is added to clarify the limitations of the newly reorganized article scope.

Revised 2008 NEC Text:

800.1 SCOPE.   This Article covers ~~telephone, telegraph (except radio), outside wiring for fire alarm and burglar alarm, and similar central station systems ; and telephone systems not connected to a central station system but using similar type of equipment, method of installation, and maintenance~~ communications circuits and equipment.

FPN No. 1: For installation requirements for equipment and circuits in an information technology equipment room, see Article 645.

FPN No. 2: For further information on remote-control, signaling, and powerlimited circuits, see Article 725.

FPN No. ~~1~~3: For further information for fire alarm, sprinkler waterflow, and sprinkler supervisory systems, see Article 760.

FPN No. ~~2~~4: For installation requirements of optical fiber cables, see Article 770.

FPN No. ~~3~~5: For installation requirements for network-powered broadband communications circuits, see Article 830.

800.2 DEFINITIONS

COMMUNICATIONS CIRCUITS.   The circuit that extends voice, audio, video, data, interactive services, telegraph (except radio), outside wiring for fire alarm and burglar alarm from the communications utility to the customer's communications equipment up to and including terminal equipment such as a telephone, fax machine, or answering machine.

Change Significance:

This revision is intended to resolve a scope/correlation issue between Articles 725 and 800. Although both articles may use similar wiring methods, the scope of each article is very different. The revised scope simplifies and clarifies that Article 800 covers communications circuits and equipment. Modern telecommunications are not limited to telephone or telegraph.

The term *telephone* has evolved to the point where it is a communications system transporting information in various forms (including voice, data, audio, video, and interactive services) and using varied technologies, including copper wire, coaxial cable, optical fiber, and radio links, as well as high-frequency carrier systems and advanced data processing and switching techniques. The addition of the definition of *communications circuit* helps clarify the scope as covering communications services and equipment provided by a communications utility, including associated services.

Notes:

_____

_____

The scope of Article 800 includes only communications circuits and equipment.

Change Summary:
- The requirements for the installation of unlisted plant cable where the cable enters a building are clarified in the positive text of a new section.

Revised 2008 NEC Text:
800.48. UNLISTED CABLES ENTERING BUILDINGS. Unlisted outside plant communications cables shall be permitted to be installed in locations as described in 800.154(C) where the length of the cable within the building, measured from its point of entrance, does not exceed 15 m (50 ft) and the cable enters the building from the outside and is terminated in an enclosure or on a listed primary protector.

FPN No. 1: Splice cases or terminal boxes, both metallic and plastic types, are typically used as enclosures for splicing or terminating telephone cables.

FPN No. 2: This section limits the length of unlisted outside plant cable to 15 m (50 ft), while 800.90(B) requires that the primary protector be located as close as practicable to the point at which the cable enters the building. Therefore, in installations requiring a primary protector, the outside plant cable may not be permitted to extend 15 m (50 ft) into the building if it is practicable to place the primary protector closer than 15 m (50 ft) to the entrance point.

FPN No, 3: See 800.2 for the definition of *Point of Entrance.*

Change Significance:
The requirements for unlisted plant cable where it enters a building existed in previous editions of the NEC in Part V, Communications Wires and Cables Within Buildings. This text was located in Exception No. 2 to 800.113. The requirement limited the length of the cable to 50 feet but did not provide installation requirements.

This revision locates this requirement in the proper part of Article 800, Part II, Wires and Cables Outside and Entering Buildings. Installation requirements are now provided with a reference to 800.154(C). Placing this requirement in a new section provides increased clarity and usability for the code user.

Similar revisions have occurred in NEC Articles 770 and 820, including 770.48 in Proposal 16-52 and 820.48 in Proposal 16-284. Note the parallel numbering format, which is being implemented across Articles 770, 800, 820, and 830 where applicable.

800.48
UNLISTED CABLES ENTERING BUILDINGS

**Chapter 8
Communications
Systems**

**Article 800
Communications
Circuits**

**Part II Wires and
Cables Outside and
Entering Buildings**

**800.48 Unlisted Cables
Entering Buildings**

Change Type: New

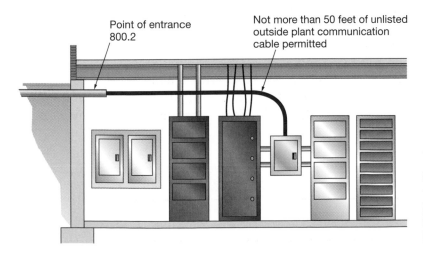

Point of entrance
800.2

Not more than 50 feet of unlisted
outside plant communication
cable permitted

The length of unlisted outside plant cable is generally limited to 50 feet. However, 800.90(B) requires that the primary protector be located as close as practicable to the point at which the cable enters the building.

800.100(B)
ELECTRODE

Chapter 8 Communications Systems

Article 800 Communications Circuits

Part IV Grounding Methods

800.100 Cable and Primary Protector Grounding

(B) Electrode

(1) In Buildings or Structures With an Intersystem Grounding Termination

(2) In Buildings or Structures With Grounding Means

(3) In Buildings or Structures Without Intersystem Grounding Termination or Grounding

Change Type: Revision

Change Summary:

- The term *intersystem bonding jumper* is incorporated into 800.100(B) to correlate with changes in 250.94 and a new definition in Article 100 of *intersystem bonding termination*.

Intersystem bonding requirements have been significantly revised to provide termination points that are readily accessible.

Revised 2008 NEC Text:

800.100 CABLE AND PRIMARY PROTECTOR GROUNDING

(B) ELECTRODE.   The grounding conductor shall be connected in accordance with 800.100(B)(1), ~~and~~ (B)(2) ~~and~~ <u>or (B)(3).</u>

(1) IN BUILDINGS OR STRUCTURES WITH AN INTERSYSTEM ~~GROUNDING MEANS~~ <u>BONDING</u> TERMINATION. <u>If the building or structure served has an intersystem bonding termination the bonding conductor shall be connected to the intersystem grounding termination.</u>

~~(1)~~<u>(2)</u> IN BUILDINGS OR STRUCTURES WITH GROUNDING MEANS. <u>If the building or structure served has no intersystem bonding termination, the grounding conductor shall be connected</u> ~~T~~to the nearest accessible location on the following:

   ((1) through(7) No changes)

   <u>A bonding device intended to provide a termination point for the grounding conductor (intersystem bonding) shall not interfere with the opening of an equipment enclosure.</u>

(continued)

A bonding device shall be mounted on non-removable parts. A bonding device shall not be mounted on a door or cover even if the door or cover is nonremovable.

For purposes of this section, the mobile home service equipment or the mobile home disconnecting means, as described in 800.90(B), shall be considered accessible.

~~(2)~~(3) IN BUILDINGS OR STRUCTURES WITHOUT <u>INTERSYSTEM BONDING TERMINATION OR</u> GROUNDING MEANS.

If the building or structure served has no <u>intersystem bonding termination or</u> grounding means, as described in 800.100(B)(2)~~(1)~~, the grounding conductor shall be connected to either of the following:.

(1) To any one of the individual electrodes described in 250.52(A)(1), (A)(2), (A)(3), or (A)(4)

(2) If the building or structure served has no <u>intersystem grounding termination or has no</u> grounding means, as described in 800.100(B)(2)~~(1)~~ or (B)(3)~~(2)~~(1), to an effectively grounded metal structure or to a ground rod or pipe not less than 1.5 m (5 ft) in length and 12.7 mm (½ in.) in diameter, driven, where practicable, into permanently damp earth and separated from lightning conductors as covered in 800.53 and at least 1.8 m (6 ft) from electrodes of other systems. Steam or hot water pipes or air terminal conductors (lightning-rod conductors) shall not be employed as electrodes for protectors.

Change Significance:

This revision correlates changes in requirements for bonding other systems as required in 250.94 with communications systems. The new requirements of 250.94 outlines three options for bonding at (1) terminals secured to the meter enclosure, (2) a bonding bar near the service equipment or meter, or (3) a bonding bar near the grounding electrode conductor. An intersystem bonding termination is defined in Article 100 as a device for bonding communications equipment at the service disconnect. Similar revisions have occurred in other articles, including 810.21, 820.100, and 830.100.

Notes:

_____

_____

_____

_____

_____

800.156
DWELLING UNIT COMMUNICATIONS OUTLET

Chapter 8 Communications Systems

Article 800 Communications Circuits

Part V Installation Methods Within Buildings

800.156 Dwelling Unit Communications Outlet

Change Type: New

Change Summary:
- A new requirement is added to require a minimum of one communications outlet installed in all new dwelling units.

Revised 2008 NEC Text:
800.156 DWELLING UNIT COMMUNICATIONS OUTLET.   For new construction, a minimum of one communications outlet shall be installed within the dwelling and cabled to the service provider demarcation point.

Change Significance:
This new section in Article 800 requires for the first time that a communications outlet be installed in all new dwelling units. This is extremely significant in that the enforcement community will now be required to ensure that this communications outlet is installed in all new construction of dwelling units. In many areas and jurisdictions, there is no enforcement of the requirements of Chapter 8 articles. This new requirement will aid in achieving uniform enforcement of the entire communications installation.

The original proposal sought to require two communications outlets in all dwelling units being newly constructed. The submitter identified a number of reasons to require these outlets, including the capability to contact emergency services, installation problems, and Americans with Disabilities Act (ADA) requirements.

At least one communications outlet is now required in a dwelling unit. There is no direction given on which room to locate the outlet.

Change Summary:

- Requirements for power limitation on cable television (CATV) and network-powered broadband systems are clarified.

Revised 2008 NEC Text:

820.15 ~~ENERGY~~ POWER LIMITATIONS.   Coaxial cable shall be permitted to deliver ~~low energy~~ power to equipment that is directly associated with the radio frequency distribution system if the voltage is not over 60 volts and if the current ~~supply~~ is <u>supplied by</u> ~~from~~ a transformer or other device that has ~~energy~~ <u>power</u>-limiting characteristics.

<u>Power shall be blocked from premises devices on the network that are not intended to be powered via the coaxial cable.</u>

830.15 POWER LIMITATIONS.   Network-powered broadband communications systems shall be classified as having low or medium power sources as ~~defined~~ <u>specified</u> in ~~Table 830.15~~ <u>830.15(1) or 830.15(2)</u>.

<u>(1) Sources shall be classified as defined in Table 830.15.</u>

<u>(2) DC power sources exceeding 150 volts to ground, but no more than 200 volts to ground, with the current to ground limited to 10 mA dc, that meet the current and power limitation for medium power sources in Table 830.15 shall be classified as medium power sources.</u>

<u>FPN: One way to determine compliance with 830.15(2) is listed information technology equpment intended to supply power via a telecommunication network that complies with the requirements for RFT-V circuits as defined in UL 60950-21, Standard for Safety for Information Technology Equipment–Safety–Part 21: Remote Power Feeding.</u>

Change Significance:

820.15 is renamed for clarity and correlation with 830.15. The text is editorially modified to "power" limitation instead of "energy" limitation. A new last sentence requires that power be blocked from premises devices (on a network) not intended to be powered.

830.15 is reorganized to address existing power sources as Low or Medium, as outlined in Table 830.15. New text is added to recognize DC power sources that exceed 150 volts (but not more than 200 volts) to ground as medium-power sources provided they are limited to 10 milliamps DC. A new informational FPN is added to reference a UL standard to assist the code user in identifying a DC source that would comply with 830.15(2).

820.15 and 830.15
POWER LIMITATIONS

Chapter 8 Communications Systems

Article 820 Community Antenna Television and Radio Distribution Systems

Part I General

820.15 Power Limitations

Article 830 Network-Powered Broadband Communications Systems

Part I General

830.15 Power Limitations

Change Type: Revision

Article 820 permits voltages up to 60 volts on coaxial cables, and Article 830 permits voltages up to 200 volts on coaxial cables.

APPENDIX

Change Title	Proposal #	Comment #
90.2 Scope	1-4 & 1-5	1-1, 1-2, & 1-3
Article 100 Definition of Bonded (Bonding)	5-2	none
Article 100 Definition of Clothes Closet	1-20	1-22
Article 100 Definition of Device	9-7	none
Article 100 Definition of Electric Power Production and Distribution Network	1-28	13-2
Article 100 Definition of Equipment	1-31	none
Article 100 Definition of Grounded (Grounding)	5-9, 10, 12, & 13	none
Article 100 Definition of Grounding Conductor, Equipment (EGC)	5-6	5-3
Article 100 Definition of Grounding Electrode Conductor	5-18	5-6
Article 100 Definition of Intersystem Bonding Termination	5-20	5-11
Article 100 Definition of Kitchen	1-36	none
Article 100 Definition of Luminaire	18-4, 4b, 5, 42, & 44	none
Article 100 Definition of Metal-Enclosed Power Switchgear	9-5	none
Article 100 Definitions of Neutral Conductor and Neutral Point	5-36	none
Article 100 Definition of Premises Wiring (System)	1-43	none
Article 100 Definition of Qualified Person	1-45	none
Article 100 Definitions of Surge Arrester and Surge Protective Device (SPD)	5-340	none

Change Title	Proposal #	Comment #
Article 100 Definition of *Short-Circuit Current Rating*	10-2	none
Article 100 Definitions of *Utility-Interactive Inverter* and *Interactive System*	13-3 & 184	none
110.11 Deteriorating Agents	1-61	none
110.12 Mechanical Execution of Work, (A) Unused Openings	1-71 & 1-72	1-50
110.16 Flash Protection	1-82, 84, & 87	1-54
110.20 Enclosure Types	1-93, 94, & 95	1-64
110.22 Identification of Disconnecting Means	1-98	1-70 & 10-22
110.26 Spaces About Electrical Equipment, (C) Entrance to and Egress from Working Space	1-119, 120, 121, 123, & 127	1-84 & 1-92
110.26(G) Locked Electrical Equipment Rooms or Enclosures	1-100	1-71
110.31(C) Outdoor Installations	1-145	none
110.33 Entrance to Enclosures and Access to Working Space	1-147 & 148	1-99
200.2(B) Continuity	5-90	5-47
210.4 Multiwire Branch Circuits	2-10 & 14	2-8 & 2-10
210.4(D) Grouping	2-17	2-10 & 2-13
210.5(C) Ungrounded Conductors	2-19, 22, 23, 24, & 25	2-18
210.6(D) 600 Volts Between Conductors	2-32	2-20
210.8(A) Dwelling Units	2-40, 41, 50, 51, 56, & 57	2-34 & 2-35
210.8(B) Other Than Dwelling Units, (2) Kitchens	2-72, 73, & 82	2-5 & 2-50
210.8(B) Other Than Dwelling Units, (4) Outdoors	2-70	2-39 & 2-41
210.8(B) Other Than Dwelling Units, (5) Sinks	2-71, 77, 81,& 85	2-44 & 2-57
210.12 Arc-Fault Circuit-Interrupter Protection	2-105, 118a, 119, 142, 143, & 147	2-89, 2-95, 2-126, 2-129, & 2-137
210.25 Branch Circuits in Buildings With More Than One Occupancy	2-184	none
210.52 Dwelling Unit Receptacle Outlets	2-189 & 190	2-198

Change Title	Proposal #	Comment #
210.52 Dwelling Unit Receptacle Outlets, (C) Countertops	2-206a, 207, & 211	2-216, 2-218, & 2-221
210.52 Dwelling Unit Receptacle Outlets, (E) Outdoor Outlets	2-228 & 229	2-225, 2-227, & 2-230
210.52 Dwelling Unit Receptacle Outlets, (G) Basements and Garages	2-240	2-237
210.60 Guest Rooms, Guest Suites, Dormitories, and Similar Occupancies	2-242	2-238a
210.62 Show Windows	2-244	none
215.10 Ground-Fault Protection of Equipment	2-286 & 287	none
215.12 Identification for Feeders	2-290, 291, 292, & 293	2-254
220.52 Small-Appliance and Laundry Loads—Dwelling Unit	2-319 & 320	none
225.18 Clearance for Overhead Conductors and Cables	4-4, 12	none
225.33 Maximum Number of Disconnects	4-20	none
225.39 Rating of Disconnect	4-22 & 25	none
230.40 Number of Service-Entrance Conductor Sets	4-42	none
230.44 Cable Trays	4-52	none
230.53 Raceways to Drain	4-58	4-27
230.54 Overhead Service Locations	4-59 & 4-60	4-29 & 4-30
230.82 Equipment Connected to the Supply Side of Service Disconnect	4-4 & 66	4-35 & 4-39a
230.95 Ground-Fault Protection of Equipment	4-4	none
240.4(D) Small Conductors	10-10	10-2 & 10-10
240.21 Location in Circuit	10-18a	none
240.21(C) Transformer Secondary Conductors	10-26	none
240.21(C)(2) Transformer Secondary Conductors Not Over 3 m (10 ft) Long	10-27	10-11 & 9-7b
240.21(H) Battery Conductors	10-33	10-13
240.24(B) Occupancy	10-37	none

Change Title	Proposal #	Comment #
240.24(F) Not Located Over Steps	10-40	10-15
240.86(A) Selected Under Engineering Supervision in Existing Installations	10-50a & 52	none
240.92(B) Feeder Taps	10-21, 22, & 23	none
250.4(B) Ungrounded Systems	5-69, 71, 72, & 73	5-40
250.8 Connection of Grounding and Bonding Equipment	5-84	none
250.20(D) Separately Derived Systems	5-95	none
250.22 Circuits Not to Be Grounded	5-100	none
250.28 Main Bonding Jumper and System Bonding Jumper	5-107	5-53
250.30(A) Grounded Systems	5-77 & 107a	none
250.30(A)(4) Grounding Electrode Conductor, Multiple Separately Derived Systems	5-110	none
250.32 Buildings or Structures Supplied by Feeder(s) or Branch Circuit(s)	5-76 & 119	5-58
250.35 Permanently Installed Generators	5-128	5-67
250.36 High-Impedance Grounded Neutral Systems	5-129	none
250.52(A)(2) Metal Frame of the Building or Structure	5-148	5-85
250.52(A)(3) Concrete-Encased Electrode	5-137	5-86
250.52(A)(5) Rod and Pipe Electrodes and (6) Other Listed Electrodes	5-160	5-97
250.54 Auxiliary Grounding Electrodes	5-170	none
250.56 Resistance of Rod, Pipe, and Plate Electrodes	5-174	5-100
250.64 Grounding Electrode Conductor Installation	5-16,	5-103, 5-116, & 5-119
250.64(D) Service with Multiple Disconnecting Means Enclosures	5-192	none
250.64(F) Installation to Electrode(s)	5-186 & 203	5-113

Change Title	Proposal #	Comment #
250.68 Grounding Electrode Conductor and Bonding Jumper Connection to Grounding Electrodes	5-213	none
250.94 Bonding for Other Systems	5-220	5-122
250.104(A)(2) Buildings of Multiple Occupancy	5-229	none
250.112(I) Remote-Control, Signaling, and Fire Alarm Circuits	5-252	none
250.119 Identification of Equipment Grounding Conductors	5-266	5-141
250.122(C) Multiple Circuits	5-282a	none
250.122(D) Motor Circuits	5-284	5-153
250.122(F) Conductors in Parallel	5-287	none
250.146(A) Surface Mounted Box	5-300	none
250.166 Size of the Direct-Current Grounding Electrode Conductor	5-310	none
250.168 Direct-Current System Bonding Jumper	5-313	5-159
Article 280 Surge Arresters, Over 1 kV	5-335	5-162 & 5-163
Article 285 Surge Protective Devices (SPDs) 1 kV or less	5-335	5-168
300.4(E) Cables and Raceways Installed Under Roof Decking	3-31	3-10
300.5(B)(C) Wet Locations and Underground Cables Under Buildings	3-42 & 3-45	3-20a & 3-22
300.5(D) Protection From Damage	3-46	none
300.6(B) Aluminum Metal Equipment	3-59 & 60	none
300.7(B) Expansion Fittings	3-61	3-50
300.9 Raceways in Wet Locations Above Grade	3-63	3-52
300.11(A) Secured in Place	3-67a	none
300.12 Mechanical Continuity—Raceways and Cables	3-65	3-53

Change Title	Proposal #	Comment #
300.19(B) Fire-Rated Cables and Conductors	3-85	3-67
300.20(A) Conductors Grouped Together	3-85a & 86	none
300.22(C) Other Spaces Used for Environmental Air	3-95	3-69
Table 300.50 Minimum Cover Requirements	3-105	none
310.4 Conductors in Parallel	6-7a	6-2 & 6-3
310.15(B)(2) Adjustment Factors, (a) More Than Three Current-Carrying Conductors in a Raceway or Cable	6-47	6-46
310.15(B)(2) Adjustment Factors, (c) Conduits Exposed to Sunlight on Rooftops	6-51	none
314.24 Minimum Depth of Boxes for Outlets, Devices, and Utilization Equipment	9-52	9-20
314.27(A), (B) Outlet Boxes	9-56	none
314.27(E) Utilization Equipment	9-63	9-35
314.28(A) Minimum Size	9-66 & 67	none
314.30 Handhole Enclosures	9-71 & 77	9-38
320.10 Uses Permitted	7-2	none
330.10 Uses Permitted	7-23 & 27	none
334.12 Uses Not Permitted	7-50 & 51	7-15
334.15 Exposed Work	7-1, 58, 61, 62, & 63	none
334.80 Ampacity	7-70, 72, & 73	7-40
336.10 Uses Permitted	7-80	none
338.12 Uses Not Permitted	7-84 & 93	none
342.30(C), 344.30(C), 352.30(C), and 358.30(C) Unsupported Raceways (IMC, RMC, PVC, and EMT)	8-9, 23, 65, & 104	8-7, 8-20, 8-38, & 8-57
344.10 Uses Permitted	8-15	8-14 & 8-15
3XX.12 Uses Not Permitted	8-29, 59, 70, 76, 84, 101, 113, 143, 160, 167, & 174	none
348.12 Uses Not Permitted	8-26, 27, & 28	none
348.60 and 350.60 Grounding and Bonding	8-40 & 49	none
350.30 Securing and Supporting	8-44 & 46	8-29

Change Title	Proposal #	Comment #
Article 352 Rigid Polyvinyl Chloride Conduit: Type PVC	8-53	none
352.10(F) Exposed	8-57	none
Article 353 High-Density Polyethylene Conduit: Type HDPE Conduit	8-68, 68a, 69, & 74	none
Article 355 Reinforced Thermosetting Resin Conduit: Type RTRC	8-78	8-49b
362.30 Securing and Supporting	8-119	none
366.2 Definitions	8-124	none
Article 376 Metal Wireways	8-151, 155, 157, & 157a	8-73
Article 382 Nonmetallic Extensions	7-98	7-56
388.30 and 388.56 Securing and Supporting and Splices and Taps	8-175 & 176	8-76
392.3(A) Wiring Methods	8-181	8-81 & 82
392.9 Number of Multiconductor Cables, Rated 2000 Volts or Less, in Cable Trays	8-180 & 194	none
392.11(C) Combinations of Multiconductor and Single-Conductor Cables	8-180 & 197	none
404.4 Damp or Wet Locations	9-88	none
404.8(C) Multipole Snap Switches	9-92	none
406.4(G) Voltage Between Adjacent Devices	18-24	18-13
406.8 Receptacles in Damp or Wet Locations	18-28, 33, & 35	18-16, 18-18, 18-19, 18-20, 18-23, & 18-24
406.11 Tamper-Resistant Receptacles in Dwelling Units	18-40	none
408.3(F) High-Leg Identification	9-109	none
408.4 Circuit Directory or Circuit Identification	9-101 & 105	none
408.36 Overcurrent Protection	9-117	9-70
409.2 Definition of *Industrial Control Panel*	11-3	none
409.110 Marking	11-12, 14, & 15	none

Change Title	Proposal #	Comment #
410.6 Listing Required	18-40a	18-65
410.16 Luminaires in Clothes Closets	18-57	18-66
410.130(G) Disconnecting Means	18-90b	18-79
410.141(B) Within Sight or Locked Type	18-100	none
411.2 Definition of *Lighting Systems Operating at 30 Volts or Less*	18-105	18-91
411.3 Listing Required	18-106	18-92
411.4 Specific Location Requirements	18-107	18-94
422.51 Cord-and-Plug-Connected Vending Machines	17-27	none
422.52 Electric Drinking Fountains	17-28	17-15
424.19 Disconnecting Means	17-31 & 33	17-50
427.13 Identification	17-52a	17-50
430.32(C) Selection of Overload Device	11-37	none
430.73 and 430.74 Protection of Conductor from Physical Damage and Electrical Arrangement of Control Circuits	11-52 & 53	11-19
430.102(A) Controller	11-81	11-23
430.102(B) Motor	11-67 & 68	11-27
430.103 Operation	11-71	none
430.110(A) and 440.12 General and Rating and Interrupting Capacity	11-75 & 92	11-30,
430.126 Motor Overtemperature Protection	11-77 & 79	none
430.227 Disconnecting Means	11-83	none
445.18 Disconnecting Means Required for Generators	13-19	none
445.19 Generators Supplying Multiple Loads	13-8	none
480.5 Disconnecting Means	13-16	13-21
490.44 and 490.46 Fused Interrupter Switches and Circuit Breaker Locking	9-155 & 156	none

Change Title	Proposal #	Comment #
500.1 Scope—Articles 500 through 504	14-1a	none
500.7(K) Combustible Gas Detection System	14-17	none
500.8(A) Suitability	14-18a	none
501.10(B)(1) General	14-33a	14-8
501.30(B) Types of Equipment-Grounding Conductors	14-44,52a, 138, & 155a	none
502.115(A) Class II, Division 1	14-63	none
502.120 Control Transformers and Resistors	14-64	14-28
502.130(B)(2) Fixed Lighting	14-69	none
502.150(B) Class II, Division 2	14-73 & 74	14-33a
505.7(A) Implementation of Zone Classification System	14-109, 110, & 111	none
506.2 Definitions	14-140 & 14-143	14-60 & 14-61
511.2 and 511.3 Definitions and Area Classification, General	14-156	14-77
513.2 and 513.3(C) Definition of *Aircraft Painting Hangar* and Vicinity of Aircraft	14-165a & 165b	none
517.2 Definitions	15-1, 11, 18, 20, 21, & 23	15-4, 15-16, 15-8, 15-11, 15-15, & 15-16
517.2 Definition of *Patient Care*	15-22	15-16
517.13(B) Insulated Equipment Grounding Conductor	15-28	15-19
517.32 Life Safety Branch	15-61, 63, & 64	none
517.34 and 517.43 Equipment System Connection to Alternate Power Source and Connection to Critical Branch	15-66, 67, 70, 81, & 82	15-45
517.40(B) Inpatient Hospital Care Facilities	15-73	15-49
517.80 Patient Care Areas	15-103	none
517.160(A)(5) Conductor Identification	15-106	none
518.4(A) General	15-109	none
518.5 Supply	15-116	none
520.2 and 520.27(B) Definitions and Neutral Conductor	15-125, 128, & 129	none
Article 522 Control Systems for Permanent Amusement Attractions	15-116	15-64 & 15-66

Change Title	Proposal #	Comment #
Article 525 Carnivals, Circuses, Fairs, and Similar Events	15-146, 148, 150, 152, & 156	15-71 & 15-74
525.2 Definitions	15-144a & 145	15-70
Article 547 Agricultural Buildings	19-4 & 9	none
547.9 Electrical Supply to Building(s) or Structure(s) from a Distribution Point	19-27 & 29a	none
547.9(B)(3) Grounding and Bonding	19-26	19-16
550.33(A) Feeder Conductors	19-69	19-28
551.4 General Requirements	19-74	none
555.9 Electrical Connections	19-117	19-46
555.21 Motor Fuel Dispensing Stations—Hazardous (Classified) Locations	19-127	19-48
590.4(D) Receptacles	3-115a	none
590.6 Ground-Fault Protection for Personnel	3-125	none
600.2 and 600.4(C) Definition of *Section Sign* and Section Signs	18-111 & 113	none
600.6 Disconnects	18-118 & 120	none
600.7 Grounding and Bonding	18-123, 124, & 125	18-95a
600.21(E) Attic and Soffit Locations	18-136, 137, & 138	none
600.32(K) Splices	18-154	none
600.41(D) Protection	18-157	none
600.42(A) Points of Transition	18-158	none
Articles 600, 610, 620, 625, 647, 665, and 675 Special Equipment Disconnects	multiple	none
604.2 and 604.6(A)(4) Definition of *Manufactured Wiring System and Busways*	19-129 & 132	none
620.2 Definitions of *Remote Machine Room and Control Room (for Elevator, Dumbwaiter)* and *Remote Machinery Space and Control Space (for Elevator, Dumbwaiter)*	12-16a, 17, & 20a	none
Article 626 Electrified Truck Parking Space	12-81	12-44
640.6 Mechanical Execution of Work	12-94	12-67 & 12-69

Change Title	Proposal #	Comment #
645.2 Definition of *Abandoned Supply Circuits and Interconnecting Cables*	12-105 & 106	none
645.5(D) Under Raised Floors	12-112	12-82
645.5(F)(G) Abandoned Supply Circuits and Interconnecting Cables and Installed Supply Circuits and Interconnecting Cables Identified for Future Use	12-116	12-80
645.10 Disconnecting Means	12-120	12-87
680.2 Definitions	17-60	none
680.10 Underground Wiring Location	17-76 & 77	none
680.12 Maintenance Disconnecting Means	17-80	17-70 & 17-71
680.21(A) Wiring Methods	17-81	17-65 & 17-72
680.22 Area Lighting, Receptacles, and Equipment	17-85	17-75
680.22(E) Other Outlets	17-96	none
680.23(A)(6) and 680.51(C) Bottom-Lined Luminaires and Luminaire Lenses	17-101 & 154	none
680.23(B) Wet-Niche Luminaires	17-103	none
680.25(A) Wiring Methods	17-113	17-65 & 17-90
680.26 Equipotential Bonding	17-114a, 115, & 115a	17-92
680.26(C) Pool Water	17-122	17-98
680.31 Pumps	17-135	none
680.43(D) Bonding	17-144 & 147	none
680.50 General	17-152	none
680.71 Protection	17-164 & 165	17-106
680.74 Bonding	17-166, 167, & 168	12-110
Article 680 Receptacle Locations, Parts II, III, IV, VI, and VII	17-85a, 86, 88, & 89	none
682.13 Wiring Methods and Installation	17-175	none
682.30 Grounding	17-178	none
Article 690 Solar Photovoltaic Systems	multiple	multiple
690.5 Ground-Fault Protection	13-22, 23, 25, & 26	13-20 & 13-29
690.10 Stand-Alone Systems	13-29, 30, & 31	13-35

Change Title	Proposal #	Comment #
690.31 Methods Permitted	13-37, 38, 39, & 40	13-46
690.33 Connectors	13-41 & 42	13-51
690.43 Equipment Grounding	13-47 & 48	13-57, 13-58, & 13-59
690.45 Size of Equipment-Grounding Conductor	13-22	13-29
692.41 System Grounding	13-72 & 73	none
695.6(B) Circuit Conductors	13-99, 13-100, & 13-101	none
700.6(C) and 701.7(C) Automatic Transfer Switches and Automatic Transfer Switches	13-117 & 150	none
700.9(B) Wiring	13-118 & 120	13-156
700.9(D) Fire Protection	13-121, 122, 124, &126	13-163
700.12(B)(6), 701.11(B)(5), and 702.11 *Generator Disconnecting Means Located Outdoors*	13-131, 13-155, & 13-181	13-169, 13-219, & 13-259
700.23 Dimmer Systems	13-134	13-176
700.27 and 701.18 Coordination	13-135 & 13-161	13-185 & 13-238
702.5 Capacity and Rating	13-168	13-255
Article 705 Interconnected Electric Power Production Sources	13-184	13-262
Article 708 Critical Operations Power Systems (COPS)	20-1	20-2, 20-8a, 20-9, 20-11, 20-12, 20-69, & 20-99
725.3(B) Spread of Fire or Products of Combustion	3-148	13-96
Chapters 7 and 8 Abandoned Cables	3-140	3-111
725.48(B) Class 1, Circuits with Power Supply Circuits	3-160	3-112
725.154 Applications of Listed Class 2, Class 3, and PLTC Cables	3-175 & 176	none
725.154(D) Hazardous (Classified) Locations	3-179	none
725.179(E) Type PLTC Cables	3-192	none
Articles 725 and 760 Class 1, Class 2, and Class 3 Remote-Control, Signaling, and Power-Limited Circuits and Fire Alarm Systems	3-137 & 211	none
727.4 Uses Permitted	3-205	none

Change Title	Proposal #	Comment #
760.3(G) Installation of Conductors With Other Systems	3-221	none
760.41 NPLFA Circuit Power Source Requirements	3-239	none
760.46 and 760.130(B) NPLFA Circuit Wiring Methods and PLFA Wiring Methods and Materials	3-240 & 246	3-159
760.121 Power Sources for PLFA Circuits	3-245	none
760.176 and 760.179 Listing and Marking of NPLFA Cables and Listing and Marking of PLFA Cables and Insulated Continuous Line-Type Fire Detectors	3-258 & 264	none
Article 770 Optical Fiber Cable and Raceways	16-2, 25, & 38	none
770.2 Definitions	16-12, 18, & 36	none
800.1 and 800.2 Scope and Definitions	16-98	16-79 & 16-80
800.48 Unlisted Cables Entering Buildings	16-145	none
800.100(B) Electrode	16-167 & 168	16-128, 16-131, & 16-129
800.156 Dwelling Unit Communications Outlet	16-207	none
820.15 and 830.15 Power Limitations	16-268 & 369	none